SYNTHETIC ENGINEERING MATERIALS AND NANOTECHNOLOGY

SYNTHETIC ENGINEERING MATERIALS AND NANOTECHNOLOGY

IBRAHIM M. ALARIFI

Department of Mechanical and Industrial Engineering, College of Engineering, Majmaah University, Al-Majmaah, Riyadh, Saudi Arabia

Engineering and Applied Science Research Center, Majmaah University, Al-Majmaah, Riyadh, Saudi Arabia

ELSEVIER

Elsevier
Radarweg 29, PO Box 211, 1000 AE Amsterdam, Netherlands
The Boulevard, Langford Lane, Kidlington, Oxford OX5 1GB, United Kingdom
50 Hampshire Street, 5th Floor, Cambridge, MA 02139, United States

Library of Congress Cataloging-in-Publication Data
A catalog record for this book is available from the Library of Congress

British Library Cataloguing-in-Publication Data
A catalogue record for this book is available from the British Library

ISBN: 978-0-12-824001-4

Publisher: Matthew Deans
Acquisitions Editor: Kayla Dos Santos
Editorial Project Manager: Isabella C. Silva
Production Project Manager:
Vijayaraj Purushothaman
Cover Designer: Vicky Pearson Esser

Typeset by STRAIVE, India

Contents

About the Author

Dr. Ibrahim M. Alarifi received his PhD in 2017 from the Department of Mechanical Science and Engineering at Wichita State University. He then joined the Department of Mechanical Engineering at Majmaah University in 2012 as a lecturer and received his tenure and was promoted to the role of Assistant Professor in 2017 and became the Head of the Department. He is currently the Director of Engineering & Applied Sciences Research Centre. Throughout his research experience, he has published 50 journal papers and 18 conference proceedings; edited 2 books; authored 4 book chapters and 1 laboratory manual; received 4 funded proposals, 1 patent, and 17 honors/awards; presented 17 presentations; chaired many international conferences; and reviewed several manuscripts in international journals and conference proceedings.

Preface

In the era of immense information, writing an academic book for any level is challenging. As we have been living in the information age, new developments are ever-increasing at a fast pace. Synthesis is a fundamental area of engineering and chemistry that deals with inventing and developing new materials. The synthesis is the process of making chemical reactions to manufacture more advanced and useful products.

The present book aims to serve the students as an academic book to present necessary and updated knowledge of synthetic materials and progressively encompass a detailed account of nanotechnology. This book disseminates essential information that will be useful for chemistry and engineering students at all levels. This book will be a single source of information as it provides knowledge about materials and discusses the methods of synthesizing engineering and nanomaterials. This book is not affluent with verbose and inapprehensible languages in which usually chemistry is written as a laboratory oriented language. Preferably the book contains apprehensible language about properties, materials and methods, and applications of synthesis material and nonmaterial-based products. Most importantly, real-life applications and solved exercises, hints, and objective type questions with answers are some of the book's unique features. These features will enhance the understanding of readers and will serve as an interactive element.

The book has been set out in two significant parts. From Chapters 1–7, it covers synthetic engineering materials. From Chapters 8–14, the book serves a detailed account of knowledge related to nanomaterials. Both the sections cover properties, methods, techniques, and applications of advanced ceramic materials, metals, biomaterials, advanced ceramic materials, polymer composite, and optical materials. This book is extremely useful for academia, scientists, engineers, researchers, students, and industry people. The book is interdisciplinary and has been written for readers with a background in physical science. It carries the basic concepts of the relevant fields that will facilitate as a primer for engineering and science students.

I have tried my best to illustrate and present the essential concepts, for instance, based on contemporary literature. My motivation behind writing this book is to provide the students with foundations by which they can comprehend the concepts and knowledge and apply the acquired knowledge in the emerging field of synthetic chemistry and engineering.

Acknowledgments

To my loving parents, Mr. Mohammed and Modhi Alarifi, for giving me their full support during my overseas study; to my wife Helen Alarifi and my family who openhandedly provided me with a healthy environment to proceed with my studies here in the United States; to my dear brothers and sisters back home, who have supported me during my absence. I want to express my true appreciation to my supervisor, Dr. Ramazan Asmatulu, for his support and for providing me with clear guidance, inspiration, suggestions, criticism, and financial assistance on my writing book journey. I could not have asked for a better mentor, and I owe him much for his support over the years. His leadership consistently challenged me to be a better researcher and provided the right environment for me to achieve my goals.

Engagement Ring

Ibrahim Alarifi

Condolences

My condolences are on the passing of my father, Mohammed Alarifi. I am deeply saddened by the loss that my family has encountered—my condolences. My deepest sympathies go out to my family. I offer you my thoughts, prayers, and well-wishes during this dark time in my life.

Mohammed Alarifi (Father)

CHAPTER 1

Introduction, properties, and application of synthetic engineering materials

Abbreviations

CHDA	cyclohexanedicarboxylic acid
EC	electrocoagulation
EF	electroflotation
EMT	electrochemical monitoring technique
FRP	fiber-reinforced plastics
GAG	glycosaminoglycan
GRP	glass reinforced polymers
ICTAC	International Confederation for Thermal Analysis and Calorimetry
NGC	nerve guide conduit
PCL	polycaprolactone
PEMs	permeation rates
PGA	polyglycolic acid
PLA	poly-lactic acid
PLGA	poly lactic-*co*-glycolic acid
PMMA	methyl-methacrylate
PU	polyurethane
TA	thermal analysis
TMA	thermomechanical analysis

1.1 Introduction

Biological materials have been considered as both blessing and envy to material science. Wood, bone, nacre, silk, and muscles encompass extraordinary properties, which include durability, strength, and flexibility. They are produced through widely used feedstock methods, which are sustainable and environmentally friendly. However, lots of efforts are gathered to execute structures and characteristics of the naturally found materials. Production and important development are in a continuous process, due to the large-scale developments in the field of biomaterial. Consider the example, as shown in Fig. 1.1, where the mechanical benefits of silk are acquired by attaining full control on their intricate chemical composition of protein

Synthetic Engineering Materials and Nanotechnology
https://doi.org/10.1016/B978-0-12-824001-4.00006-5

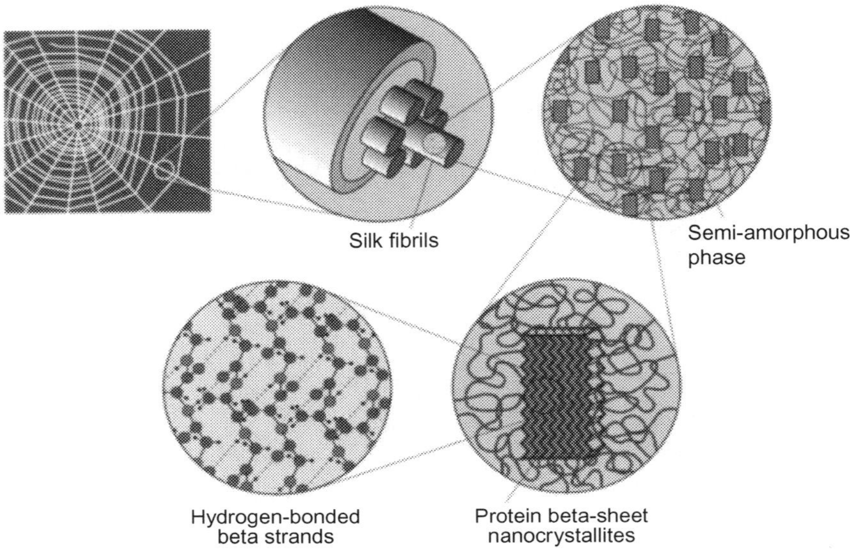

Fig. 1.1 The classified formation of silk: microns, nanometers, and molecule-scale [1].

built-in polymers along with fiber positioning and hydration through the process of gyrating to introduce the hierarchical structure in the final product. Arrangement of anigurd treating is altered, which results in diverse forms of silk fiber yield that are "engineered" for multifold purposes [1].

1.1.1 Genesis of synthetic biology and synthetic engineering

Synthetic biology is the new field of science, but the idea was generated several years ago. German American biologist Jacques Loeb, in The Mechanistic Conception of Life, provided the idea that a living organism could be viewed with an engineering perspective. They are analyzed as organisms that are engineered by progression through the tools of human involvement, handling, and modification. The author asserted the idea that man can play a role as creator in nature and may shape it according to his own wish [1]. Many of the bioprocesses are now using microbial, plants, or even animal cells to acquire useful material from them, such as enzymes, in order to obtain industrial biochemical, pharmaceuticals, and other materials. Economic benefits have been outnumbered in the past two decades from the hike in the use of different bioprocesses. Economic growth was brought by various findings and innovations in the manufacturing of products [2].

$$m = \frac{Mn}{DP} \tag{1.1}$$

$$m = \frac{m - fs\,ms}{fx} \tag{1.2}$$

The use of fermentation is one such method to produce chemicals and natural products which were commenced a long time ago. It was the time when research laboratories, pharmaceutical, and food companies separated microorganisms (from different sources such as soil, garden, and forests) to produce useful compounds or materials [2]. A recurring field in synthetic biology is the role of engineering. This notion is popular worldwide, as different scholars, as well as biologists, are studying and working on this field [3]. Synthetic organisms are very dissimilar to the drugs and have a more complicated process and impact to a greater extent. It stipulates enhanced specialized solutions as synthetic biology encompasses various subfields, numerous techniques, and more specialized methods [4].

Hint statement

Polymer and polymer-ceramics compounds materials have been exploited for manufacturing synthetic bone scaffolds owing to their tunable degradability, biocompatibility, ability to process, and adaptability.

Polymer and polymer-ceramics compound constituents are exploited for engineering the synthetic bone scaffoldings as they are biologically well-matched, dissolvable, and possess the ability to process as well as possess adaptability [5]. There is extensive use of polymers in electrospinning, which can be further classified as naturally found polymer, human-made polymer, and polymer composites. Synthetic or human-made polymers possess improved perfunctory characteristics in contrast to the naturally found polymers. Blending two synthetic polymers or the duo of natural polymers or, most preferably, the combination of synthetic and natural polymers can produce better results in the manufacturing of electrospun mat with increased quality and properties such as mechanical durability or biocompatibility, as illustrated in Fig. 1.2 [7, 8].

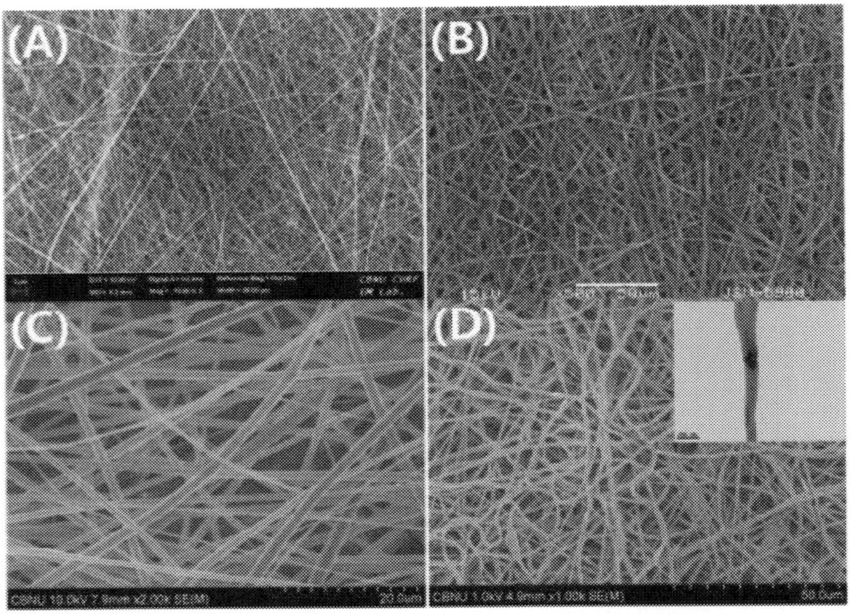

Fig. 1.2 (A) Silk nanofiber; human-made material (B) polycaprolactone (PCL); co-polymer composites (C) PCL-cellulose acetate along with polymer-ceramic composite (D) PCL-cellulose acetate-β-TCP with the nanofiber [6].

Example 1.1

An alternating co-polymer is known to have a number average molecular weight of 250,000 g/mol, and a degree of polymerization of 3420 molecular weight of styrene, ethylene, propylene are 104.14 g/mol, 28.05 g/mol, and 42.08 g/mol respectively. What is the other repeat unit? Why?

Noted the fraction are $fs = fx = 0.5$.

Solution:

$$m = \frac{M\,n}{DP} = \frac{250,000\ \text{g/mol}}{3420} = 73.10\ \text{g/mol}$$

$$m = \frac{m - fs\,ms}{fx}$$

$$= \frac{73.10\ \text{g/mol} - (0.5)(104.14\ \text{g/mol})}{0.5} = 42.06\ \text{g/mol}$$

$$m = 42.06\ \text{g/mol}$$

Therefore, propylene is the other repeat unit type since its "m" value is almost the same as the calculated mx.

Some of the most commonly used polymers for manufacturing bone tissue include collagen, silk retains, alginate, elastin, chitosan, glycosaminoglycan (GAG), etc. [9, 10]. In contrast with synthetic polymers, natural polymers do not possess good biocompatibility and have low mechanical properties [5]. Synthetic polymers that are abundantly required to manufacture bone tissue include poly-lactic acid (PLA), poly (lactic-*co*-glycolic acid) (PLGA), polyurethane (PU), polycaprolactone (PCL), Polypropylene, poly (glycolic acid) (PGA), etc. while abundantly required natural polymers that are used for bone tissue engineering include: collagen, alginate, chitosan, silk, gelatin, elastin, glycosaminoglycan (GAG), etc. [9, 10]. Although natural polymers are well-suited biologically, but possess little mechanical characteristics in comparison to synthetic polymers [5]. However, the common synthetic polymers required in manufacturing bone tissue include polycaprolactone (PCL), poly-lactic acid (PLA), poly (glycolic acid) (PGA), poly (lactic-*co*-glycolic acid) (PLGA), polyurethane (PU), Polypropylene, etc. [9, 10].

Hint statement

The most commonly used polymers for the purpose of bone tissue engineering are alginate, silk retains, glycosaminoglycan (GAG), collagen, elastin, chitosan, etc.

1.1.2 Properties of synthetic materials

Most of the polymers are synthetic in nature. Although, the basic molecular organization of all plants and animals is parallel to the structure of a synthetic polymer. Most of the polymers are required to bring about designs that have synthetic properties as especially manufactured by chemists or chemical engineers. Engineers of other fields usually manufacture engineering components through common materials. They also work in direct collaboration of chemists or chemical engineers to manufacture a polymer having firm features [11]. The major valuable characteristics possessed by polymers are due to the nature of uniqueness and due to their protracted chain molecular structure [11].

Synthetic materials are used to mimic the ECM (Extracellular Matrix). Due to their manageable surface and structural characteristics, they make it possible to exploit their critical advantages for NGC (Nerve Guide Conduit) fabrication. Biodegradable synthetic materials are more favorable in comparison to nonbiodegradable, since the necessity for a supplementary operation is excluded from averting the guidance conduit [12]. The use of polymers is in the automotive engineering industry, computer industry, aerospace industry, construction, and numerous other applications. Such as, polymer composites nowadays manufacture the shock absorbers of automobiles as possess an adequate degree of tenderness, which is compatible with state and federal standards [11]. It is beneficial to reduce weight. This alteration of metal is regarded as cost-effective owing to the decline in fuel cost as well as the capability of convenient recycling of the polymer composite either they are wrecked automobile or shock absorber for new automobiles.

Hint statement

The polymers are used in the automotive industry, the computer industry, building trades, aerospace industry, and many other applications is common.

The bumpers of an automobile are now manufactured by polymer blend that possesses an adequate degree of tenderness, which is compatible with state and federal standards.

In the manufacturing of sports goods, automotive, aerospace, construction, and other applications, fiber reinforced plastics (FRP) or polymer matrix compounds are essential. They are known by various terms. For example, the FRP constituents are composed of crystal fibers, which are known as glass-reinforced polymers (GRP) or fiberglass [11]. One of the earlier adopted applications of polymeric materials was for the insulating of electric ropes for the power connectivity as they were low in conductivity. Moreover, the polymers are also employed as thermal insulating tools in houses, vehicles, and in many other fields [11]. The polymers (such as polybenzimidazole) who have a higher degree of thermal resistance are used as fabrics for outfits of firefighters who encounter intense heat. The insulation characteristic of polyurethane foam and polystyrene foam in comparison with block and timber are presented in Fig. 1.3.

Mechanical characteristics and rate of degradation augment the procedure of synthetic components for the application of neural regeneration in terms of productivity. Due to their enhanced features, they are the core contenders for NTE tubes. They are found helpful in overcoming the issues of immune response and rapid downgrading [12]. Synthetic materials are more beneficial in comparison to natural materials as they can be manufactured with thoroughly defined composition and well-developed elements and assemblies [13].

$$E_{cl} = E_m V_m + E_{f1} V_{f1} + E_{f2} V_{f2} \tag{1.3}$$

$$E_m V_m = E_{m(1-Vf)} \tag{1.4}$$

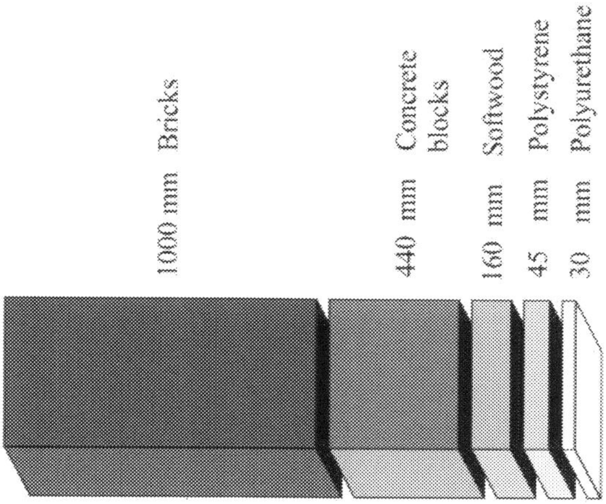

Fig. 1.3 Thermal insulation characteristics of different materials [11].

Example 1.2

The elastic moduli of aramid, glass fibers, and polyester resin matrix are 131 GPa and 72.5 GPa, and 4.0 GPa, respectively. The volume fractions of aramid and glass fibers 0.25 and 0.35, respectively. Explain an expression for the modulus of elasticity for a hybrid composite in which all fibers of both types are oriented in the same direction and compute the longitudinal modulus of elasticity of a hybrid composite.

The solution I: For a hybrid composite having all fibers aligned in the same direction:

$$E_{cl} = E_m V_m + E_{f1} V_{f1} + E_{f2} V_{f2}$$

The expression for the modulus of elasticity for a hybrid composite.

Solution II: For a hybrid composite (glass + aramid fibers) having all fibers aligned in the same direction:

$$E_m V_m = E_m \left(1 - V_f\right) \; V_f = V_{aramid} + V_{glass}$$

$$V_f = 0.25 + 0.35 = 0.60$$

$$E_{cl} = (4\,\text{GPa})(1.0 - 0.25 - 0.35) + (131\,\text{GPa})(0.25) + (72.5\,\text{GPa})(0.35)$$

$$\text{E}_{cl} = 59.7\,\text{GPa}$$

The longitudinal modulus is for the hybrid composite elasticity.

1.2 Techniques for acquiring synthetic engineering materials

Synthetic polymers are acquired through different processes of polymerization, which is homogeneous in nature. It includes mass polymerization or solution. They can also be treated through heterogeneous processes such as emulsion polymerization, dispersion, suspension, and mini-emulsion. Mostly in the industrial practices, polymerization of heterogeneous processes is engaged owing to its benefits, i.e., betterment in heat transmission resistor and a lesser degree of viscidness in terms of retorting [14]. Some of the other techniques are as follows:

Hint statement
Synthetic polymers possess enhanced mechanical properties in contrast to natural polymers, and they can be used to design several useful materials.

1.2.1 Polymerization emulsion

Polymerization emulsion is a simple process with regard to practice and the stages of emulsion polymerization, as shown in Fig. 1.4, which functions through a reaction system where the presence of surfactants makes the

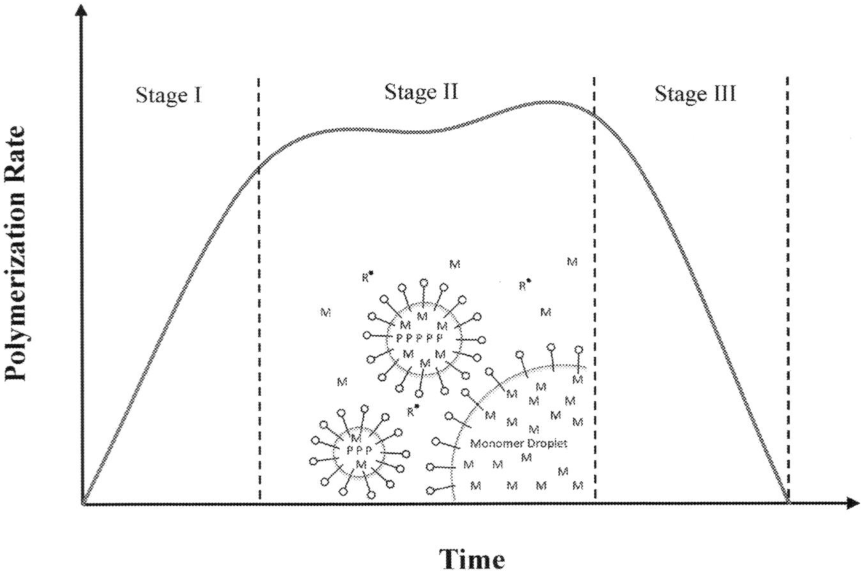

Fig. 1.4 The typical stages of emulsion polymerization.

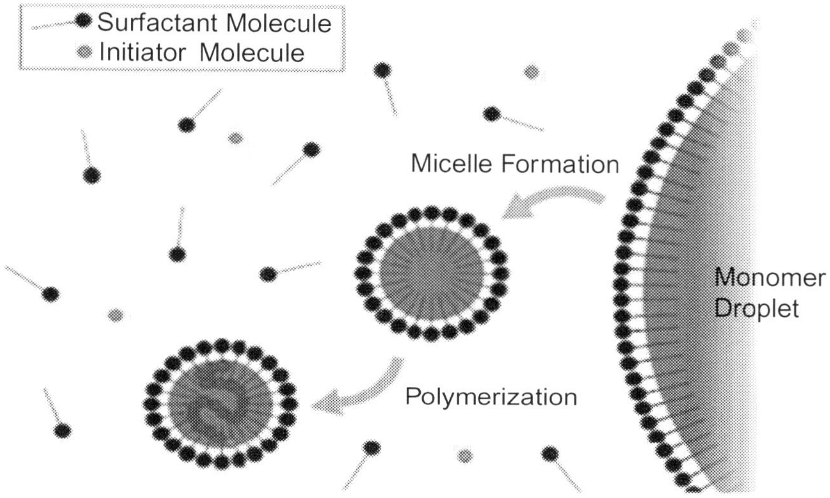

Fig. 1.5 Schematic representation of emulsion polymerization [15].

monomers constant and is disseminated in diluted form, as shown in Fig. 1.5. The application of emulsification is held to blend the polymers which are hydrophobic by diluting amphipathic emulgent. Afterward, free radicals composed of oil-soluble water originators are generated [15].

When diluting a more concentrated solution must be considered, such the volume increases V_1 and V_2. The number of moles stays the same M_1 and $M_{2,}$ as shown in Fig. 1.6.

$$V_1 M_1 = M_2 V_2 \tag{1.5}$$

Initial and final concentration and volumes must have the same units.

Example 1.3

If 45.0 mL of a 6.00 M HCL solution is diluted to a final volume of 0.250 L, what is the final concentration?

Given data;

$$M_1 = 6.00\,\text{m}\ M_2 = ?$$

$$V_1 = 45.0\,\text{mL}\ V_2 = 0.250\,\text{L} \times \frac{1000\,\text{mL}}{1} = 2.50 \times 10^2\,\text{mL}$$

Solution:

$$V_1 M_1 = M_2 V_2$$

Continued

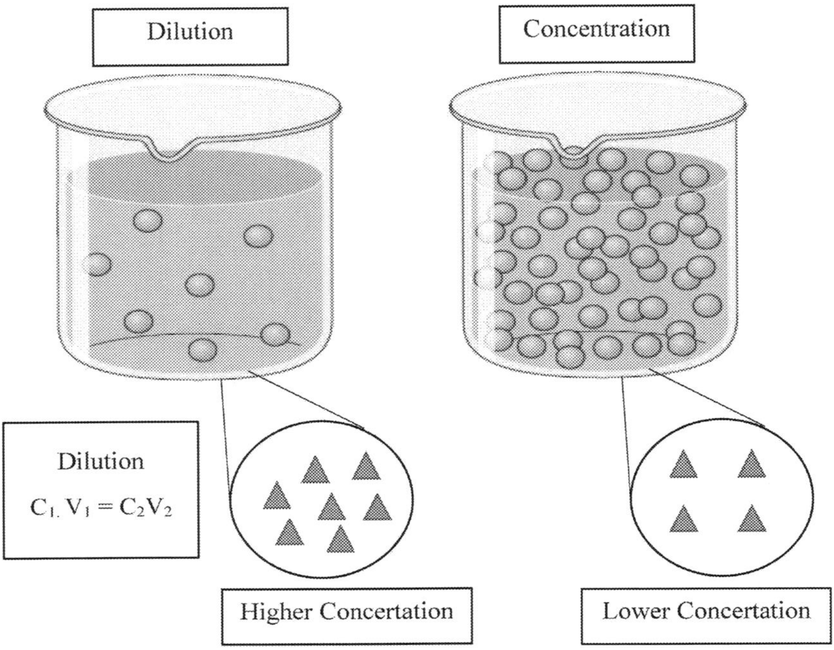

Fig. 1.6 The comparison between the dilution and concertation synthetic.

Example 1.3—cont'd

$$(6.00\,\mathrm{m})\,(45.0\,\mathrm{mL}) = (M_2)\,(0.250\,\mathrm{L})$$

$$\frac{(6.00\,\mathrm{m})\,(45.0\,\mathrm{mL})}{(0.250\,\mathrm{L})} = \frac{(M_2)\,(0.250\,\mathrm{L})}{(0.250\,\mathrm{L})}$$

$$\frac{(6.00\,\mathrm{m})\,(45.0\,\mathrm{mL})}{(2.50\times 102\,\mathrm{mL})} = \frac{(M_2)\,(0.250\,\mathrm{L})}{(2.50\times 102\,\mathrm{mL})}$$

$$M_2 = 1.08\,\mathrm{m}$$

1.2.2 Cross-linked polymer synthesis technique

Polymers can be cross-linked either due to their strong chemical bonding or physical interaction. Synthetic techniques for the polymers which are chemically linked are categorized into two types. First is the technique of cross-linking in the course of polymerization. Second is the postcross-linking of polymer chains. The former technique is characterized by polymerization of chain-growth, such as; step-growth polymerization and radical polymerization, which deals with monomers either bi-functional or multifunctional (like poly-condensation and poly-addition) and they are employed in order to produce of cross-linked polymers. The process is made by the reaction between reactive agents on the polymer chain and a cross-linker, having two or more volatile groups [16]. Cross-linking can be suitable for increased physical properties, heat resistance in aggregate materials, and long-term durability, which is important in the making of pipe, foam, cable wires, and other common PE (polyethylene) applications [17]. Methyl-methacrylate or PMMA compounds with diamine are manufactured through condensation reactions in the carbonyl group (>C=O) along with the Amine group (—NH). The usability of these reactions is observed to enhance PMMA characteristics, particularly thermal stability and quality of texture. It is processed through an ordinary chemical reaction by applying aliphatic diamines and aromatic diamines in varying quantities as cross-linking agents (Fig. 1.7) [19].

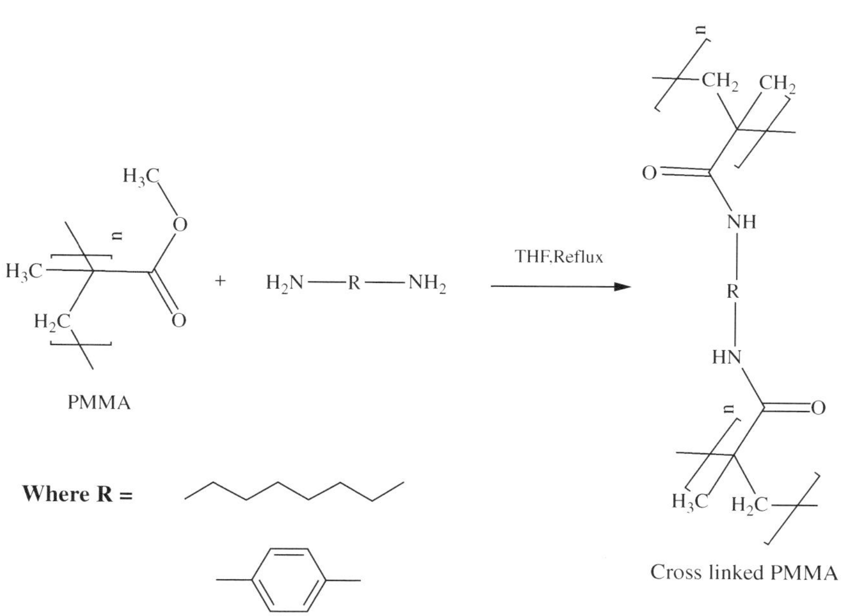

Fig. 1.7 Graphic representation of cross-linking of (methyl-methacrylate) PMMA polymers with aliphatic and aromatic diamines [18].

1.2.3 Reinforcing-filler technique

Reinforcing filler method is a well-developed method to augment mechanical, electrical, and thermal properties of plastics. New multifunctional materials are developed through the formulation and compounding of polymer matrices with organic and inorganic fillers. Organic fillers are beneficial as they decrease the cost of plastics. They are not as expensive as polymers. Other physical characteristics include friction coefficient, molding shrinkage, and weather ability, which are considered to be affected by filler addition [20].

Generally, reinforcing filaments are found as vulnerable and are more prone to get damaged amidst the process of compounding operation. Hence, the contemporary method is employed wherein the compounds are feed into the molded melt, and no compounding tools are employed (Fig. 1.8). This technique brings about a reduction in the abrasion of fibers when the melts perform as a lubricant. More importantly, wear considerations are needed to be taken into account when mostly abrasives are involved as the reinforcing materials [21].

1.2.4 Intercalation technique

The term "intercalation" denotes a procedure of implanting a molecule or an Inorganic into a host lattice, as shown in Fig. 1.9. It is alterable to some extent to assemble the guest-host or intercalation composite, which are perturbed from the host configuration and the reaction to form the compound

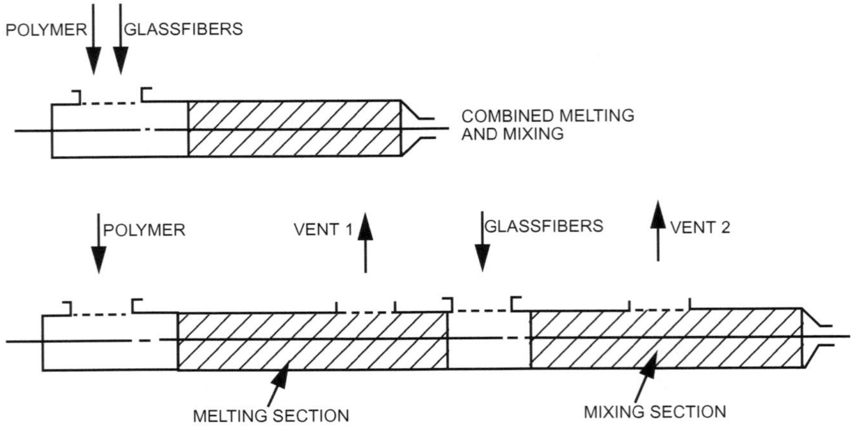

Fig. 1.8 Fundamental feeding and screw mechanism for single and double-stage compounders to manufacture reinforced plastics [21].

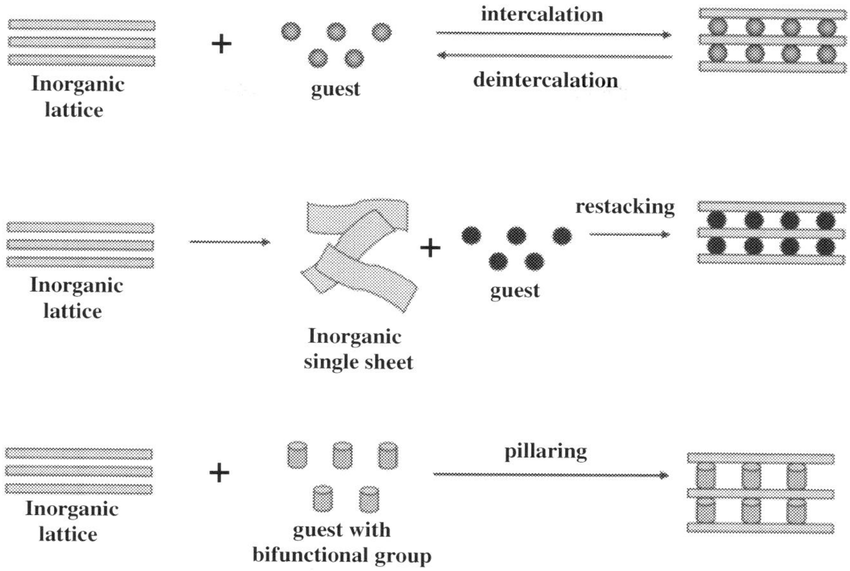

Fig. 1.9 Model of intercalation of inorganic lattice reaction.

[22]. The intercalation reactions are described as topo tactic solid-state reactions, and the product's intercalation compounds are evidently eminent from the enclosure and interstitial composites [23].

1.2.5 In situ polymerization technique

In situ polymerization was widely employed earlier to make coated silicates/polymer nanocomposites. When the sheets of clay are disseminated into the monomer suspension, through polymerization. The pieces of silicates are inflated into the liquefied monomer, which forms layered silicates consequent by polymerizing the monomer. The method for the mixture of nanocomposites has been practiced extensively. For the formation of thermoset-based clay nanocomposites, in situ intercalative polymerization is the only appropriate technique, despite some of its demerits [24].

Fig. 1.10 exhibits the phases to initiate mineral dispersal amid polyesterification. Firstly, a standardized mixture of MTEtOH and diethylene glycol is prepared. Subsequently, supplementary monomers (1,6-hexanediol, 1,4-CHDA, and maleic anhydride) are blended with the dispersal of diethylene glycol/MTEtOH. The clay mineral (CloisiteVR Na1), which is not modified, is combined when the temperature reaches

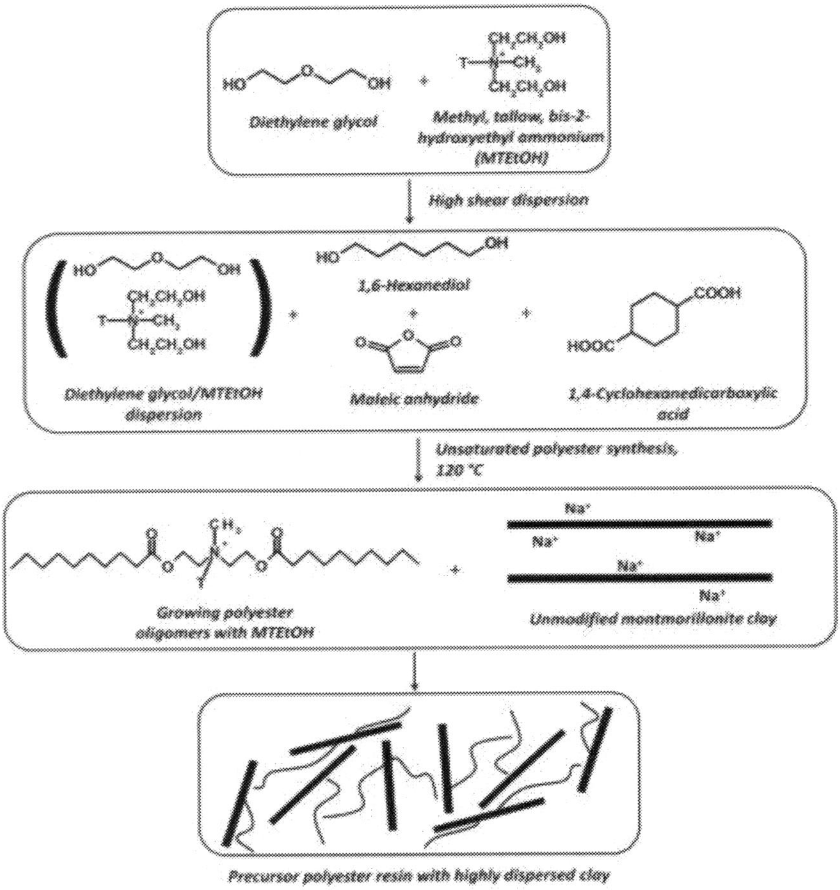

Fig. 1.10 The schematic exhibit of in situ polymerization to manufacture more scattered clay mineral with polyester resin through exchange reaction of in situ ion [25].

at 1208°C during polyesterification. This method is employed to proceed with an insitu ion exchange reaction to ascribe the organic modifier wit unmodified clay mineral [25].

1.2.6 Melt compounding technique

Fig. 1.11 showed the melting process changes a molten into a firm composite form while giving it a fixed structure and shape. Several chemical and physical properties of liquid melts have been found. The melt compounding process consists of three phases. The first phase is a flow where the melt pours,

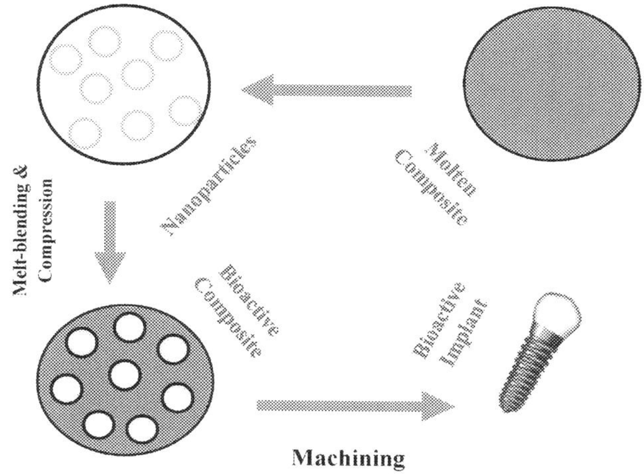

Fig. 1.11 The melting process changes a molten into a machined bio-composite.

either by gravity or an exterior force. The second step takes place and usually intersects the former. It is called the shape definition. The melt streams into a mold to make a three-dimensional (3D) shape, which is known as shape retention. It takes place to firm up and solidify through crystallization or glass conversion by making it cold [26].

1.3 Methods of producing synthetic material

1.3.1 Thermo-mechanical methods

The thermochemical methods are often restricted to solids and are used to measure the change in length with temperature. After the calibration, the coefficient of expansion is inferred. This process is defined as "thermodilatometry." When the expansion of solid takes place by the application of load, an amalgamation of expansion impact and modulus alteration are recorded. The process is called "thermomechanical analysis" (TMA) [27].

Thermal analysis (TA) is the study of the relationship between a sample property and its temperature when provided heat. As the sample is either given the heat or ventilated in an exact manner is common. The International Confederation explains the Thermo-mechanical technique for Thermal Analysis, and Calorimetry (ICTAC) is a procedure in which the distortion of the illustration is measured under the constant load [28].

Thermomechanical analysis (TMA) was performed to investigate the linear length change of the composite sample [29]. A Seiko Instruments TMA/SSC120C automatic cooling dynamic load thermomechanical analyzer was used for this study. In this set of experiments, a constant load of 10 mg was applied on the quartz probe to maintain contact with the top of the sample. The dimensions of all samples were approximately $20 \times 6 \times 1.5$ mm. The tests were performed to measure the length change along the longest dimension over the temperature range from 30°C to 230°C. The investigations were performed based on length change, which can be found using the following equation;

$$S_i = \frac{L_0 - L_i}{L_0} \times 100\% \tag{1.6}$$

Where S_i is the instantaneous length change, L_0 is the initial length before the start of the experiment, and L_i is the immediate length of the sample at any temperature, as shown in Fig. 1.12.

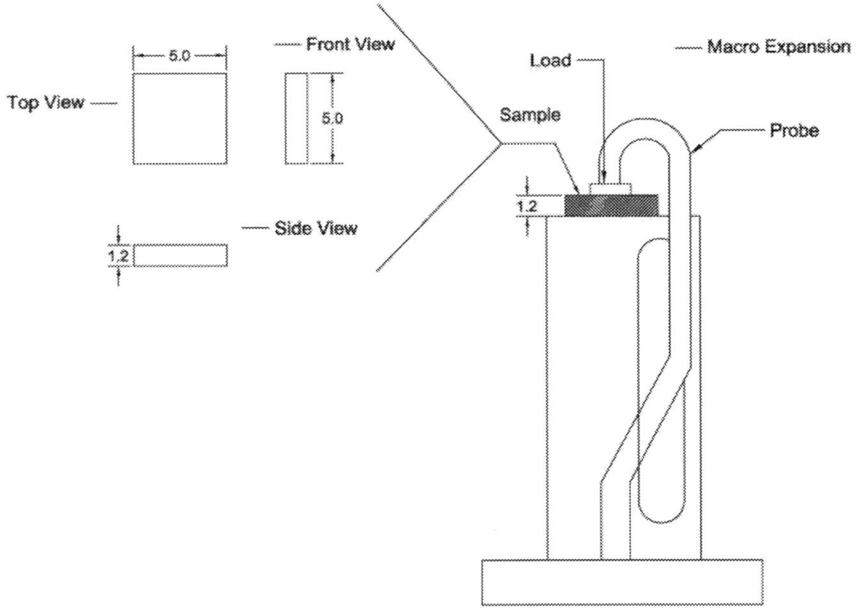

Fig. 1.12 Thermomechanical analysis (TMA) technique setup [30].

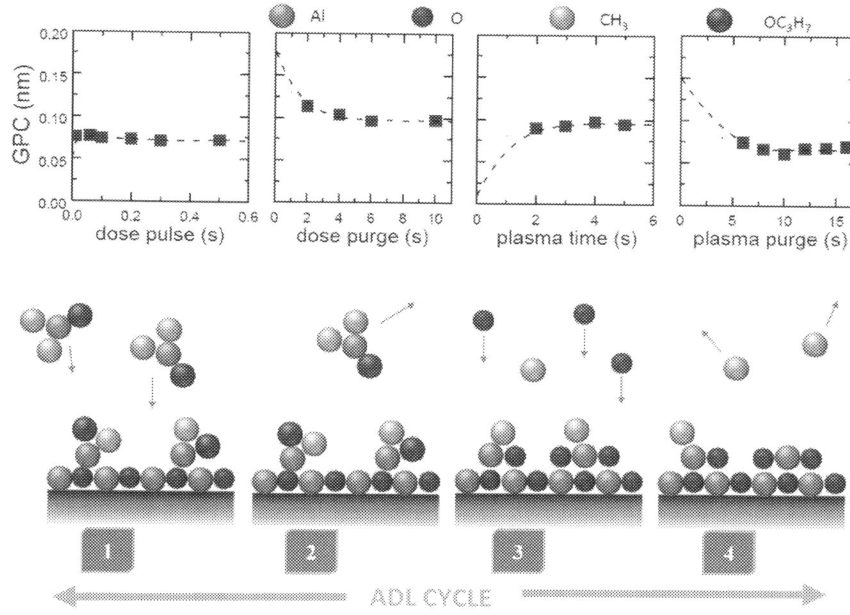

Fig. 1.13 Chemical method ADL cycle atomic layer deposition.

1.3.2 Chemical methods

Fig. 1.13 showed the chemical methods are transforming enzymes, implanting various monomer, action with dissimilar substances with the use of supercritical carbon dioxide, sol-gel technique, layer-by-layer deposition such O, CH_3, and OC_3H_7. High pH treatment, saline combination, saline-alkaline action, maleic anhydride coupling, and maleic anhydride coupling over alkaline treatment are the other chemical methods [3, 31].

1.3.3 Electrochemical methods

The term electrochemical denotes the relationship between chemicals, their reactions, and electricity. Ascertain chemical reactions can create electricity that can galvanize particular chemical reactions that are possible to stimulate otherwise.

When electricity is produced through a battery, two metals are involved in a solution that is chemical-based (Fig. 1.14). When the metals react chemically, electrons are discharged from one metal to another. The battery is linked to the metals from both sides. The end of the battery, which discharges more electrons, gains the charge of protons (positive) while the

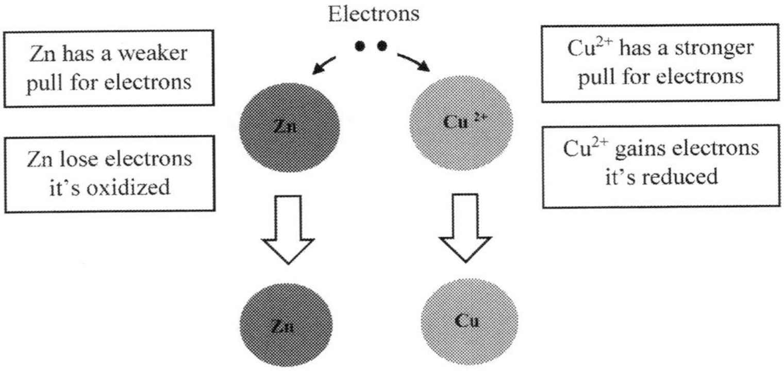

Fig. 1.14 Example of Cu and Zn as certain chemical reactions can create electricity.

other end gets a negative charge on it. The flow of electrons is processed through the wire, if both the ends of the battery are linked together (Fig. 1.15). Thus, a balance of electrical charge is maintained.

Once a metal is dipped into the chemical solution, i.e., suspension of sulfuric acid, it begins to dispense electrons. Moreover, all the metals are possessed with varying levels (more or less) of dispensing electrons. For instance, if the strips of copper and zinc metals are dipped into a watery solution that contains sulfuric acid, the metal of coper will show a greater tendency to dispense more electrons [32] (Fig. 1.16).

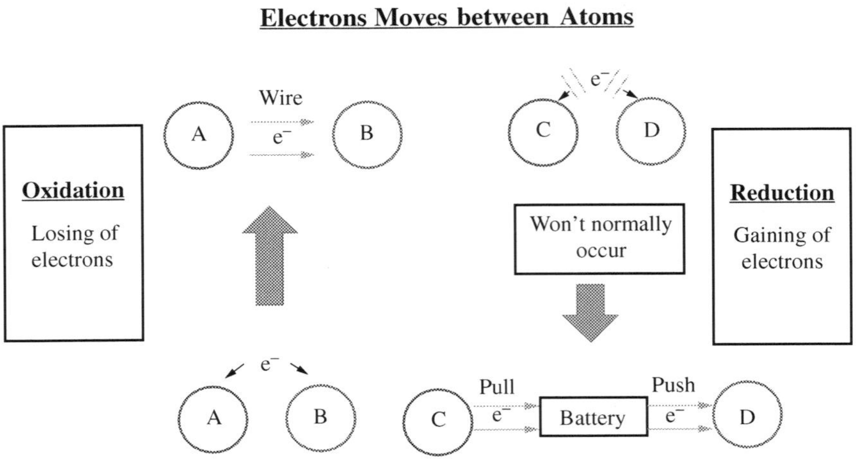

Fig. 1.15 The relationship between chemical reactions and electricity.

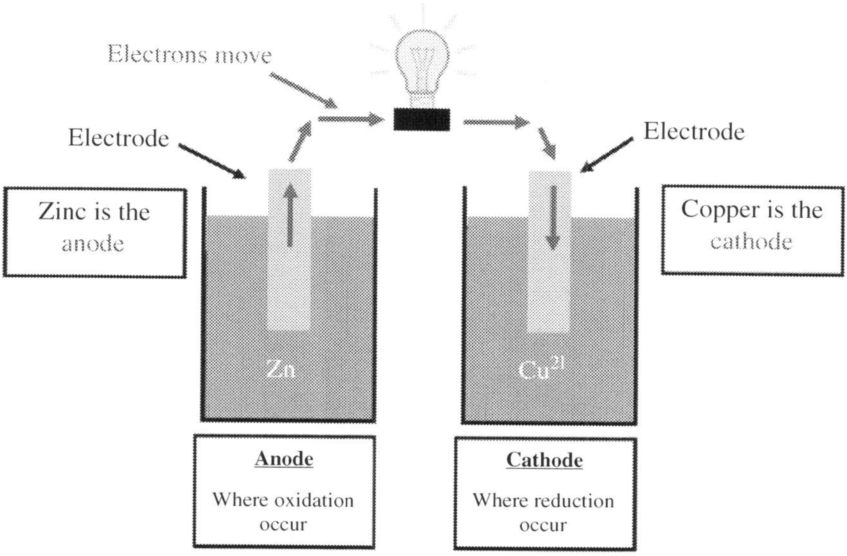

Fig. 1.16 Schematic certain chemical reactions can create electricity.

Standard electrode potential (E°) [33], erstwhile recognized as "standard reduction potential" denotes the half-reaction is presented by:

$$Ox + ne^{-} = Red \tag{1.7}$$

Where n is an integer, which is denoted as 1, the standard state is identically standardized (activity coefficient = 1) as 1 M concentration for alluding aqueous species. Hence, when H^{+} takes place in the half-reaction, the standard state is reported as pH 0. The species that are in the form of gases at 298 K, the rate of the standard is fractional coercion, is determined as 105 Pa [33]. The electrode potentials that are nonstandard are symbolized at $E^{\circ\prime}$ [34].

Electrochemical thermodynamics are central to the chemical potential (μ), which is equivalent to the molar Gibbs energy (G) for a pristine material and for the energy of unfettered fractional molar Gibbs for preparing a solution component [35]. In order to prepare a solution of species A:

$$\mu_A = \mu_A^{o} + RT \ \ln\left(\frac{\gamma C}{C_o}\right) \tag{1.8}$$

Wherein C alludes concentration. Moreover, with a circle in superscript, it refers to standard state's concentration. While a and γ nominate the activity and its coefficient. Usually, the standard state of the gas phase has 1 atm or 1 bar

(0.987 atm) pressure. In contrast with standard sates for liquid phase have a quantity of 1 M or 1 molal [35]. Notably, the activity which has been referred, has no dimension of activity coefficients γ_i [36] which are represented by:

$$a_{i=\gamma_i} \frac{C_i}{C_i^o} \tag{1.9}$$

While applying Eq. (1.10) on the reaction:

$$Ox + e^- \rightarrow Red \tag{1.10}$$

Ox is construed as oxidant, and Red is represented as reductant. Herein, the reaction of free molar energy is presented by

$$\Delta G = G^\circ + RT \ \ln Q = G^\circ + RT \ \ln Q\left(\frac{a_{Red}}{a_{0x}}\right) \tag{1.11}$$

Q alludes the reaction of quotient, which as dimensionless. The relation of excessive electric work and free energy is exhibited as electrode potential *E* of a half cell [37] is

$$\Delta G = -nFE \tag{1.12}$$

Here *F* denotes to the Faraday constant (96,485 C mol^{-1}) while *n* nominates the quantity of electrons, involved in half-reaction. While integrating it with Eq. (1.12), Nernst equation is obtained [38].

$$E = E^\circ \frac{RT}{F} \ \ln \left(\frac{a_{Red}}{a_{0x}}\right) \tag{1.13}$$

Where E° is the standard electrode potential, interchangeably known as half-cell potential, in the case of the equal proportion of all species, *E* equals E°. While this standard state is not found usually in practice, thus repositioned by formal potentials, $E^{\circ\prime}$. More specifically, the formal potential is termed as conditional potentials for alluding the unspecified circumstances in spite of under standard conditions [36].

Conspicuously, the measurement of this quantity is the half-cell where the proportion of the accumulated quantity of oxidizing and diminished species is uniformed. On the other hand, the rest of the specified materials are positioned at a selected concentration. Probably, they are described to match the half-cell potentials at the time of concentrating the quotients (Q_c) in the equation of Nernst [35].

$$E = E^{\circ\prime} \frac{RT}{F} \ \ln Q_c E^{\circ\prime} \frac{RT}{F} \ \ln \left(\frac{C_{ox}}{C_{Red}}\right) \tag{1.14}$$

Subsequently, the formal potential ($E^{\circ\prime}$) is linked to the standard reduction potential (E°) which is demonstrated as [35]:

$$E = E^{\circ\prime} \frac{RT}{F} \ln\left(\frac{\gamma_{ox}}{\gamma_{Red}}\right) \tag{1.15}$$

Fig. 1.17 shows the standard reduction potentials in an aqueous solution at 25°C.

Electrochemical methods are used to eliminate and recover heavy metals. The process is applied to the principle, according to which metals transform

Stronger oxidizing agent ↑ / Weaker reducing agent ↓

Half-reaction			E°(V)
$F_2(g) + 2e^-$	→	$2F^-$	2.87
$Co^{3+} + e^-$	→	Co^{2+}	1.82
$Au^{3+} + 3e^-$	→	$Au(s)$	1.50
$Cl_2(g) + 2e^-$	→	$2Cl^-$	1.36
$O_2(g) + 4H^+ + 4e^-$	→	$2H_2O(l)$	1.23
$Br_2(l) + 2e^-$	→	$2Br^-$	1.07
$2Hg^{2+} + 2e^-$	→	Hg_2^{2+}	0.92
$Hg^{2+} + 2e^-$	→	$Hg(l)$	0.85
$Ag^+ + e^-$	→	$Ag(s)$	0.80
$Hg_2^{2+} + 2e^-$	→	$2Hg(l)$	0.79
$Fe^{3+} + e^-$	→	Fe^{2+}	0.77
$I_2(s) + 2e^-$	→	$2I^-$	0.53
$Cu^+ + e^-$	→	$Cu(s)$	0.52
$Cu^{2+} + 2e^-$	→	$Cu(s)$	0.34
$Cu^{2+} + e^-$	→	Cu^+	0.15
$Sn^{4+} + 2e^-$	→	Sn^{2+}	0.15
$S(s) + 2H^+ + 2e^-$	→	$H_2S(g)$	0.14
$2H^+ + 2e^-$	→	$H_2(g)$	0.00
$Pb^{2+} + 2e^-$	→	$Pb(s)$	-0.13
$Sn^{2+} + 2e^-$	→	$Sn(s)$	-0.14
$Ni^{2+} + 2e^-$	→	$Ni(s)$	-0.25
$Co^{2+} + 2e^-$	→	$Co(s)$	-0.28
$Tl^+ + e^-$	→	$Tl(s)$	-0.34
$Cd^{2+} + 2e^-$	→	$Cd(s)$	-0.40
$Cr^{3+} + e^-$	→	Cr^{2+}	-0.41
$Fe^{2+} + 2e^-$	→	$Fe(s)$	-0.44
$Cr^{3+} + 3e^-$	→	$Cr(s)$	-0.74
$Zn^{2+} + 2e^-$	→	$Zn(s)$	-0.76
$Mn^{2+} + 2e^-$	→	$Mn(s)$	-1.18
$Al^{3+} + 3e^-$	→	$Al(s)$	-1.66
$Be^{2+} + 2e^-$	→	$Be(s)$	-1.70
$Mg^{2+} + 2e^-$	→	$Mg(s)$	-2.37
$Na^+ + e^-$	→	$Na(s)$	-2.71
$Ca^{2+} + 2e^-$	→	$Ca(s)$	-2.87
$Sr^{2+} + 2e^-$	→	$Sr(s)$	-2.89
$Ba^{2+} + 2e^-$	→	$Ba(s)$	-2.90
$Rb^+ + e^-$	→	$Rb(s)$	-2.92
$K^+ + e^-$	→	$K(s)$	-2.92
$Cs^+ + e^-$	→	$Cs(s)$	-2.92
$Li^+ + e^-$	→	$Li(s)$	-3.05

Weaker oxidizing agent / Stronger reducing agent

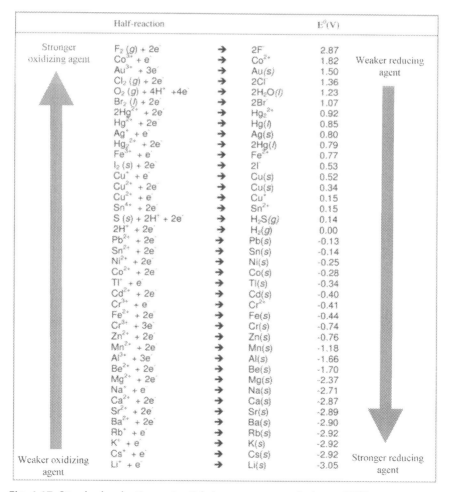

Fig. 1.17 Standard reduction potentials in an aqueous solution at 25°C.

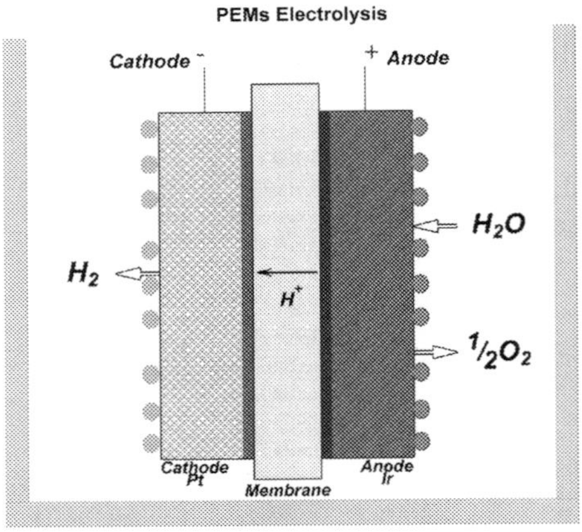

Fig. 1.18 Schematic diagram of PEMs electrolysis cell membrane.

in their fundamental method on the dense conductors once electrodes are applied with electricity. The reduction is divided into three types according to their roles; electro flotation (EF), electrocoagulation (EC), and electrodeposition [39]. The detection of the small anodic current that is developed by oxidizing hydrogen, which permeates from the side of the membrane to the other, is referred to as the classical electrochemical methods. The diffusion of limiting conditions through the membrane is monitored continuously through the electrochemical monitoring technique and is widely used in the characterization of gas permeation rates for PEMs, as shown in Fig. 1.18.

1.4 Application of materials

For various medical applications, different hybrid, synthetic, and biological polymers are used. Numerous polymers are adjustable to meet the requirements of specific applications in terms of physical, chemical, and biological features [40]. The core principle of polymers involves several assemblies of simple structural units to shape a 3D construct and thus includes a widespread distribution of all biological systems. This spans over intracellular filaments and cytoskeleton via essential proteins. They are made of soft extracellular matrix and matrices with mechanical functions in cartilage or ligament to the keratin of hair and skin at the human exterior interface with the environment. Moreover, silk polymers can also be formed by insects for constructing external structures [40].

Human-made synthetic polymers are found abundantly as natural polymers. However, the breakthrough happened in the Second World War. Freshly manufactured polymers are also used in medical applications. For example, the polyesters and polyamides are used as synthetic suture constituents. The demand for synthetic polymers has increased due to its medical as well as technical application for several reasons. Depending upon the monomer units, several physical and chemical properties can be attained through polymerization reaction and development of co-polymers with varied components at modifiable applications [41].

1.4.1 Uses of synthetic materials in medicine

Table 1.1 presents some examples regarding the uses of synthetic polymers in the field of medicine.

Table 1.1 Application of synthetic polymers in medicine.

Synthetic polymers	Full form	Application	Reference
BTHC	Butyryl-trihexyl-citrate	The alternative plasticizer of PVC in blood bags	[42]
EVAL	Ethylene-vinyl alcohol copolymer	Hemodialysis membrane component	[43]
HXPE	Highly cross-linked PE	Obtained by gamma sterilization of UHMWPE	[44]
DEHP	Di(2-ethylhexyl) phthalate	The most frequent plasticizer of PVC	[45]
ePTFE	Expanded PTFE Gore-Tex	This type of polymer is generally utilized for repairing surgical meshes, vascular grafts, tendon, and ligament mending	[46]
EVAL	Ethylene-vinyl alcohol copolymer	Hemodialysis membrane component	[43]
HDI	Hexamethylene diisocyanate	Diisocyanate for polyurethane formation	[47]
HDPE	High-density PE Stiff polyolefin	It is utilized for packing, as an implant for craniofacial contour augmentation, or internal coating of catheters	[48]

Continued

Table 1.1 Application of synthetic polymers in medicine—cont'd

Synthetic polymers	Full form	Application	Reference
PCL	Poly(caprolactone diol)	Diol for polyurethane formation	[47]
IPDI	Isophorone diisocyanate	Diisocyanate for polyurethane formation	[47]
PA	Poly(amide) Nylon	Used as suture material, the balloon of tubes, tendon and ligament mending, etc.	[49]
PEG	Poly(ethylene glycol)	They are used as an antifoaming coating on tubes or as pore former material in dialysis sheaths	[43]
PDMS	Poly (dimethylsiloxane)	Silicones are highly torpid elastomer, used for catheters, plastic surgical treatment, intraocular lenses, as devices for glaucoma drainage and dialysis membranes	[50]
PEEK	Polyether ether ketone	Solid polymer to use for orthopedic purposes or internal lining of tubes	[51]

Most of the polymer devices are placed outside the body as they are used for the drug packaging, while PTFE or HDPE are generally placed as an inner coating of interventional tubes to develop appropriate sliding on the guide cable. In the application of general surgery, suture constituents and staples are used as common polymers [40]. The key factors for the selection of suture materials include its flexible strength, resistance to tissue, degradability of knots, as shown in Fig. 1.19.

Advanced active polymer wound dressings are generally developed by the inclusion of high adsorption properties to support physiological processes while eliminating detrimental influences. Modern surgical techniques that are used in either robotic surgery or the existing surgical methods of laparoscopy or those held for different body organs such as liver or lungs have a wider application of adhesives, specifically in cases where puncture defects of the needle are problematic [40]. Bone cement acts as the anchorage associated to the bone joint prosthesis that must provide a uniform transfer of load from implant to the bone. This type of application is majorly done

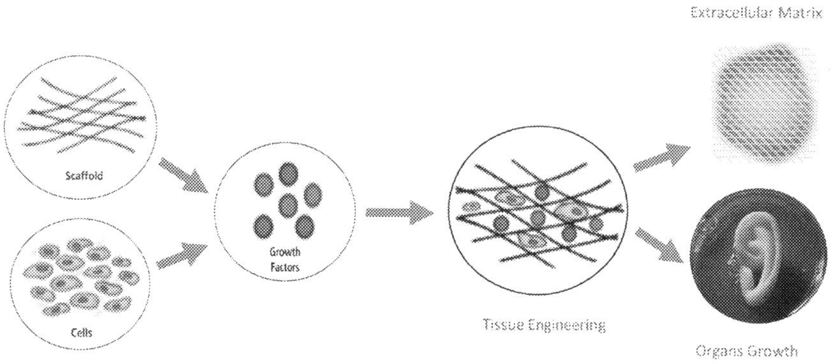

Fig. 1.19 Synthetic materials in tissue engineering applications.

by PMMA [52]. Several materials are incorporated to meet the tendon and ligament defects, specifically in cases where autologous material absent or strong enough [53].

1.4.2 Applications of synthetic materials in construction

Synthetic materials have been widely used in the construction industry on a larger scale for long, as shown in Fig. 1.20. The use of synthetic as well as natural fibers has generated many advantages in terms of mechanical and physical properties of those materials [54]. One of its use is held in the process of making concrete. Fibers augment the properties of concrete. Most fibers are used in preparing concrete slabs-on-ground [55], due to their ability to strengthen the low-tension capacity of concrete [56, 57]. The most used macro plastic fiber includes Polypropylene [57], which is commonly used in mortar and concrete applications to smooth construction while minimizing the shrinkage and cracking of plastic [43]. Asphalt concrete is also a chief element that is highly used in the construction of road infrastructure

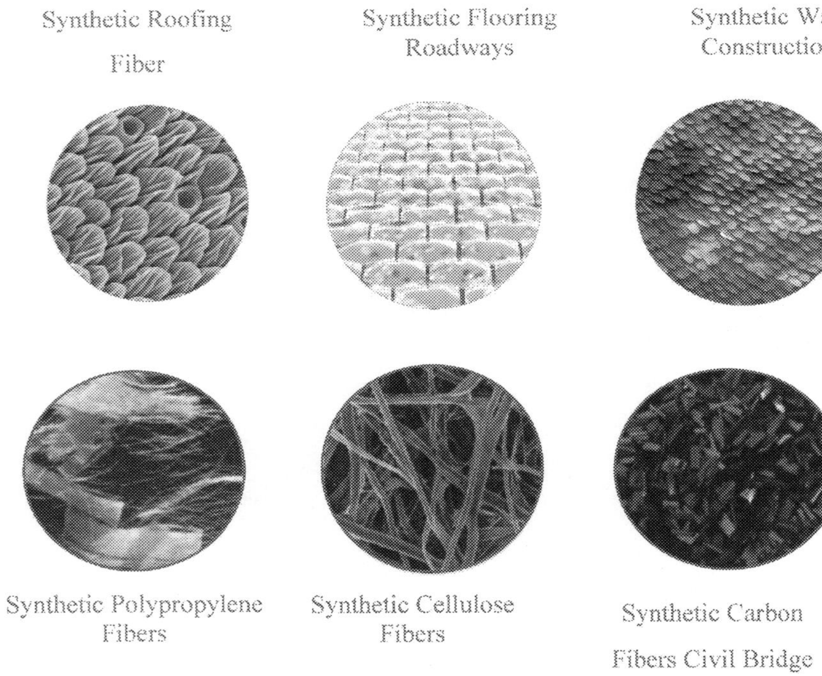

Fig. 1.20 Schematic of synthetic materials in construction applications.

and roadways globally [54]. In addition, stone mastic asphalt (SMA) is a type of mixture [58], which is commonly used to develop resistance toward rutting [59].

Moreover, in similar grades of aggregate and gap-graded SMA mixtures [60], fibers have brought about various advantages, including the lessening of corrosion and enhancement in terms of the stabilization of asphalt [58]. Fibers can further be utilized to enhance the properties of poor soil in terms of ductile capacity, shear strength, compressibility, mass, and hydraulic conductivity [61]. The quality of certain earth material composites that can also be enhanced by fibers includes adobe blocks and compressed earth blocks. The first type is the synthetic sundried blocks, which are chiefly composed of water and earth materials [62, 63]. This also includes carbon fiber composites that possess exceptional mechanical properties and thus have been profoundly used in the automotive industry.

In cases where carbon fibers are utilized as reinforcement, lightweight material is designed. Nevertheless, this is not cost-effective. For instance, the weight of a vehicle body can be minimalized by approximately 40%–60%. Moreover, in space technology, aircraft, defense, and sporting industries, where explicit and highly advanced applications or technology are prerequisites, carbon fibers are used [64]. Also, the Glass fiber-reinforced polypropylene composites have been progressively utilized in the automotive industry as they possess unique mechanical properties. They are easy to construct, lightweight, and cost-effective. These compounds have been used in the preparation of bus bumpers and car seats etc. [65, 66].

1.5 Conclusion

The development of a greater number of synthetic materials has introduced an improvement in the areas of manufacturing and industries with cost-effective techniques. Despite the plethora of new catalysts and development of new synthetic routes through joining small molecules into long polymers chains, the demand for enhancing the quality of life with products that are energy-efficient, sustainable, and quite capable of reducing global pollution is increasing. The above chapter provided a detailed account of the properties, outnumbered application, and significance of synthetic materials as an indispensable part of the era of technological innovation. Various methods and techniques have been described which are used in the preparation of synthetic materials such as emulsion polymerization, cross-linked polymer synthesis technique, reinforcing-filler technique, intercalation technique, in situ polymerization technique, melt compounding technique. It also covers different methods such as; thermo-mechanical methods, chemical methods, and electrochemical methods and discusses the multifold application of synthetic materials in different fields and industries.

References

[1] P. Ball, Synthetic biology—engineering nature to make materials, MRS Bull. 43 (7) (2018) 477–484.
[2] W.S. Hu, Engineering Principles in Biotechnology, John Wiley & Sons, Incorporated, 2018.
[3] M. Simons, The diversity of engineering in synthetic biology, NanoEthics (2020) 1–21.
[4] B. Kolodziejczyk, A. Kagansky, Consolidated G20 synthetic biology policies and their role in the 2030 Agenda for Sustainable Development, G20 Insights (2017).
[5] D.P. Bhattarai, L.E. Aguilar, C.H. Park, C.S. Kim, A review on properties of natural and synthetic based electrospun fibrous materials for bone tissue engineering, Membranes 8 (3) (2018) 62.
[6] A.I. Rezk, A.R. Unnithan, C.H. Park, C.S. Kim, Rational design of bone extracellular matrix mimicking tri-layered composite nanofibers for bone tissue regeneration, Chem. Eng. J. 350 (2018) 812–823.
[7] A.P. Tiwari, M.K. Joshi, C.H. Park, C.S. Kim, Nano-nets covered composite nanofibers with enhanced biocompatibility and mechanical properties for bone tissue engineering, J. Nanosci. Nanotechnol. 18 (1) (2018) 529–537.
[8] I.M. Alarifi, W.S. Khan, R. Asmatulu, Synthesis of electrospun polyacrylonitrile-derived carbon fibers and comparison of properties with bulk form, PLoS ONE 13 (8) (2018) e0201345.
[9] M. Liu, X. Zeng, C. Ma, H. Yi, Z. Ali, X. Mou, Y. Deng, N. He, Injectable hydrogels for cartilage and bone tissue engineering, Bone Res. 5 (1) (2017) 1–20.
[10] K. Gulati, M.K. Meher, K.M. Poluri, Glycosaminoglycan-based resorbable polymer composites in tissue refurbishment, Regen. Med. 12 (4) (2017) 431–457.
[11] H.F. Brinson, L.C. Brinson, Characteristics, applications and properties of polymers, in: Polymer Engineering Science and Viscoelasticity, Springer, Boston, MA, 2015, pp. 57–100.
[12] M.E. Marti, A.D. Sharma, D.S. Sakaguchi, S.K. Mallapragada, Nanomaterials for neural tissue engineering, in: Nanomaterials in Tissue Engineering, Woodhead Publishing, 2013, pp. 275–301.
[13] R.L. Williams, H.J. Levis, R. Lace, K.G. Doherty, S.M. Kennedy, V.R. Kearns, Biomaterials in Ophthalmology, (2019).
[14] D. Distler, W.S. Neto, F.M. Silva, Emulsion polymerization. in: Reference Module in Materials Science and Materials Engineering, 2017. https://doi.org/10.1016/B978-0-12-803581-8.03746-2.
[15] A.N.M.B. El-hoshoudy, Emulsion polymerization mechanism. in: Recent Research in Polymerization, vol. 1, 2018. https://doi.org/10.5772/intechopen.72143.
[16] T. Oyama, Cross-linked polymer synthesis. in: Encyclopedia of Polymeric Nanomaterials, 2014. https://doi.org/10.1007/978-3-642-36199-9_181-1.
[17] M. Tolinski, Additives for Polyolefins: Getting the Most Out of Polypropylene, Polyethylene and TPO. William Andrew, 2015. https://doi.org/10.1016/B978-0-323-35884-2.00015-6.
[18] H.K. Albeladi, A.N. Al-Romaizan, M.A. Hussein, Role of cross-linking process on the performance of PMMA, Int. J. Biosens. Bioelectron. 3 (2017) 279–284.

[19] M.A. Hussein, R.M. El-Shishtawy, B.M. Abu-Zied, A.M. Asiri, The impact of cross-linking degree on the thermal and texture behavior of poly (methyl methacrylate), J. Therm. Anal. Calorim. 124 (2) (2016) 709–717.
[20] A. Pegoretti, A. Dorigato, Polymer composites: reinforcing fillers, in: Encyclopedia of Polymer Science and Technology, 2002, pp. 1–72.
[21] S. Jakopin, Compounding techniques for fiber reinforced properties, Adv. Polym. Technol. 3 (4) (1984) 365–381.
[22] A.J. Jacobson, L.F. Nazar, Intercalation chemistry. in: Encyclopedia of Inorganic and Bioinorganic Chemistry, 2011. https://doi.org/10.1002/9781119951438.eibc0093.
[23] A. Lerf, Storylines in intercalation chemistry, Dalton Trans. 43 (27) (2014) 10276–10291.
[24] A. Malas, Rubber nanocomposites with graphene as the nanofiller, in: Progress in Rubber Nanocomposites, Woodhead Publishing, 2017, pp. 179–229.
[25] E. Pavlacky, D.C. Webster, An in situ intercalative polymerization method for preparing UV curable clay–polymer nanocomposites, J. Appl. Polym. Sci. 132 (39) (2015).
[26] P. Hornsby, Compounding of particulate-filled thermoplastics, in: S. Palsule (Ed.), Polymers and Polymeric Composites: A Reference Series, 2013, pp. 1–16.
[27] P.K. Gallagher, M.E. Brown, R.B. Kemp, Handbook of Thermal Analysis and Calorimetry, Elsevier, New York, 1998.
[28] T. Zhang, C.S. Wu, G.L. Qin, X.Y. Wang, S.Y. Lin, Thermomechanical analysis for laser+ GMAW-P hybrid welding process, Comput. Mater. Sci. 47 (3) (2010) 848–856.
[29] I.M. Alarifi, Fabrication and Characterization of Electrospun Polyacrylonitrile Carbonized Fibers as Strain Gauges in Composites for Structural Health Monitoring Applications, (Doctoral dissertation) Wichita State University, 2017.
[30] I.M. Alarifi, A. Alharbi, W.S. Khan, A.S. Rahman, R. Asmatulu, Mechanical and thermal properties of carbonized PAN nanofibers cohesively attached to surface of carbon fiber reinforced composites, Macromol. Symp. 365 (1) (2016, July) 140–150.
[31] M.S. Bodur, M. Bakkal, H.E. Sonmez, The effects of different chemical treatment methods on the mechanical and thermal properties of textile fiber reinforced polymer composites, J. Compos. Mater. 50 (27) (2016) 3817–3830.
[32] M. Nogueira, A. Black, Basics of solar electricity: photovoltaics (PV), in: Northern California Solar Energy Resource Guide, 2003, pp. 1–4.
[33] T. Renner, Quantities, Units and Symbols in Physical Chemistry, Royal Society of Chemistry, 2007.
[34] D.A. Armstrong, R.E. Huie, S. Lymar, W.H. Koppenol, G. Merényi, P. Neta, … P. Wardman, Standard electrode potentials involving radicals in aqueous solution: inorganic radicals, BioInorg. React. Mech. 9 (1–4) (2013) 59–61.
[35] J. Ho, M.L. Coote, C.J. Cramer, D.G. Truhlar, Theoretical calculation of reduction potentials, Org. Electrochem. (2015) 229–259.
[36] F. Scholz, Electroanalytical Methods, vol. 1, Springer, Berlin, 2010.
[37] G.N. Lewis, M. Randall, K.S. Pitzer, L. Brewer, Thermodynamics, McGrawHill, New York, 1961, 353.
[38] R.S. Berry, A. Stuart, Rice, J. Ross (Eds.), Physical Chemistry, 2000.
[39] J. Koelmel, M.N.V. Prasad, G. Velvizhi, S.K. Butti, S.V. Mohan, Metalliferous waste in India and knowledge explosion in metal recovery techniques and processes for the prevention of pollution, in: Environmental Materials and Waste, Academic Press, 2016, pp. 339–390.
[40] M.F. Maitz, Applications of synthetic polymers in clinical medicine, Biosurf. Biotribol. 1 (3) (2015) 161–176.
[41] A. Lendlein, Polymers in biomedicine. Macromol. Biosci. 10 (2010) 993–997, https://doi.org/10.1002/mabi.201000300.

[42] C.F. Högman, L. Eriksson, Å. Ericson, A.J. Reppucci, Storage of saline-adenine-glucose-mannitol-suspended red cells in a new plastic container: polyvinylchloride plasticized with butyryl-n-trihexyl-citrate, Transfusion 31 (1) (1991) 26–29.
[43] H. Klinkmann, J. Vienken, Membranes for dialysis, Nephrol. Dial. Transplant. 10 (supp3) (1995) 39–45.
[44] M. Slouf, H. Synkova, J. Baldrian, A. Marek, J. Kovarova, P. Schmidt, U. Gohs, Structural changes of UHMWPE after e-beam irradiation and thermal treatment, J. Biomed. Mater. Res. B Appl. Biomater. 85 (1) (2008) 240–251.
[45] J. Sampson, D. De Korte, DEHP-plasticised PVC: relevance to blood services, Transfus. Med. 21 (2) (2011) 73–83.
[46] U. Klinge, J.K. Park, B. Klosterhalfen, The ideal mesh? Pathobiology 80 (4) (2013) 169–175.
[47] P. Ferreira, A.F. Silva, M.I. Pinto, M.H. Gil, Development of a biodegradable bioadhesive containing urethane groups, J. Mater. Sci. Mater. Med. 19 (1) (2008) 111–120.
[48] D. Jenke, Evaluation of the chemical compatibility of plastic contact materials and pharmaceutical products; safety considerations related to extractables and leachables, J. Pharm. Sci. 96 (10) (2007) 2566–2581.
[49] L. Pruitt, J. Furmanski, Polymeric biomaterials for load-bearing medical devices, JOM 61 (9) (2009) 14–20.
[50] P.J. Mackenzie, R.M. Schertzer, C.M. Isbister, Comparison of silicone and polypropylene Ahmed glaucoma valves: two-year follow-up, Can. J. Ophthalmol. 42 (2) (2007) 227–232.
[51] S.M. Kurtz, J.N. Devine, PEEK biomaterials in trauma, orthopedic, and spinal implants, Biomaterials 28 (32) (2007) 4845–4869.
[52] S.M. Kenny, M. Buggy, Bone cements and fillers: a review, J. Mater. Sci. Mater. Med. 14 (11) (2003) 923–938.
[53] U.G. Longo, A. Lamberti, N. Maffulli, V. Denaro, Tendon augmentation grafts: a systematic review, Br. Med. Bull. 94 (1) (2010) 165–188.
[54] A. Mohajerani, S.Q. Hui, M. Mirzababaei, A. Arulrajah, S. Horpibulsuk, A. Abdul Kadir, F. Maghool, Amazing types, properties, and applications of fibres in construction materials, Materials 12 (16) (2019) 2513.
[55] J.R. Roesler, S.A. Altoubat, D.A. Lange, K.A. Rieder, G.R. Ulreich, Effect of synthetic fibers on structural behavior of concrete slabs-on-ground, ACI Mater. J. 103 (1) (2006) 3.
[56] I.M. Alarifi, W.S. Khan, A.S. Rahman, Y. Kostogorova-Beller, R. Asmatulu, Synthesis, analysis and simulation of carbonized electrospun nanofibers infused carbon prepreg composites for improved mechanical and thermal properties, Fibers Polym. 17 (9) (2016) 1449–1455.
[57] S. Yin, R. Tuladhar, F. Shi, M. Combe, T. Collister, N. Sivakugan, Use of macro plastic fibres in concrete: a review, Constr. Build. Mater. 93 (2015) 180–188.
[58] S. Serin, N. Morova, M. Saltan, S. Terzi, Investigation of usability of steel fibers in asphalt concrete mixtures, Constr. Build. Mater. 36 (2012) 238–244.
[59] A.K. Arshad, S. Mansor, E. Shafie, W. Hashim, Performance of stone mastic asphalt mix using selected fibres, Jurnal Teknologi 78 (7-2) (2016).
[60] Y. Huang, R.N. Bird, O. Heidrich, A review of the use of recycled solid waste materials in asphalt pavements, Resour. Conserv. Recycl. 52 (1) (2007) 58–73.
[61] S.M. Hejazi, M. Sheikhzadeh, S.M. Abtahi, A. Zadhoush, A simple review of soil reinforcement by using natural and synthetic fibers, Constr. Build. Mater. 30 (2012) 100–116.
[62] A. Laborel-Preneron, J.E. Aubert, C. Magniont, C. Tribout, A. Bertron, Plant aggregates and fibers in earth construction materials: a review, Constr. Build. Mater. 111 (2016) 719–734.

[63] A. Alharbi, I.M. Alarifi, W.S. Khan, R. Asmatulu, Synthesis and analysis of electrospun $SrTiO_3$ nanofibers with NiO nanoparticles shells as photocatalysts for water splitting, Macromol. Symp. 365 (1) (2016, July) 246–257.
[64] V. Dhand, G. Mittal, K.Y. Rhee, S.J. Park, D. Hui, A short review on basalt fiber reinforced polymer composites, Compos. Part B 73 (2015) 166–180.
[65] S. Suresh, V.S. Kumar, Experimental determination of the mechanical behavior of glass fiber reinforced polypropylene composites, Procedia Eng. 97 (2014) 632–641.
[66] M. Akermi, N. Jaballah, I.M. Alarifi, M. Rahimi-Gorji, R.B. Chaabane, H.B. Ouada, M. Majdoub, Synthesis and characterization of a novel hydride polymer P-DSBT/ZnO nano-composite for optoelectronic applications, J. Mol. Liq. 287 (2019) 110963.

CHAPTER 2

Synthetic polymers

Abbreviations

CDLP	continuous direct light processing
CRP	controlled radical polymerization
DLP	digital light processing
FKM	fluorocarbon elastomers
IARC	International Agency for Research on Cancer
MW	molecular weight
NIOSH	National Institute of Occupational Safety and Health
NLM	National Library of Medicine
NR	natural rubber
OSHA	Occupational Safety and Health Administration
PE	polyethylene
PS	polystyrene
PVC	poly vinyl chloride
ROP	ring-opening polymerization
SLA	stereolithography

2.1 Introduction

The end of the nineteenth century has been marked as the genesis and development of the science of polymer in organic chemistry as a result of discovering several characteristics of various materials when undergone to the process of dispersion and was related to the volume of the molecules. Earlier to 1920, Hermann Staudinger (1881–1965) discovered the elements which are acquired through the process of poly-reactions such as cellulose proteins, natural rubber, as well as various synthetic resins comprised of infinite molecules. These molecules are known as "macromolecules," a term coined by Staudinger. These macromolecular materials or polymers owing to their increased weight of molecules encompass such characteristics whose presence has not been marked into other existing materials.

Moreover, chemical propensity, volume, and the organization of these molecules make them widely known to embrace exceptional mechanical and technical characteristics. These polymers have the ability to exhibit

Synthetic Engineering Materials and Nanotechnology
https://doi.org/10.1016/B978-0-12-824001-4.00008-9

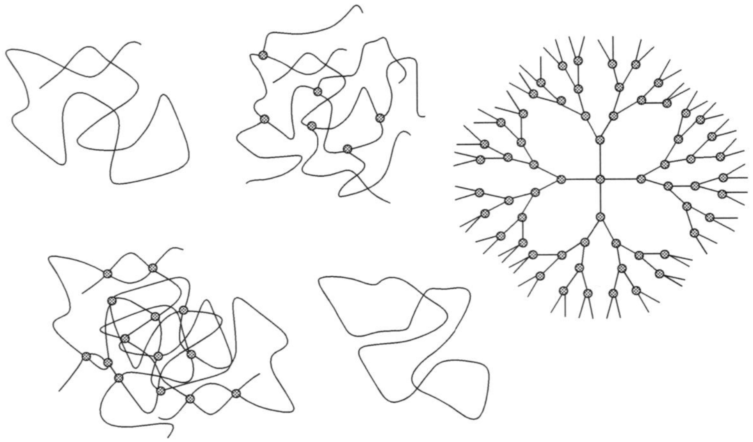

Fig. 2.1 The architecture of macromolecular [4].

the cyclically extended line cuffs, crosslinking, branched as well as hyper-branched and dendritic structural designs [1–3]. As a result of the thermoplastic tendency and the likelihood of crosslinking, the molecules of polymers ease the process of manufacturing numerous commodities, such as: synthetic rubber, the flicks, plastics, smears, and fibers, which have been illustrated in Fig. 2.1 [4].

Hint statement
The uses of synthetic polymers have been augmented for recent years, which has surpassed the use of naturally occurring fibers.

2.1.1 Fundamental structure of polymer

The term polymer refers to the meaning of many units (poly = many; mer = units) where simply structured units that are composed of chemicals reoccur the process for a prolonged period in the molecular structure. A chain polymer molecule is acquired through simplified molecular species, which is denoted as a monomer [5]. Fig. 2.2 exhibits the fundamental structure of polymers.

Furthermore, many units are combined together to form larger units or large molecular structures, and they are called polymers. Thus, the process from monomer to form the polymers is termed as polymerization.

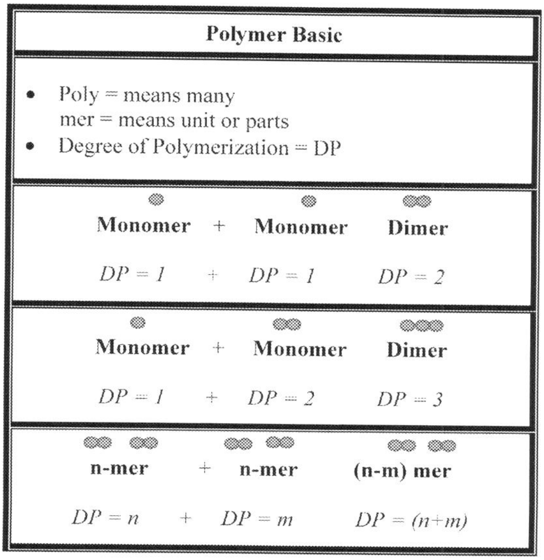
Polymer Basic
Poly = means many
mer = means unit or parts
Degree of Polymerization = DP
Monomer + Monomer Dimer
DP = 1 + DP = 1 DP = 2
Monomer + Monomer Dimer
DP = 1 + DP = 2 DP = 3
n-mer + n-mer (n-m) mer
DP = n + DP = m DP = (n+m)

Fig. 2.2 The basic structure of the polymer.

$$CH_2{=}CH_2$$

$$-CH_2-CH_2-$$

$$n(CH_2{=}CH_2) - (CH_2-CH_2)n$$

Monomer ethane – Polyphone

The foremost molecular materials which have been identified with technical perfunctory had improved chemical features, and cellulose nitrate (Celluloid) or crosslinked casein (Galalith) are to name a few of them. The era of manufacturing the polymers took its roots since 1910 in the form of phenol and formaldehyde (Bakelite). In contrast, the use of styrene or vinyl chloride polymers began from 1930 to the present era. Nowadays, the use of polymers has been outnumbered in terms of production, which is more than 260 million tons annually [4].

Hint statement

The macromolecular materials or polymers owing to their higher weight of molecules encompass such characteristics, whose presence has not been marked into other materials. That exist.

Synthetic polymers denote to the polymers that are manufactured at laboratories by human-made methods. Some of the examples of synthetic polymers are epoxy, synthetic rubber, Teflon, synthetic rubber, polyamides (nylon), Polystyrene (PS), polyvinyl chloride (PVC), and Polyethylene (PE). Usually, the petroleum oil plays an indispensable role in producing synthetic polymers into organized environs. Carbon-carbon bonds are the basic building blocks of synthetic polymers. The pressure is asserted on chemical bonds at high temperatures with a catalytic agent that composes a bond of all the monomers [6].

Catalytic agent increases the speed of the chemical reaction of monomers. Synthetic polymers have now been regarded as a principal tool in millions of uses, which are in the form of thermoplastics, elastomers, thermosets, and synthetic fibers [6]. Synthetic polymers are the principal element of manufacturing the elastomers, fibers, and plastics. The augmented demand and use have attracted the researchers to discover more new dimensions in the field of polymers. Through the databanks of National Library of Medicine (NLM), National Institute of Occupational Safety and Health (NIOSH), International Agency for Research on Cancer (IARC), and Occupational Safety and Health Administration (OSHA), and core principles, basic concepts, features, venomousness and impact, and observation on health have been identified. According to the databank, 1-butene monomers, propylene, and ethylene react as harmful as asphyxiated gases. In addition, environmental hazards have been observed due to polyethylene and polypropylene, which are vulnerable to oxygen. Polyethylene gets crosslinked with oxygen, while polypropylene resists the increase in the weight of molecular compounds. On the other hand, monomer filtrate has not been witnessed to be hazardous to human health [4].

Hint statement

Synthetic polymers have several advantageous characteristics such as higher modulus strengths, tenderness ability of stretching and elasticity, reliance on chemicals, and recyclability.

In order to examine and characterize synthetic polymers, various methods are involved. Some of these methods are also employed in nuclear magnetic resonance, electromagnetic spectroscopy, liquefied chromatography, and gas chromatography. While others are used in the field flow fractionation, mass exclusion chromatography, light sprinkling, and osmometric [4].

2.1.2 Physical characteristics of synthetic polymers

Different kinds of intermolecular forces are the determining factors to generate the physical characteristics of synthetic polymers. The formation of perfunctory groups is based on their composite monomers. For the interface of polymer chains, London forces, dipole-dipole forces, and hydrogen bonds have a significant part to play. The nature of the bond determines the formation of the basic structure of the polymer. A combination of polymers is obtained from polymerization, which varies in terms of their molecular mass and characteristics, where the length of the chains has a principal role.

Crosslinking reactions are employed to alter the characteristics of polymers, and they also constitute bonds between chains. London forces are considered as the feeblest intermolecular forces that are indispensable to the polymer in terms of the length of the chains as they provide initial contact between adjacent chains [4].

2.1.3 Structure of polymers

The structure of molecular chains is determined by the comprehensive procedure of the reoccurrence of the units that is held individually. This reoccurrence is held on the scale of atoms that affect all the characteristics of polymers. Polymers are classified in different ways. Conspicuously, polymers are classified as crosslinked, linear, and branched polymers on the basis of intermolecular connectivity, which is presented in Table 2.1 [7].

Table 2.1 The structure of polymers.

Polymers structure		
Linear polymer	**Branched chain**	**Crosslinking polymer**
Long strength chains	Long chains with branches	3D structures
Close packing	Irregularly packing	Cross linking
High melting points	Low melting points	Insoluble
High tensile strength	Low tensile strength	And infusible
High density	Low density	Various position by covalent bond
Example: Polythene	Example: Nylons	Example: Bakelite

2.1.4 Classification of synthetic polymers

The composition of the polymers laid the foundations to classify them as homopolymers and copolymers. The polymers that contain one kind of unit that continues to occur are called a homopolymer. On the contrary, the polymers that comprise more than a single monomer are known as copolymers [8]. The classification of polymers can be in various ways on the basis of their mechanism, variation in reaction, physical characteristics, and their technological advancements [9]. However, Table 2.2 presents the classification of polymers based on structure, and Fig. 2.3 presents the categorization of polymers in line with their origin.

2.1.5 Modes of polymerization

The methods that are employed to transform polymers from smaller molecules are addition polymerization and condensation polymerization. Addition polymerization is interchangeably known as chain-growth or chain-polymerization, while condensation polymerization termed as step-growth or step-reaction polymerization [10].

2.1.5.1 Additional polymerization

Additional polymers are manufactured through monomers. This process does not involve the loss of even a single molecule, which is smaller

Table 2.2 Classification of synthetic polymers based on structure.

Classification of synthetic polymers			
Elastomers	**Thermoplastic**	**Thermosetting polymers**	**Fibers**
• Weak in the molecular forces • A high degree of elasticity • Rough, tough, and stretchable • Ex. Rubber	• Softens on heating • Hardens on cooling • Remain soluble and fusible • Ex. Polyethylene	• Undergoes permeant changes on heating • Harden, and infusible • Creep resistance • Ex. Bakelite	• Strong in the molecular forces • High tensile strength • Stiff and tough and resistance • Ex. Nylon

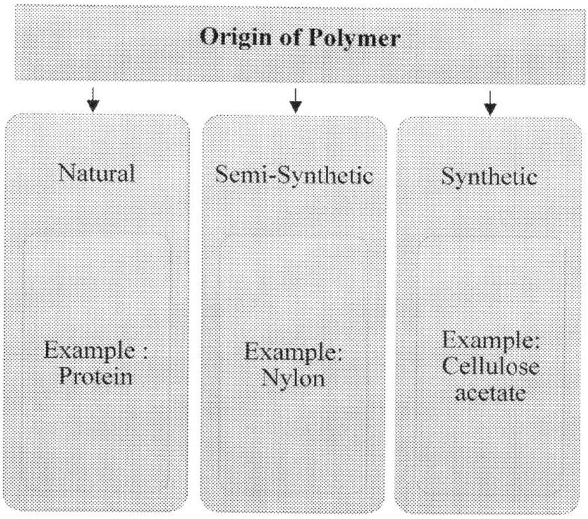

Fig. 2.3 Classification of polymers based on origin.

due to bond opening. Chief addition polymers are produced by polymerization that encompasses monomers which have a carbon and carbon-double bond. These monomers are known as substituted ethylenes and vinyl monomers [11].

Example 2.1

Polythene

$$-CH_2-CH_2- + -CH2-CH_2- = -CH_2-CH_2-CH_2- + -CH_2-CH_2-CH_2- + -CH_2-CH_2-CH_2-$$

Addition Monomers + Addition Monomers + Addition Monomers

A homopolymer is a polymer that is composed of a single kind of monomer and comprises of a unique type of reoccurring unit is a homopolymer. Usually, they are produced through addition polymerization. The monomers must possess single or double bonds. The reoccurring unit is encircled with braces that reflects the chemical mechanism of a homopolymer. For instance, to present "X" as the reoccurring unit of a particular polymer, its structure will be exhibited as –[X]– [12].

Example 2.2

Buna (s) Styrene Butadiene

$n\mathrm{CH_2 = CH\text{-}CH{=}CH_2{+}XC_6CH_5\text{-}CH{=}CH_2 \text{ - } (CH_2\text{-}CH{=}CH\text{-}CH_2)}_n\,\mathrm{(CH_2\text{-}CH\text{-}C_6CH_5)}_x$

Butadiene Styrene co-polymers

Example 2.3

Nylon 6, 6

$n\ \mathrm{H_2\,N(CH_2)_6NH2{+}}n\mathrm{COOH\ (CH_2)4COOH - (NH(CH_2)_6\ NH\text{-}CO(CH_2)_4\ CO)\ n + n\ H_2O}$

Hexamethylene diamine Acetic Acid Nylon 6, 6

Co-polymer: Unlike the homopolymer, a polymer that comprises two or more than two kinds of polymer, is known as copolymer. They have more than two kinds of reoccurring units. Generally, they are produced through condensation polymerization [12].

2.1.5.2 Condensation polymerization

IUPAC described condensation polymerization as a method of polymerization which contains reoccurrence of condensation while emitting molecules of simple size. The materials that are manufacture through the condensation polymerization method are called condensation polymers. An apparent condition process is a determining factor of condensation polymers with cyclic compounds molecular units. It can be illustrated from the example of polyimides which are manufactured by cyclic lactams with no effect on water (Fig. 2.4) [13].

Fig. 2.4 Process of acquiring polyimides [13].

2.1.6 Degree of polymerization

It is an accumulated quantity of monomer units in the polymer, which is exhibited as DP or X_n. It measures the proportion of the molecular mass of a polymer and molecular volume of reoccurring units, as shown in Table. 2.3. The two major kinds of measurement are number average DP and weight average DP as shown in Fig. 2.5. A higher degree of polymerization is required for enhanced mechanical characteristics [14].

$$\text{Degree of Polymerization (DP)} = \frac{M_n}{m} \tag{2.1}$$

Where, M_n = number of average molecular weight, m = repeat unit molecular weight (number of atoms wt.).

Table. 2.3 The molecular mass of molecules in different types.

Number of molecules (N_1)	Mass of each molecule (M_1) (g/mol)	The total mass of each type of molecule (N_1 M_1) in (g/mol)	$M^2{}_1 N_1$
1	800,000	800,000	6.4×10^{11}
3	750,000	2250,000	16.8×10^{11}
5	700,000	3,500,000	24.5×10^{11}
8	650,000	5,200,000	33.8×10^{11}
1	600,000	6,000,000	36×10^{11}
13	550,000	7,150,000	39.3×10^{11}
20	500,000	10,000,000	50×10^{11}
13	450,000	5,850,000	26.3×10^{11}
10	400,000	4,000,000	16×10^{11}
8	350,000	2,800,000	98×10^{11}
5	300,000	1,500,000	4.5×10^{11}
3	250,000	750,000	18.7510^{11}
1	200,000	200,000	0.4×10^{11}
$\Sigma N_1 = 100$		$\Sigma N_1 M_1 = 50{,}000{,}000$ (total masses)	$\Sigma M^2{}_1 N_1 = 370.75 \times 10^{11}$

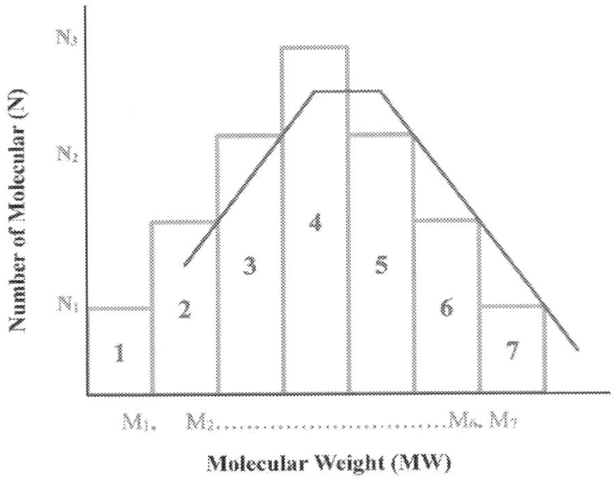

Fig. 2.5 Schematic diagram of the variation of molecular mass.

Example 2.4

Suppose for synthetic Polyvinyl chloride, $M_n = 21.150\,\mathrm{g/mol}$ and then repeat unit molecular weight, $m = 2 \times 12 + 1 + 35 = 60\,\mathrm{g/mol}$.

Solution:

$$DP = \frac{M_n}{m}$$

$$DP = \frac{21,150\,\mathrm{g/mol}}{62\,\mathrm{g/mol}} = 341$$

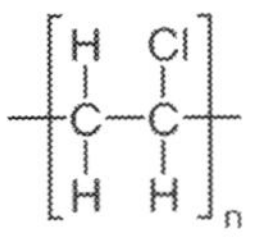

PVC

Example 2.5

$$\text{The number of average molecular weight, } M_n = \frac{N_1 M_1 + N_2 M_2 + N_3 M_3}{N_1 + N_2 + N_3}$$

$$\text{The number of average molecular weight, } M_n = \frac{\Sigma N_1 M_1}{\Sigma N_1} \frac{50,000,000}{100}$$

$$= 500,000 \text{ g/mol}$$

$$\text{The weighted of average molecular weight } M_w = \frac{\Sigma M_1^2 N_1}{\Sigma N_1 M_1}$$

$$= \frac{370.75 \times 10^{11}}{50,000,000} = 741.500 \text{g/mol}$$

$$\text{Polydispersity index} = \text{PI} = \frac{M_w}{M_n} = \frac{741,500 \text{ g/mol}}{500,000 \text{ g/mol}} = 1.48$$

2.1.7 Types of polymers

There are three types of polymers which are described below:

2.1.7.1 Elastomers

Elastomers are the polymers which are connected slackly through crosslinking. They are more flexible and elastic, likewise rubber. Elastomers comprise materials that are twisted and bonded with crosslinking. Therefore, they can be strained and have the ability to revert on their prior shape, when the applied pressure is lifted. Manifold crosslinks cause the enhancement of the firmness and implant the thermoset characteristics in compounds (Fig. 2.6) [15].

These are the polymers that can recover their actual shape if they are slanted amid processing that is physical in nature. While distending the compounds,

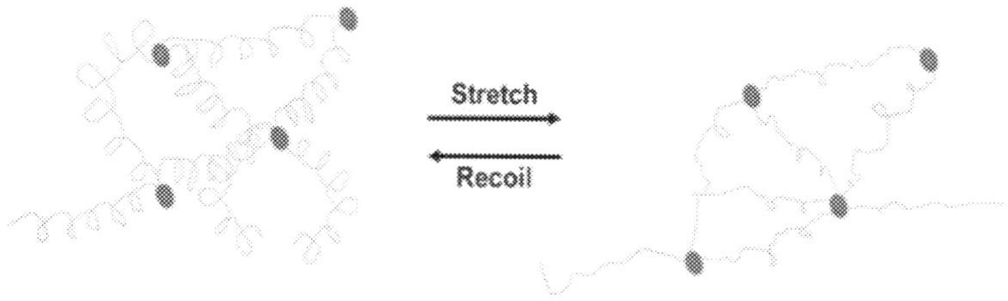

Fig. 2.6 Elastomers of stretch polymers [15].

single chains are warped. Once the compression is unfettered, the molecule regains the actual twisted structure of it. Rubber is an example of an elastomer that comprises mono-saturated units that are isolated through sp-hybridized atoms of carbon, which increases the elasticity of the polymer. Elastomers recover their pristine form once they are slanted while proceeding physically. The single chains are twisted and extended through enlarging the compounds. When the pressure is freed, it causes the return of the original state of the molecules in a twisted form. The most commonly found elastomer is a rubber. Its units are saturated that is detached through sp3-hybridized atoms of carbon, which bring flexibility in polymers [16]. Elastomers are distinct polymers and have excellent alterable latitude along with hysteresis and lesser permanence. Due to their capacity of molecular reactions, crystallinity, and firm restrictions of chains, they are called ideal polymers. The characteristics of elastomers include limited flexibility, reduced scrape resilience to chemicals [16].

It is unique from other compounds that show an elastic response. In the context of Hooke's law, an elastic response refers to the stain, which makes proportionate stress. However, the volume of this strain is very low. For instance, silicate glass has a 0.001% approximate amount of strain. Fig. 2.7 illustrates the comparative curve of strain and stress between natural rubber and polypropylene, a thermoplastic. The cause of alteration in terms of the strength of polymers is evident from the crystallinity of the polypropylene [17].

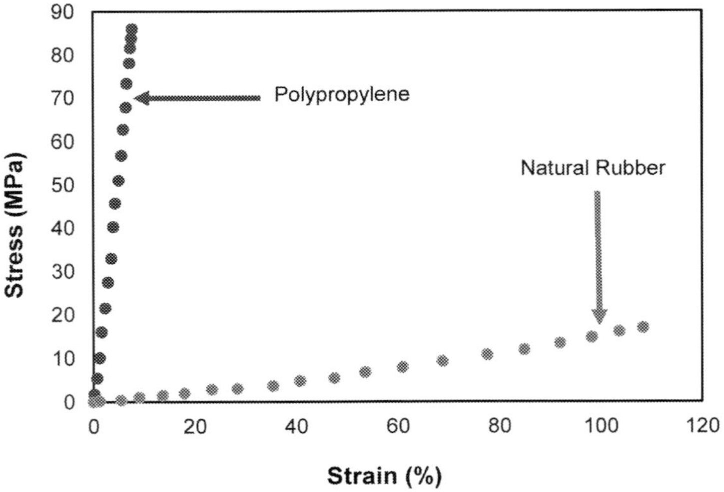

Fig. 2.7 The stress-strain curve of comparing elasticity between a thermoplastic and elastomer.

Elastomers are mostly manufactured with chain-growth polymerization. Since the high weight of molecules is the core element of this method, which is a prerequisite for high chain latitude [17]. Some of the specific elastomers are stated [17] below:

- Natural rubber (NR): This is an elastomer that comprises of *cis*-1,4-isoprene monomer. It is composed of the fluid that excretes from a tree (*Hevea brasiliensis*), which is condensed and combined with the fillers, catalysts of vulcanization, and antioxidants.
- Halogen and Nitrile Substituted Elastomers Polychloroprene: This elastomer is considered as first to be used for commercial purposes. Its composition is based on halogen sodium chloride. The development of Polychloroprene was marked by 1932 amid the technique of emulsion polymerization [17].
- Sulfide Elastomers Polysulfide: It is used for commercial purposes, and its fundamental constituents are the bonds of sulfur atoms. Nathan Mnookin and Joseph C. Patrick manufactured and patented this polymer through the experiment, which developed low-cost antifreeze. The name of Thiokol coined it. The extraordinary chemical resilience for oil and lubricants, along with brilliant nonconductor features [17].
- Fluorocarbon Elastomers (FKM): They are categorized as synthetic copolymers to produce these copolymers, through the atoms of fluorine, hydrogen atoms are detached from hydrocarbons. The chief characteristics include their resilience toward chemicals and heat resistance [17].

2.1.7.2 Plastics

Plastics are the polymers that harden upon cooling. They are further divided into thermosetting plastics and thermosets, which are described below (Fig. 2.8):

(1) Thermosetting plastics or thermosets: These are the plastics that cannot be unstiffened upon high temperature. They are synthetic materials that strengthen upon heating and cannot be reshaped or warmed again, once they encounter the foremost heat-forming, which is known as thermoset materials [19]. Polyethylene is one of the common examples of thermoplastic. These plastics usually have extensively crosslinked chains and can be explained as the polymers which it get harden and cannot be reversed to their actual state upon heating and reacting with some chemicals [20].

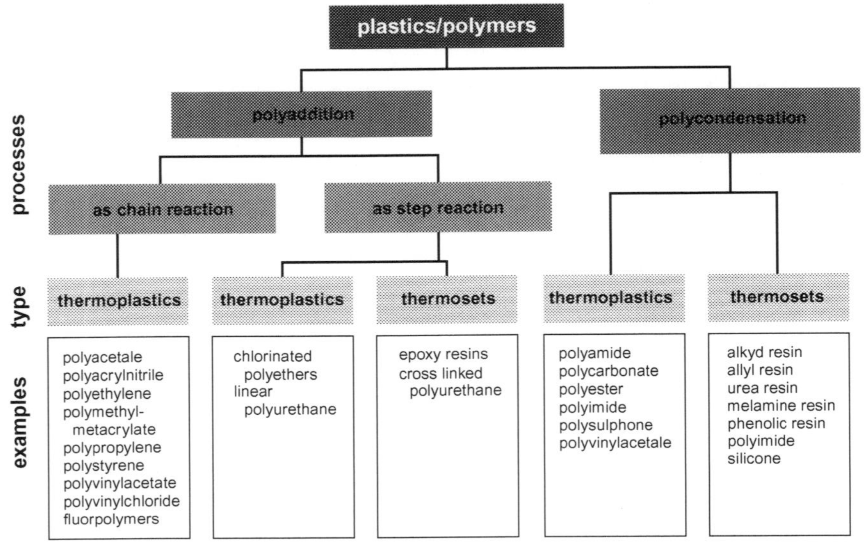

Fig. 2.8 Types of plastic and polymers [18].

Some general properties of the thermosetting or thermosets plastics include:

- Upon heating, it gets hardened and strengthened.
- They are generally found in a liquid state when at room temperature.
- They are irreversible to a prior state, while shape can only be reformed upon reheating.
- The service temperature of materials is 300°C [20].

(2) Thermoplastics: These are the polymers which get softened upon heating. Unlike, thermosets, they are found as solid at room temperature. The thermoplastic compounds can be shaped at a specific temperature while get solidified when chilled [20]. Some of the important properties of thermoplastics are described as under:

- They are normally found in a solid-state when at room temperature.
- They are often reversible to their prior state.
- They are sensitive to high temperatures and loses strength when exposed to heat.
- Upon cooling only, the shape can be changed.
- The service temperature of materials is 150°C.
- They have linear structure Celsius [20].

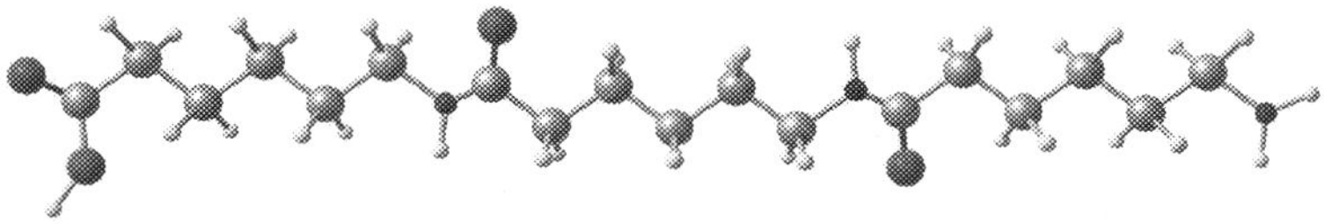

Fig. 2.9 Nylon 6, 6 synthetic fibers chemical structure.

2.1.7.3 Fibers

Fibers are categorized as thermoplastics that are transformed into compounds, i.e., natural fibers. Fibers are produced by passing the melt thermoplastics through minute pores, which makes it hard, as shown in Fig. 2.9. Another method is about dissolving the thermoplastics in an unstable solvent when the solvent begins to vaporize. Despite, the synthetic materials are manufactured from synthetic compounds like petrochemical, some kinds of synthetic fibers are produced from naturally found cellulose which includes rayon, modal, and the newly developed Lyocell. Production of synthetic fibers is cost-effective when produced on a large scale in comparison to other natural fibers [21]. Synthetic fibers are used extensively in manufacturing valued commercial goods such as nylon (2.8) [22], polyester, acrylic, and polyolefin [23].

Hint statement

Synthetic polymers denote to the polymers that are manufactured at laboratories by human-made methods. Some of the examples of polymers are polyvinyl chloride (PVC), synthetic rubber, epoxy, Teflon, polyamides (nylon), Polyethylene (PE), and polystyrene (PS).

2.2 Methods of synthesizing the polymers

There are various methods to synthesize polymers. Some of them include free radical polymerization, condensation polymerization, graft-copolymerization ring-opening polymerization, and photo-polymerization [24].

2.2.1 Free-radical polymerization

The free-radicalization method includes the processing of vinyl monomers as the chemical actions do not stipulate highly purified reactive agents and vigorous omission of all contaminations such as moisture, air, etc. since they require substitute ionic or organometallic catalysts [25]. The process of free-radical polymerization can be in mass, emulsion, diffusion, and fluid [26].

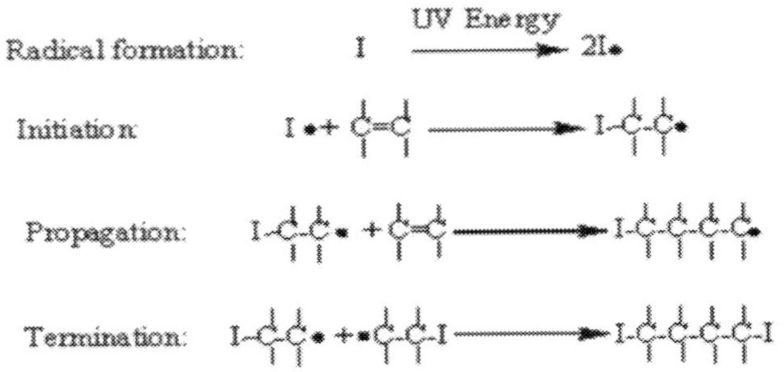

Fig. 2.10 Free radical UV-polymerization [29].

Being well-matched with clusters of function, resilience to water, and acidic solvent facilitates the processing of suspension to ease the experiment. In the meantime, through the free-radical polymerization technique, million tons of the volume of lesser compactness polyethylene, polystyrene, polyacrylate, and methacrylate are manufactured per year. Besides its numerous benefits, it also possesses some of the demerits as the compounds acquired are usually polydisperse, with a poor mechanism of molecular mass and structure [27]. The process of free-radical polymerization comprises three phases: initiation, propagation, and termination. More specifically, in the phase of initiation, radicals are produced by vinyl monomer, which involves radical reactions wherein the production of radicals undergoes rapidly in the form of a developed chain of polymers. Notably, no change in the active center is observed, typically through grouping the radicals of two developing polymer chains or through disproportionation [28]. The transition of free-radical UV-polymerization is illustrated in Fig. 2.10, where photo-initiator engrosses light energy and transform it into radical species. The initiation phase is related to the reaction of radical species where they react with carbon double bonds (—C=C—), which are associated with the acrylate groups of oligomers. Further, during the phase of propagation, the reaction of species is held with acrylate to create radical for proceeding polymerization. Thus, the phase of termination takes place where growing oligomers react that comprise of the double chain (Fig. 2.10) [29].

2.2.2 Condensation polymerization

It is the process of synthesizing polymers through the reoccurring abolition of simply structured molecules. The resultant products are called condensation polymers. The examples of simple condensation by retorting monofunctional reactive agents are esters and amides. While polyesters are the

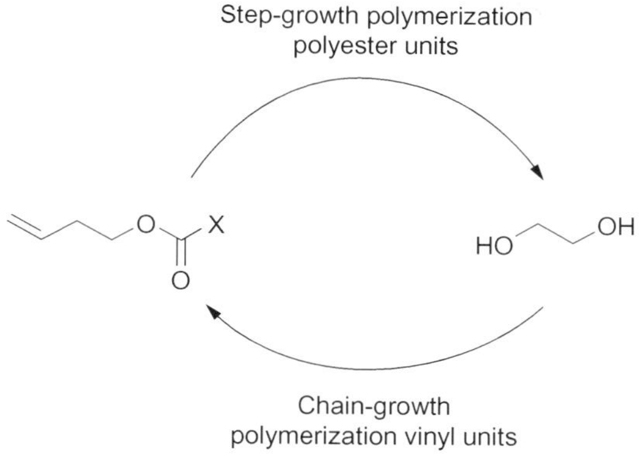

Fig. 2.11 Synthesizing of difunctional monomers by condensation and addition polymerization [32].

consequent of the condensation polymerization, it undergoes the reaction of di-functional reactive agents that also include alcohols or amines [30]. Condensation polymerization is interchangeably called as step-growth polymerization, where minor-sized molecules undergo a process and create broad operational units while emitting minor molecules as a spinoff [31].

The reactions of polymerization are divided into two phases: Chain polymerization and step-growth polymerization. What elements make them distinguished from each other are the experimental approaches and reactions and mechanisms of monomers. Formerly, step-growth polymerization was associated with the creation of polyester by the combination of free-radical and condensation techniques (Fig. 2.11) [32].

2.2.3 Graft polymerization

It is a method where monomers are united covalently and get polymerized into the actual polymer chains. It can be obtained by "grafting to" or "grafting from" methods (Fig. 2.12). The method of "grafting to" involves the perfunctory monomer. While "grafting from" is obtained through acting with a substrate to originate restrained originators. Grafted polymers are used in biomedical, textiles, vehicle manufacturing, electrolyte membranes, coatings, adhesives, laminates, commodity plastics, etc. [34–36].

2.2.4 Photo-polymerization

It is a light-tempted polymerization, which is involved in three-dimension printing when compounds such as radiation-curable resins and photopolymers liquescent placed in a container are treated into the form of sheets.

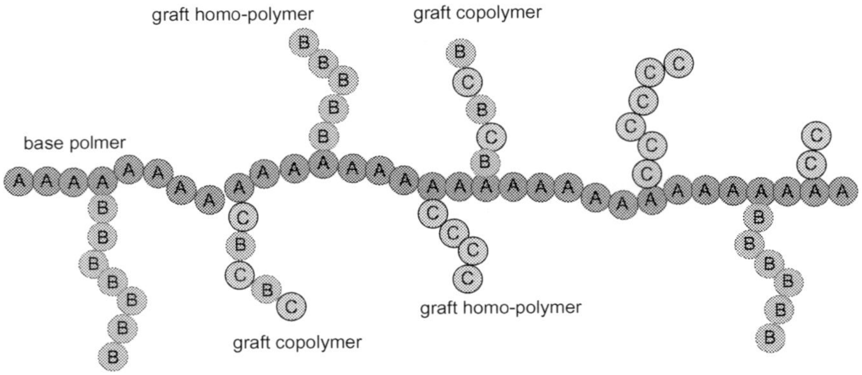

Fig. 2.12 Graft polymerization [33].

This process contains techniques, i.e., direct light processing (CDLP), digital light processing (DLP), and stereolithography (SLA) [33]. Stereolithography is used for the purpose of 3D printing. A movable laser beam is used to cure the photopolymer, which is found in a liquid state. The growth of the object takes place gradually in the form of consecutive layers inside the flask. UV radiation produces a trivial layer of photopolymer, which solidifies photopolymer. Subsequently, the liquid polymer is included to create more layers (Fig. 2.13). Moreover, Digital light processing optimizes the micro mirror projector to

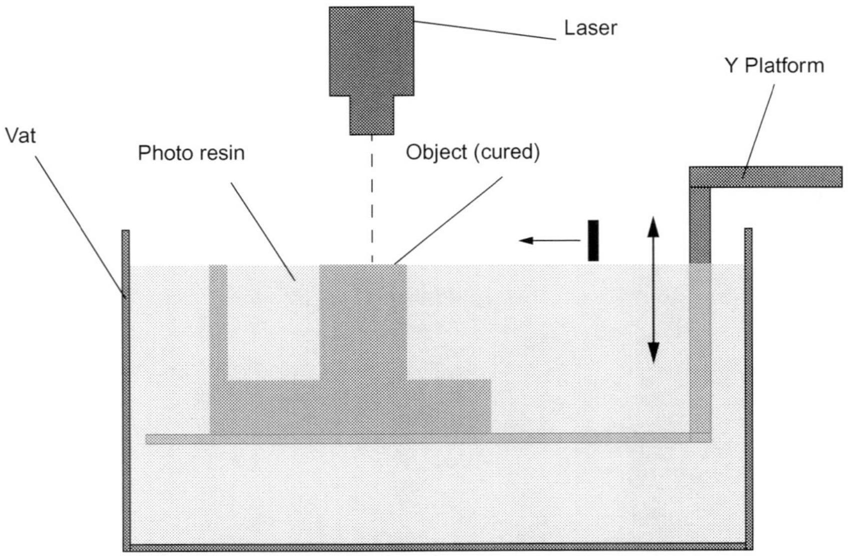

Fig. 2.13 A schematic diagram of stereo-lithography [37].

preserve the photopolymer. Likewise, Continuous digital light processing involves curing the polymer through oxygen and LED [37].

2.3 Techniques of synthesizing the polymers

2.3.1 Ring-opening polymerization (ROP)

It is one of the significant techniques for producing high molecular weight (MW) condensation polymers that are unstable spinoffs. Products that are varying in nature are produced through ring-opening polymerization, such as; epoxide thermosets, polyalkenes (via ring-opening metathesis polymerization), aliphatic polyesters, polyamides, and silicones. It is one of the prominent methods of manufacturing macromolecules, and varieties of polymers are manufactured commercially through this method (Fig. 2.14) [38, 39]. Ring-opening polymerization has been employed as a useful technique to form polymers with diverse and manageable characteristics from preparing human-made materials to naturally found materials, enhancing the biodegradability of polymers in terms of usage for agriculture and medicine [40].

2.3.2 Polymer bioconjugates

A synthetic polymer is conjugated with several biomolecules which aim to reach biohybrid compounds, which is equally useful for both natural and synthetic polymers. More precisely, "grafting to" and "grafting from" approaches are most commonly adopted in comparison with the approach of synthesizing the biomolecule that encompasses with the monomers and has conjugated with various bio-functional sets with the backbone of polymers [41]. Polymers that are reactive to ordinary alteration as the stimulus

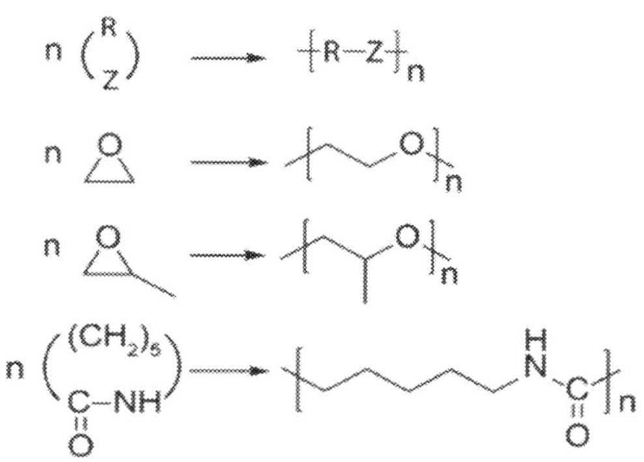

Fig. 2.14 An example of ring-opening polymerization [38].

factor of environs cause greater modifications in physical properties and are termed as intelligent polymers or smart polymers. These are conjugated with several recognition proteins such as protein A, enzymes, and antibodies. These polymers are constituted with the process of random conjugation with lysine amino groups on the exterior level of protein and site-specific conjugation of the polymer, such as groups of cysteine sulfhydryl. They are processed genetically in the form of an amino acid arrangement of protein. The smart polymers are capable of being combined in a physical and chemical manner that conjugates biomolecules to produce various groups of polymer-biomolecule that can provide responses in biological and chemical manner. Biomolecules can be conjugated with proteins and oligopeptides, simple lipids, single- and double-stranded oligonucleotides and DNA plasmids, sugar. Also, various recognition ligands and molecules of the synthetic drug are involved in it.

Probably, polyethylene glycol is conjugated to transform into an intelligent polymer backbone while implanting the stealth features (Fig. 2.15).

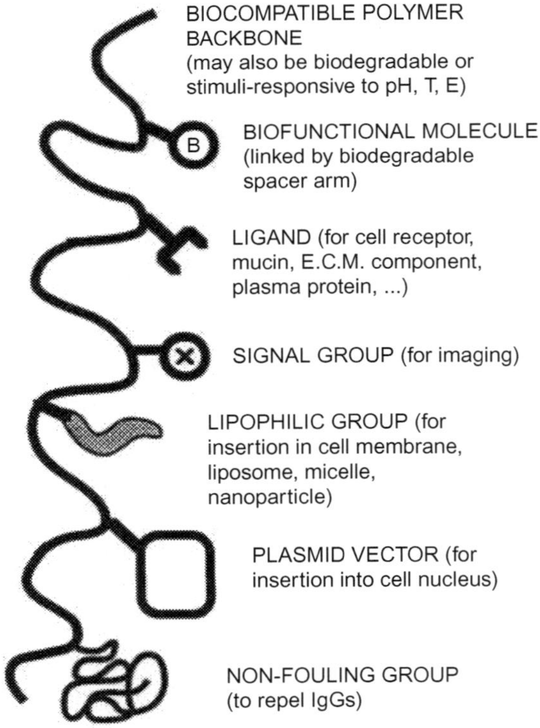

Fig. 2.15 Some of the natural and synthetic polymers that can be undergone to the conjugation to produce smart polymers [42].

The process of conjugating the biomolecules and polymers resultant to a new hybrid molecule that is capable of blending the characteristics of two compounds in a synergistic manner, in order to implant unprecedented features [42].

2.3.3 Controlled/living radical polymerization

Controlled radical polymerization assists in having control over the structure and design of the compound. It also facilitates in producing surface-tethered copolymers and organic or inorganic compounds [43]. It is further divided into three sets of techniques, which are Nitroxide-mediated Polymerization (NMP), Atom Transfer Radical Polymerization (ATRP), and Reversible Addition/Fragmentation Chain Transfer Polymerization (RAFT). CRP is related to *t* multiexecution of dimensional vinyl monomers, which are used for various purposes. In addition to it, it capacitates the new material designs and provides access to synthetic métier owing to the firmness that is implanted during polymerization [44].

The emergence of CLRP has provided convenient ways of producing the predefined polymers of prearranged narrow molecular weight distribution (MWD) and molecular weight (MW), and other several block copolymers that have intricated the macromolecular structure (Fig. 2.16). Before the progress of CLPR, these characteristics were not feasible to acquire [45].

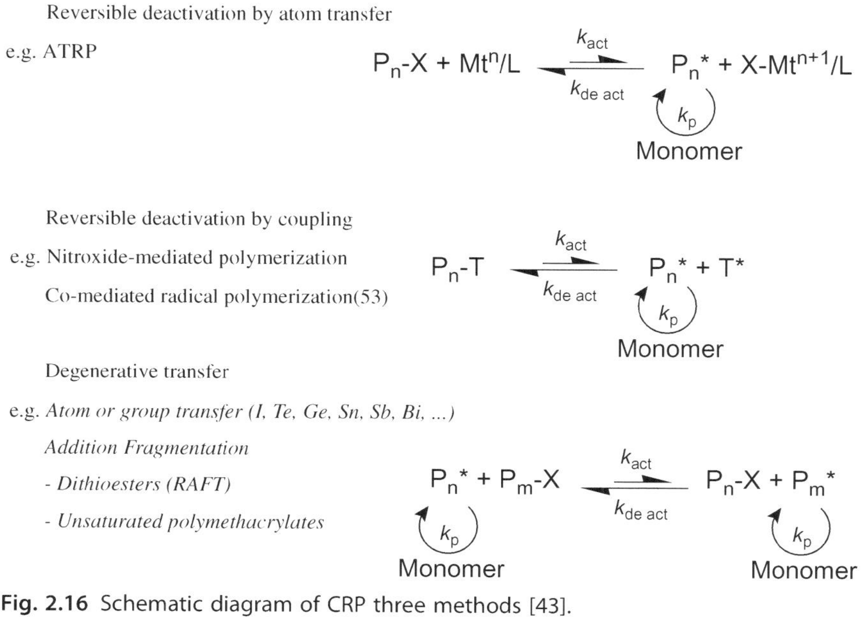

Fig. 2.16 Schematic diagram of CRP three methods [43].

2.4 Different applications of synthetic polymers

Nowadays, polymers are abundantly used all over the world, and they are the principal elements of manufacturing all the devices. For example, computer chips, laptops, all types of mobile phones, and other electronic gadgets are manufactured by synthetic polymers [46–48]. In the process of microlithography, synthetic polymers play an indispensable role and work as an organic light emitter diode. They also serve as the state-of-the-art tools used in biomedical in order to synthesize bone scaffoldings and ligaments.

Moreover, in delivering drugs through tubes, flasks for circulatory dispensation of medicines, synthetic polymers are used [49]. The internal parts of all mobile phones are composed of synthetic polymers. It is further used in the manufacturing of automobile parts. In some of the energy-saving Dreamliner and A380 aircraft (Fig. 2.17) [50], which are lighter in mass and composed of strengthened noncomposite compounds, synthetic polymers are the principal materials to form [51, 52]. Other significant uses of synthetic polymers include its usability as heat resistant materials, sealants, epoxy resin, and coats, which are related to the construction industry [49].

Smart polymers have now been manufactured to be used in sensor tools as an external stimulus. More new uses of synthetic polymers are in transforming the memory tools and the tools which are crucial in heal-up systems. Briefly, synthetic polymers occupy the greatest and undeniable importance in the present era. This can be illustrated from the fact that the volume of producing synthetic polymer products has been surpassed 200 million tons annually. However, half of the chemists in the United States, Japan, and Western Europe who are associated with polymers, have its inadequate literacy, as they have not taken most of the other distinct characteristics of the polymers and they also have no apprehension of how to monitor molecular structures [49]. Other than the biomedical and pharmacological uses of synthetic polymers, they are also used in internal prosthetic implants, suture materials, and drug carters [50].

2.5 Conclusion

Synthetic polymers have been revolutionized the several functions and processes of human development owing to the different favorable properties they encompass. Several synthetic polymers are being used in biomedical applications, construction industry automobile industry, and so on. The above chapter contains the realization and commercial adoption of synthetic polymers and their enhanced application over naturally found materials. It

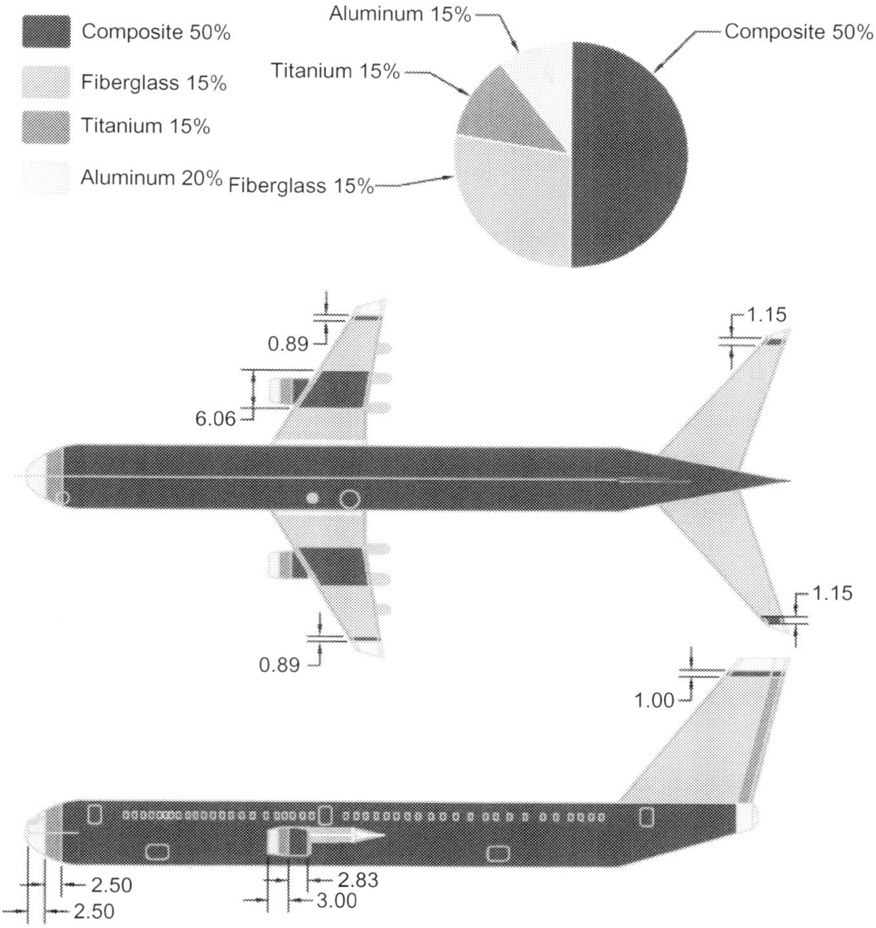

Fig. 2.17 Usage of different synthetic polymers in the Boeing [50].

also states the characteristics of synthetic materials as well as offers a brief view of the types of synthetic materials. Finally, discussion regarding different methods involved in the preparation of synthetic materials and their wide-range applications world-wide is also provided.

References

[1] I.M. Alarifi, A. Alharbi, O. Alsaiari, R. Asmatulu, Training the engineering students on nanofiber-based SHM systems, in: American Society for Engineering Education (ASEE), Zone III Conference, 2015.

[2] A.A. Abd-Elhady, H.E.D.M. Sallam, I.M. Alarifi, R.A. Malik, T.M. El-Bagory, Investigation of fatigue crack propagation in steel pipeline repaired by glass fiber reinforced polymer, Compos. Struct. 242 (2020) 112189.
[3] A.R. Alharbi, I.M. Alarifi, W.S. Khan, A. Swindle, R. Asmatulu, Synthesis and characterization of electrospun polyacrylonitrile/graphene nanofibers embedded with $SrTiO_3$/NiO nanoparticles for water splitting, J. Nanosci. Nanotechnol. 17 (8) (2017) 5294–5302.
[4] D. Braun, H. Cherdron, M. Rehahn, H. Ritter, B. Voit, Polymer Synthesis: Theory and Practice: Fundamentals, Methods, Experiments, Springer Science & Business Media, 2012.
[5] P. Ghosh, Polymer Science: Fundamentals of Polymer Science, Polymer Study Centre, Kolkata, 2006.
[6] S. Podzimek, Polymers| Synthetic, 2019, Available from: https://www.sciencedirect.com/topics/materials-science/synthetic-polymer.
[7] A. Shrivastava, Introduction to Plastics Engineering, William Andrew, 2018.
[8] S.S. Desale, J. Zhang, T.K. Bronich, Synthetic polymer-based nanomaterials, in: Nanomaterials in Pharmacology, Humana Press, New York, NY, 2016, pp. 1–26.
[9] S.V. Bhat, Synthetic polymers, in: Biomaterials, Springer, Dordrecht, 2002, pp. 51–71.
[10] F.W. Harris, Introduction to Polymer Chemistry, 1981.
[11] T.A. Saleh, V.K. Gupta, Nanomaterial and Polymer Membranes: Synthesis, Characterization, and Applications, Elsevier, 2016.
[12] Yashoda, Difference Between Homopolymer and Copolymer, 2016, Available from: https://pediaa.com/difference-between-homopolymer-and-copolymer/.
[13] Condensation Polymerization, 2011. Available from: https://onlinelibrary.wiley.com/doi/abs/10.1002/0471440264.pst072.
[14] M. Chanda, Plastics Technology Handbook Fourth Edition, 2007.
[15] V.R. Sastri, Plastics in Medical Devices: Properties, Requirements, and Applications, William Andrew, 2013.
[16] J.B. Campbell, Synthetic elastomers, Science 141 (3578) (1963) 329–334.
[17] W. Bailus Jr., D.S. Lynette, Synthetic Polymers, 2012, Available from: https://onlinelibrary.wiley.com/doi/abs/10.1002/0471435139.tox088.pub2.
[18] D. Braun, Kunststoffe: Kunststoff-Kompendium. Von A. Franck und K.-H. Biederbick. Vogel-Buchverlag Würzburg 1984. 1. Aufl., 346 S., 100 Abb., 54 Tab., DM 48,–. ISBN 3-8023-0135-8, Nachrichten Chem, Tech. Lab. 33 (5) (1985) 414–417.
[19] R.A. Shanks, General purpose elastomers: structure, chemistry, physics and performance, in: Advances in Elastomers I, Springer, Berlin, Heidelberg, 2013, pp. 11–45.
[20] M. Shafi, Types of plastics PDF-properties of thermosetting plastics, in: Thermoplastics & Glass Cutting, 2018. Available from: https://mechanicalstudents.com/properties-of-thermosetting-plastics-thermoplastics-types-of-plastics/.
[21] OpenLearn. (n.d.). Introduction to Polymers. Retrieved from: https://www.open.edu/openlearn/science-maths-technology/science/chemistry/introduction-polymers/content-section-1.2.2.
[22] E.A. Kamel, Synthetic Fibers: The Manufacturing Process and Risks to Human and Environment, 2020, Available from: https://owlcation.com/misc/Synthetic-Fibers-The-Manufacturing-Process-and-Risks-to-Human-and-Environment.
[23] A. El Nemr, From natural to synthetic fibers, in: Textiles: Types, Uses and Production Methods, Nova Science Publishers, Inc, Hauppauge, NY, 2012, pp. 1–152.
[24] S. Fibres, in: J.E. McIntyre (Ed.), Nylon, Polyester, Acrylic, Polyolefin, 2004.
[25] H. Holback, Y. Yeo, K. Park, Hydrogel swelling behavior and its biomedical applications, in: Biomedical Hydrogels, Woodhead Publishing, 2011, pp. 3–24.

[26] A. Rudin, P. Choi, The Elements of Polymer Science and Engineering, Academic press, 2012.
[27] K. Matyjaszewski, Atom Transfer Free Radical Polymerization, 2001, Retrieved from https://www.sciencedirect.com/topics/materials-science/atom-transfer-radical-polymerization.
[28] C.J. Hawker, M.E. Piotti, E. Saldívar-Guerra, Nitroxide-Mediated Free Radical Polymerization, 2016.
[29] R. Tajau, M.I. Ibrahim, N.M. Yunus, M.H. Mahmood, M.Z. Salleh, N.G.N. Salleh, Development of palm oil-based UV-curable epoxy acrylate and urethane acrylate resins for wood coating application, AIP Conf. Proc. 1584 (1) (2014) 164–169.
[30] Polymer Properties Database, 2018. Available from: https://polymerdatabase.com/polymer%20chemistry/radical%20mechanism.html.
[31] Willey Online Library, 2011. Available from: https://onlinelibrary.wiley.com/doi/abs/10.1002/0471440264.pst072.
[32] L.E. Elizalde, G. de los Santos-Villarreal, J.L. Santiago-García, M. Aguilar-Vega, Step-growth polymerization, in: Handbook of Polymer Synthesis, Characterization, and Processing, 2013, pp. 41–63.
[33] T.A. Sherazi, Graft polymerization, in: Encyclopedia of Membranes, 2014, p. 1.
[34] Polymers Properties Database, Principles of Condensation Polymerization, 2015, Available from: https://polymerdatabase.com/polymer%20chemistry/Condensation%20Polymerization.html.
[35] I. Alarifi, B. Prasad, M.K. Uddin, Conducting polymer membranes and their applications, in: Self-standing Substrates, Springer, Cham, 2020, pp. 147–176.
[36] A.R. Alharbi, I.M. Alarifi, W.S. Khan, R. Asmatulu, Highly hydrophilic electrospun polyacrylonitrile/polyvinypyrrolidone nanofibers incorporated with gentamicin as filter medium for dam water and wastewater treatment, J. Membr. Sep. Technol. 5 (2) (2016) 38–56.
[37] V. Marković, P. Živković, 3D Printing—Challenges and Perspectives, 2018, Available from: http://technoscience.ba/wp-content/uploads/2018/04/3D-PRINTING-%E2%80%93-CHALLENGES-AND-PERSPECTIVES-Viktor-Markovi%C4%87-Predrag-%C5%BDivkovi%C4%87.pdf.
[38] R.J. Young, P.A. Lovell, Introduction to Polymers, CRC press, 2011.
[39] P. Vinogradov, 3D printing in medicine: current challenges and potential applications, in: 3D Printing Technology in Nanomedicine, Elsevier Inc, Missouri, 2019, p. 1.
[40] E.J. Goethals, S. Penczek, Chain Polymerization I. Comprehensive Polymer Science and Supplements, 1989, Available from: https://www.sciencedirect.com/topics/engineering/ring-opening-polymerization.
[41] C. Chen, D.Y.W. Ng, T. Weil, Polymer bioconjugates: modern design concepts toward precision hybrid materials, Prog. Polym. Sci. 105 (2020) 101241.
[42] A.S. Hoffman, Bioconjugates of intelligent polymers and recognition proteins for use in diagnostics and affinity separations, Clin. Chem. 46 (9) (2000) 1478–1486.
[43] K. Matyjaszewski, J. Spanswick, Controlled/living radical polymerization, Mater. Today 8 (3) (2005) 26–33.
[44] Grajales, S. (n.d.). Controlled Radical Materials Science Polymerization Guide. Available from: https://www.sigmaaldrich.com/content/dam/sigma-aldrich/docs/SAJ/Brochure/1/controlled-radical-polymerization-guide.pdf.
[45] B. Le Droumaguet, J. Nicolas, Recent advances in the design of bioconjugates from controlled/living radical polymerization, Polym. Chem. 1 (5) (2010) 563–598.
[46] I.M. Alarifi, Structural analysis of hexagonal and solid carbon fibers composite, Polym. Test. 84 (2020) 106392.

[47] I.M. Alarifi, A. Alharbi, W.S. Khan, A. Swindle, R. Asmatulu, Thermal, electrical and surface hydrophobic properties of electrospun polyacrylonitrile nanofibers for structural health monitoring, Materials 8 (10) (2015) 7017–7031.
[48] V.S. Swarna, I.M. Alarifi, W.A. Khan, R. Asmatulu, Enhancing fire and mechanical strengths of epoxy nanocomposites for metal/metal bonding of aircraft aluminum alloys, Polym. Compos. 40 (9) (2019) 3691–3702.
[49] O. Nuyken, S.D. Pask, Ring-opening polymerization—an introductory review, Polymers 5 (2) (2013) 361–403.
[50] I.M. Alarifi, A. Alharbi, W. Khan, R. Asmatulu, Structural health monitoring of composite aircraft, Adv. Mater. Sci. Res. 21 (2015) 111–132.
[51] I.M. Alarifi, Investigation into the morphological and mechanical properties of date palm fiber-reinforced epoxy structural composites, J. Vinyl Addit. Technol. (2020).
[52] M.S. Butt, A. Maqbool, M.A. Umer, M. Saleem, R.A. Malik, I.M. Alarifi, H. Alrobei, Enhanced mechanical properties of surface treated AZ31 reinforced polymer composites, Crystals 10 (5) (2020) 381.

CHAPTER 3

Synthetic alloys

Abbreviation

Cu	copper
DC electricity	direct current electricity
Fe	iron
MSF	multistage flash
NaOH	sodium hydroxide
Ni	nickel

3.1 Introduction

Humans have been utilizing the metals and alloys, which has ultimately improved the quality of life [1]. An alloy is formed when two or more than two metallic components are molten, to acquire a liquid form which is then solidified. Besides specific elements, all other elements found in an alloy are metallic in nature [2–4]. Following this an alloy may consist a mixture of metal with either another nonmetal or metal [5] and is referred to as the metallic assortment of two or more than two components, which is in the solidified form. The role of alloys is critical in our daily life, such that they are used in different vehicles, mobile phones, kitchen utensils, etc. [6].

Conspicuously, an alloy can be formed with a combination of metal or components with no properties of the metal. Sometimes, the composition of an alloy is made through one level, where the alloy inclines to acquire the suppleness and flexibility like a pristine metal. It can be illustrated from the fact that the composition of the gold and silver is made through a single level. In the context of equilibrium, the preparation of an alloy has proceeded through two or more levels. Each level comprises of particular crystal organization that is dissimilar to others. For instance, the composition of copper-zinc alloy progresses through five levels that result in the formation of brass. Notably, upon melting, the alloys are similar. In contrast, they may be found in the state of heterogeneity when in the state of solid—the alloys which have mercury as the major components. The components are denoted as amalgams, while those with the combination of sodium and mercury.

Synthetic Engineering Materials and Nanotechnology
https://doi.org/10.1016/B978-0-12-824001-4.00001-6

Sodium and mercury are known as sodium amalgam. The elements of alloys may hold their chemical features, while some of the physical properties are enhanced [5]. The alloys will be called low melting if they have a melting point of lower than 232 °C. One of the examples of low meting alloy is the tin, which has lower than 232 °C of melting point [7].

The density of materials is its weight per unit volume. It is symbolized by ρ. More specifically, it is also termed as a specific weight. All the materials have different densities [8]. Fig. 3.1 exhibits the frequency of variation in the densities of materials.

> Hint statement
>
> Humans have been utilizing the metals and materials composed of metals.

Apart from mechanical characteristics, chemical and physical properties are also found in metal alloys. These characteristics of being reactive, electrical conduction, fair ductility, resilience toward deformation, and

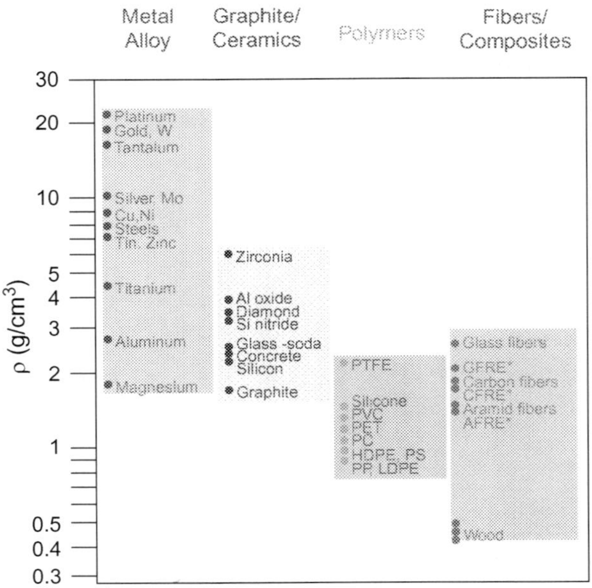

Fig. 3.1 Densities of material classes.

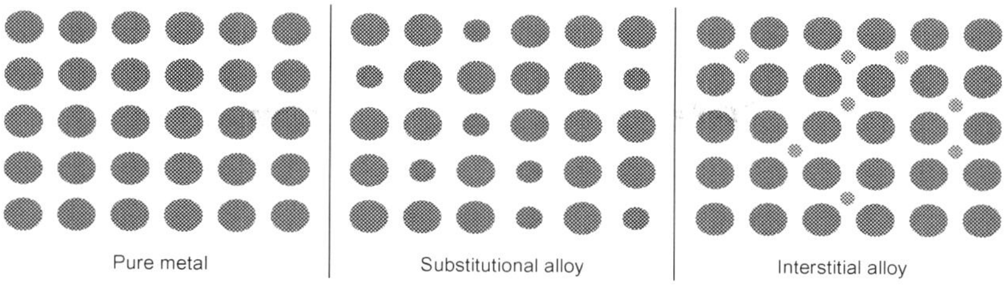

Fig. 3.2 Atomic formation of pure metal, substitutional, and interstitial allow [10].

flexibility are few names to their characteristics [7]. Fig. 3.2 shows the atomic structure of the pure metal, substitutional alloy, and interstitial alloy.

> Hint statement
>
> A metallic assortment of two or more than two components, which is in the state of solid, is called an alloy.

3.1.1 Types of synthetic alloys

Alloys can be categorized into two types. The first type includes the substitutional alloys, which are formed by mixing a molten metal with another substance. However, the formation of the substitutional alloys is held through two different mechanisms, i.e., interstitial mechanism and atom exchange. The size of each element added into the mixture of the alloy is of significant value. The exchange method in atoms generally happens when the presented atoms are smaller in size. However, some atoms that are based on the metallic crystals are generally substituted on with atoms consisting of the other common constituents and is generally termed as the substitutional alloy. Bronze and brass in which some copper atoms are replaced with wither zinc or tin atoms are the most common example of the substitutional alloys [10]. These metals commonly share covalent bonding, followed by the change in the proportion of the additional constituents. Through the formation, metals generally have a flow of electron, which serves as the prominent reason behind several characteristics found in them.

> Hint statement
>
> The alloys with the melting point of lower than 232 degrees are known as "low melting." One of the examples of low meting alloy is tin.

The second type includes the interstitial alloy, which is a compound developed by the presence of a very tiny small radius in an interstitial hole present in the metal lattice (Fig. 3.3). Some common examples of small atoms include boron, carbon, nitrogen, and hydrogen. At an industrial level, a particular transitional metal such as; nitrides and carbides are of significant value. However, within the interstitial mechanism, the atoms are relatively different in sizes such that; one atom is smaller than the other one, which makes it even more difficult to replace an atom in the crystals of the base metal (Fig. 3.4).

Mostly, the smaller atoms are trapped in the present spaced between atoms found in the crystal matrix; the phenomenon, however, is known as interstices [11]. One of the common examples of the interstitial alloy is steel, as it consists of very small carbon atoms that are found in the interstices of the iron matrix. Also, their classification can be made as homogeneous (comprise of one level) or heterogeneous (comprise of two or more stages) or intermetallic. The mechanical characteristics of the alloys are usually

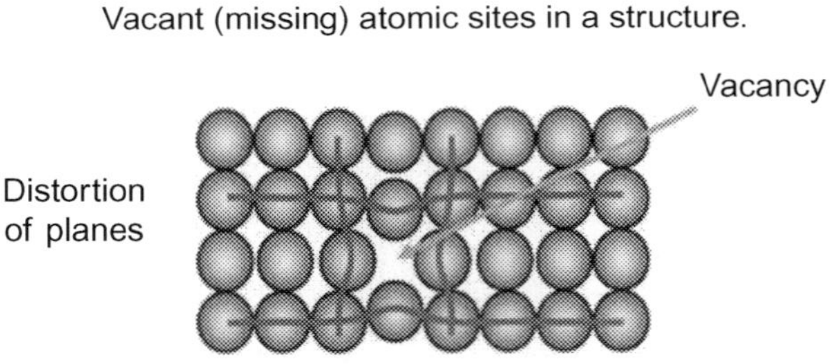

Fig. 3.3 The element missing (vacant) in atomic structure.

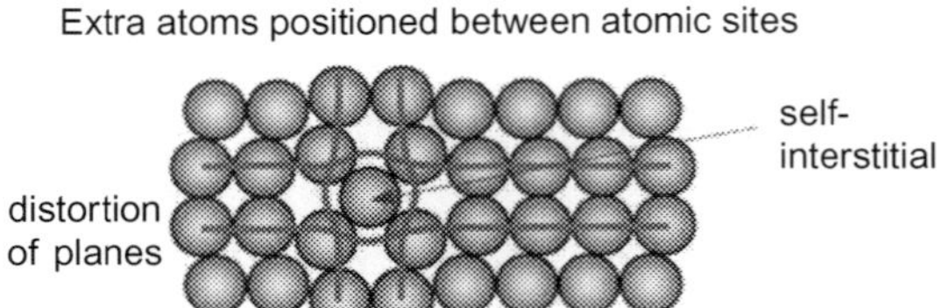

Fig. 3.4 The element self interstitial atoms positioned between atoms.

different from its components [11]. The metal which has no toughness can be transformed through alloying with other metal, having the same property, for instance, and Copper.

3.1.2 Difference between substitutional and interstitial alloys

Table 3.1 exhibits the difference between substitutional and interstitial alloys.

Fig. 3.5 displays the difference between the atomic structure and composition of substitutional and interstitial alloys.

Table 3.1 The synthetic metal alloys types.

Substitutional	Interstitial
It is formed by substitutional of metal atoms with another metal atom of the same alloy	It is formed when some metal atoms are smaller, but others are larger so, smaller atoms come in intersects of the other metal atoms
The atom exchange mechanism forms it	The insertion mechanism forms it
It has the same sized metal atoms	It has metal atoms with different sizes
For example, Brass Bronze, Copper, and Nickel	For example, Carbon, Fe, and Steel

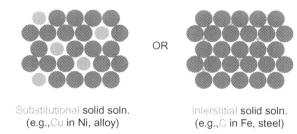

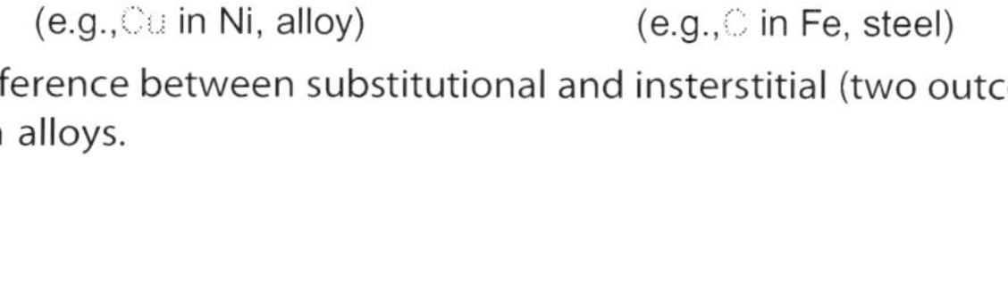

Fig. 3.5 The difference between substitutional and insterstitial (two outcomes element point defects in alloys.

Example 3.1

Analyze the degrees of freedom of Ni-alloy in 40% of Cu at the temperatures of 1200, 1250, and 1300°C.

Solution

- At the temperature of 1200 °C:

$P=1$ represents the phase where there is the presence of solid.

$C=2$ refers to the phase with the presence of copper and nickel atoms. Therefore,

$$F=2-1+1=2.$$

Now the composition and temperature will be combined to explain the solid-state entirely.

- At the temperature of 1250 °C:

$P=2$, as solid, as well as the liquid, is involved.

$C=2$, as there is the presence of the atoms of both copper and nickel. Thus,

$$F=2-2+1=1.$$

Along with fixing the temperature in the region, which is two-phased, the composition will be treated in the same manner. However, when one phase composition is determined, it causes the two-phased fixation of composition and temperature mechanically.

- At the temperature of 1300 °C:

$P=1$, represents the presence of one phase liquid.

$C=2$, refers to the presence of the atoms of nickel and copper. Subsequently,

$$F=2-1+1=2.$$

The liquid phase needs fixing the composition and temperature to elaborate on the condition of Copper-Nickel alloy in the region of liquid [12].

Since merely one degree of freedom is included in the two-phase region. The two-phase composition is determined, which is based on the specification of temperature. This is widely applicable when variation occurs in the composition of alloys [12].

Hint statement

Alloys are acquired as they can perform more effectively than metals. Usually, they are found to be stronger, resilient, and more solid than the pristine metals.

3.1.3 Purpose of the alloys

Alloys are acquired as they can perform more effectively than metals. Usually, they are found to be stronger, resilient, and more solid than the pristine metals. Besides, their favorable characteristics that are well-matched to the casting process make them useful. Other physical characteristics include the feature as magnetic and resilience toward corrosion in the particular environs [6].

In contrast with metals, usually, alloys melt at varying points but have a range of melting where the component is a combination of solid and liquid. Nevertheless, many of the alloys have a certain proportion of components, which is termed as "eutectic mixture," where the alloys are found with an exclusive point of melting [13]. Pristine metal gets molten at its melting point. However, when another metal is added to the liquefied form of metal, the combination of metals acquires the state of solid at the point of temperature, which is different from the freezing point of the pristine metal [6]. For example, Nickel has a melting point of 1452 °C, and Copper has 1084 °C, while a mixture that has an equal concentration of both the metals gets molten at the point, which is the middle value of the melting points of them (Fig. 3.6) [6].

An alloy is composed of an assortment of metal or other constituents. However, alloys that have certain stoichiometry and quartz-like organization are the intermetallic compounds. Measurement of the alloys' components is made through mass percentage for its real-life application and nuclear usage.

Steel is the mostly found alloy in today's world, as it has a higher degree of toughness and its flexibility to change upon heating. When adding the

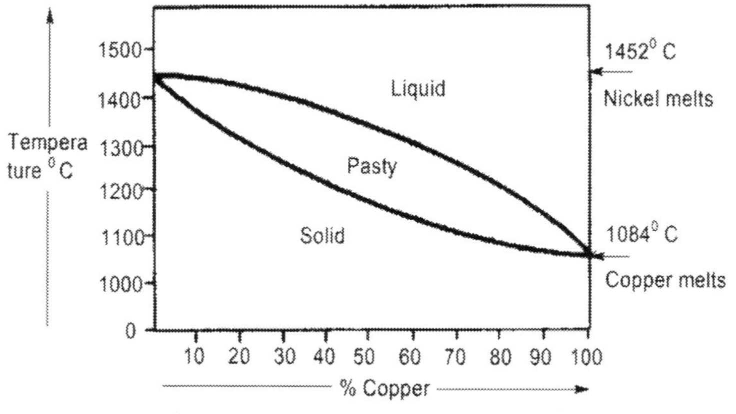

Fig. 3.6 The melting point of copper-nickel alloys [6].

Chromium with Steel, it becomes resilient toward corrosion, and it can improve further by composing the stainless steel. When Silicon is added up with steel, the process brings a change in its electrical properties and results in silicon steel [14].

A plethora of alloying elements are involved that have different properties for varying base materials. Moreover, Chromium is a metal that is mostly used in alloys to prevent corrosion and is also used to enhance its toughness. Nickel is another metal that enhances the hardness of materials. For example, austenitic stainless-steel is composed of a higher concentration of Nickel. Another metal is Copper, which acts as a significant constituent to make aluminum and makes it precipitation-harden-able. With steel, it can increase the resistance toward corrosion, which may result in the meager level of corrosion resilience of aluminum. Furthermore, carbon is an important component of manufacturing steel. It is mixed with steel and cast-iron alloys to augment the toughness [15].

3.1.4 Properties of synthetic alloys metals

Alloys are prerequisites for manufacturing high-quality, engineered materials. These materials can be implanted with the necessary characteristics. For instance, iron has significance in the advanced material industry as it possesses mechanical firmness. Moreover, stainless steel is also known for its mechanical characteristics. However, their resistance to corrosion makes them distinguished from other manufacturing materials [16]. The following are some of the properties of synthetic alloys metals:

- It has a more tensile strength.
- It has more toughness.
- It has corrosion resistance.
- It has high ductility.
- It is more durable.
- It is hard in structure in comparison to other metals.

Example 3.2

Prepare a Cu—Ni alloy that has 140 $MN\,m^{-2}$ yield strength, 420 $MN\,m^{-2}$ minimum tensile strength, and has 20% of minimum elongation. If Cu-20% Ni and pure nickel are available, project method to produce castings that possessed with desired features.

Solution:

Up to 60% Ni augments the strength of Copper. While Ni is augmented by approximately 40% of Cu, the desired composition can be analyzed in Fig. 3.7. In order to obtain the projected yield strength, the alloys must be comprised of 30%–90% Ni. On the other hand, for acquiring tensile strength, the required volume of Ni is 33%–90%. The range of the percentage of elongation must be within the range of 60%–90% of Ni. All the required concentration can be presented as:

Cu-90% (Ni)

OR

Cu-33%–60% Ni

The low nickel content is selected as it is not cost-effective in comparison with Copper. Moreover, the alloys of lower nickel comprise of lower liquids and castings that are permitted. Thus, the alloy, that is rational, can be Cu-35% Ni.

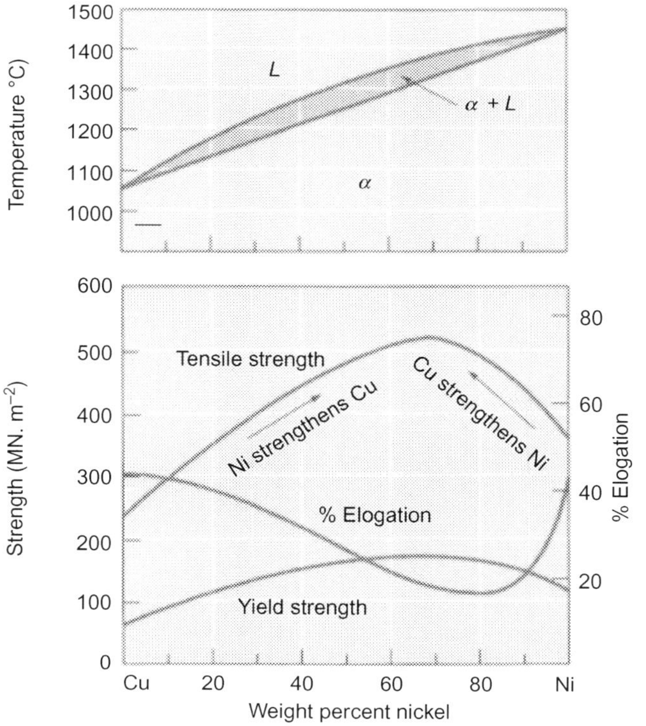

Fig. 3.7 Mechanical characteristics of copper-nickel alloys.

For the purpose of acquiring the same composition by the obtainable the melting stock, pure nickel is blended with 20% of Cu Ni ingot. To obtain the alloy that is 10 kg:

The required mass of Ni = $(10\ \text{kg})\ \frac{(35\%\text{Ni})}{100\%} = 3.5\,\text{kg}$.

Suppose,

y kg is the mandatory volume of Cu-20 ingot,

Subsequently, $(10 - y)$ kg is the required quantity of pure Ni.

$3.5\ \text{kg Ni} = y \times \frac{20}{100} + (10 - y) \times \frac{100}{100}$

$3.5 = 0.2x + 10 - y$

$0.8y = 6.5$

$y = 8.125\ \text{kg}$

Hence, 8.125 kg of Cu and 20% Ni is required to be melted with pure Nickel in the quantity of 1.875 to obtain the desired alloy. Afterwards, the alloys would be heated at the temperature of 1250 °C, prior to containing the liquid metal in the mold [12].

Example 3.3

Natural alloys

$$\text{Electrum} \rightarrow \text{Ag} + \text{Au}$$

$$\text{Meteorite} \rightarrow \text{Fe} + \text{Ni}$$

Human-made alloys

$$\text{Brass} \rightarrow \text{Cu} + \text{Zn}$$

$$\text{Bronze} \rightarrow \text{Cu} + \text{Sn}$$

$$\text{Cast Fe} \rightarrow \text{Fe} + \text{C} + \text{Si}$$

$$\text{Steel} \rightarrow \text{Fe} + \text{C} + (\text{Al, Cu})$$

3.2 Methods and techniques for obtaining synthetic alloys

Different methods that are employed in order to produce alloys synthetically are described in the following:

3.2.1 Fusion method

The fusion method carries the process of blending organized constituents of alloys, with the symmetric amount in the refractory brick-lined container. In cases where the metals of components are possessed with a dissimilar melting point, the metal with a high melting point gets molten, while other metals with low melting points bulge. In this method, component element, having a higher melting point in suitable quantity along with other component elements with lower melting points are added to it in sequined quantity. To avoid oxidation of molten mass, and it is covered with fine charcoal powder. While the molten mass is stirred with graphite rods, and the molten mass is to get uniform alloy as the final output. In the preparation of Brass, Copper (m. p. 1089 °C) is melted in the baker, and then the sequined quantity of Zinc (419 °C) is added (Fig. 3.8).

The liquefied mass is well mixed with the graphite rod. However, the metals with dissimilar compactness such as the denser constituents are assorted in last to prevent the settling, or else alloy of variable composition is acquired in brick-lined containers [6]. The components of metal are stirred well and permitted to get molten. However, the liquefied mass is protected by powered carbon to revert the process of corrosion of the liquefied alloy constituents as they are prone to have a strong reaction toward oxygen, which is present in the environment. The liquefied mass is left at room temperature to maintain its temperature [17].

3.2.2 Reduction method

The reduction method is involved in reducing the appropriate constituent (oxide) of metal through other compound metals of the alloys. For instance, aluminum bronze is acquired through plummeting alumina, where Cu is present in an electric furnace [6]. Probably, metals are present as compounds, while reduction is characterized by a chemical process where other

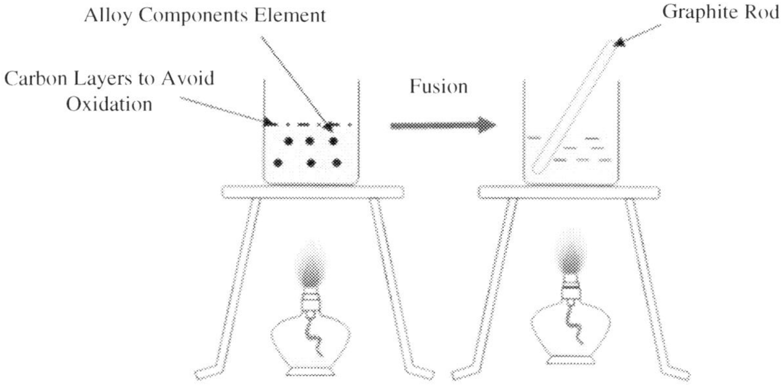

Fig. 3.8 Schematic view of fusion method.

compounds can detach a constituent of a compound in order to get a pristine form of metal. The reduction method is processed under an electrical furnace [17]. Notably, methods that are used to acquire alloys, carbon, salts, and oxides are certain residuals that might serve as the barrier in hindering the development of properties in alloy being produced [17].

The process of synthesizing the $Co_{60}Fe_{20}Ni_{20}$, $Co_{50}Fe_{25}Ni_{25}$, $Co_{40}Fe_{40}Ni_{20}$, and $Co_{33}Fe_{33}Ni_{33}$ has been illustrated in Fig. 3.9, where hydrazine ($N_2H_4{\cdot}H_2O$) was used as an agent for reduction, along with a device that could exhaust the air. The required volume of the chemicals was stirred in the watery solution, based on ethanol. The volume was 3:1. A magnetic stirrer was used for 20 min to prepare a standardized solution. Afterward, 6 mL of NaOH was put in the solution to acquire 11 pH. The prepared solution was heated at 100°C while the stirring with a magnetic stirrer was kept continued. Subsequently, hydrazine ($N_2H_4{\cdot}H_2O$) 10 mL was poured into the solution. A conical flask was used to avert air intervention. Amid the reaction, vapors and gases were dispensed. The temperature of 100 °C was kept continued for 45 min. Finally, the resultant solution was dried out in the vacuumed oven at 60°C to 4 h [18].

3.2.3 Electrodeposition method

This method takes place where the concurrent deposition of different compound metal from the electrolytic solution along with their salt's solutions by

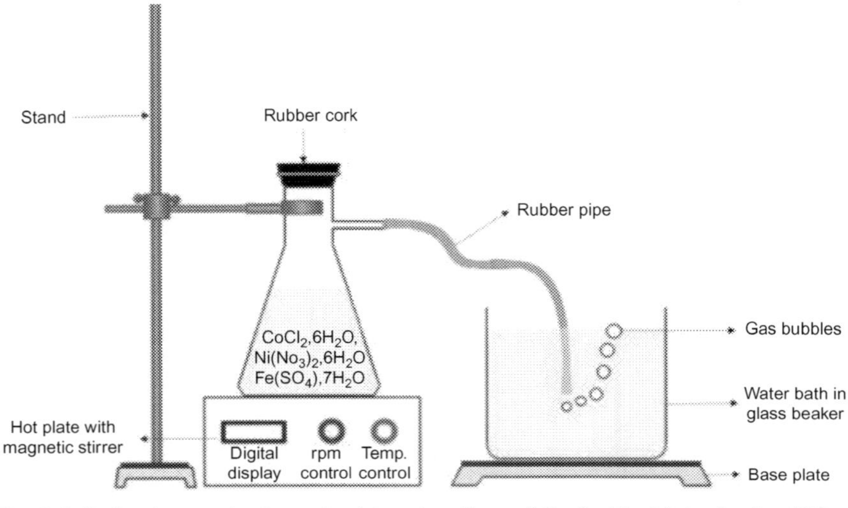

Fig. 3.9 Reduction method: synthesizing the alloys of Co-Fe-Ni with hydrazine [18].

fleeting DC electricity, i.e., brass [6–14]. The electrodeposition method augments the superficial features, embellished, and perfunctory features of numerous ranges of materials. Presently, electrodeposition is developing as a widely used multipurpose method to acquire nanomaterials. It is a film growth process that comprises the development of metallic or semi-conduction [19]. Nowadays, electrodeposition coverings conductive substrates that initiated from metal ion signs in a favorable solvent and followed by charge transmission [20].

Fig. 3.10 shows the electrodeposition of alloys in a deep eutectic solvent. In the ratio of Ni/Fe in a deep eutectic solvent, an optimized oxygen reaction was obtained by $Ni_{75}Fe_{25}$ catalyst, which was based on 316 mV to have a density of $10\,mA\,cm^{-2}$. Whereas, $62\,mV\,dec^{-1}$ was the Tafel slope [21].

3.2.4 Powder metallurgy or compression methods

This technique involves the combination of two or more powder of metal that is exposed to excessive coercion at the heat that is lesser than the melting point of alloys [6]. Powder metallurgy can be further explained as the processing of finished metal powders and subsequently preparing articles from distinct metal powders or those of alloys [17]. This method is employed to acquire superalloys of Nickel that are more alloyed and intricate for casting. However, it cannot be prepared through the traditional method of casting techniques but possesses an ability to deal with casting and restrictions as well as to manufacture a better grain size [22]. The notion of the power metallurgy encompasses the way of manufacturing substances from metal powders either through addition or no addition of nonmetallic compounds. The

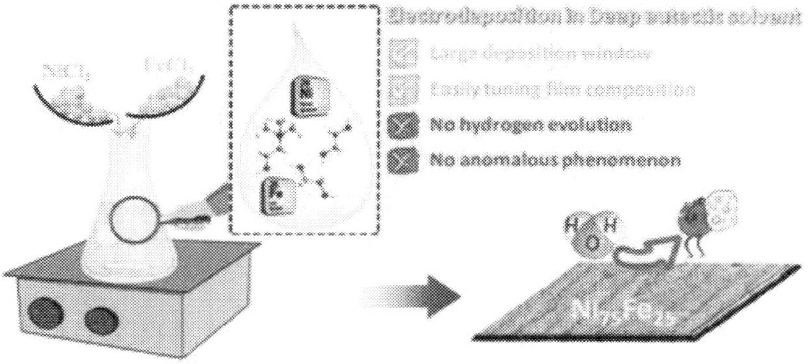

Fig. 3.10 Electrodeposition of alloys in a deep eutectic solvent [21].

powders can be compressed to a required form, which is further provided by the temperature to prepare a welded, alloyed, or blended mass. The substance is favorable in order to use instantly, or it can be functioned through outdated techniques.

The prominent advantage of this method is that it can manufacture the substances that have distinct characteristics, which is not possible by other methods, i.e., controlled porosity. This method can make parts in metals which cannot be liquefied at a commercial scale but associated with predicaments. Besides, it has an appropriate blend of metal and nonmetals and can be employed with the metals which are not alloyed. Further, it can deal with the metals that have different melting points or compactness and facilitates a mass-production method of preparing large bulk of parts which are indistinguishable in volume and quality [23]. The process of powder metallurgy comprises sequential activities, during which feedstock in the powdered form μm to nm is added, specifically in conditions where the compounds are different in mass and compositing. Fig. 3.11 displays the traditionally employed method that was used to produce the final composites [24].

3.3 Application of synthetic alloy

Alloys are used to augment the toughness of metal, alter the color, and to dwindle the melting point. It further employed to augment tensile hardness as well as to facilitate fair cast-ability. Through alloying the metal, the limpness of the metal is increased, which leads to amplify the resilience toward

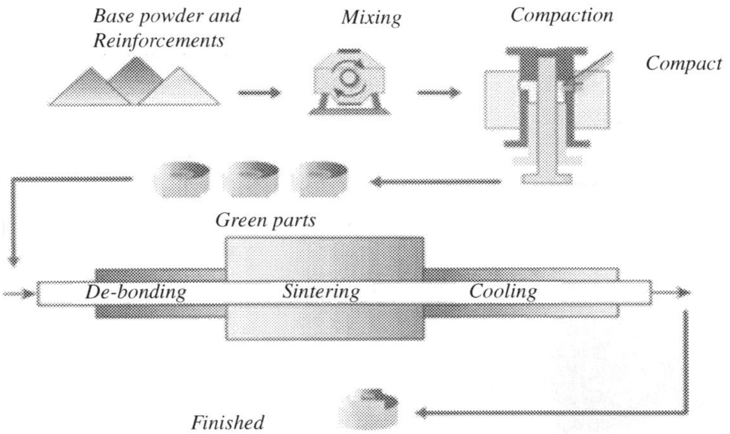

Fig. 3.11 The traditional method of powder metallurgy process [24].

corrosion [25]. Alloys have a greater role in the lives of humans and are used in multiple forms, and it could not be wrong to say that humans cannot even pass a single day without using alloys. In addition, tools of the machine and engineering equipment are also prepared by alloys [7]. Market data reveals that approximately 15% of the world's consumption of Zinc is consumed to acquire alloys that are later used to prepare automotive parts, electric systems, water taps and sanitary tools, household goods, and fashion accessories, etc. Usually, half of the consumption of Zinc is in spurring steel and averting corrosion [26]. Other uses of Zinc are involved in making alloys of brass, bronze, aluminum, and magnesium. Also, it is used as an oxide in the chemical and pharmaceutical industries in cosmetic products, paint, rubber, and agriculture industries [27]. Recently, Zinc has been found as a promising substitute for iron and magnesium due to biodegradable metal [28]. In weaponry, space technology, conveyance manufacturing, and automobile industries, there is extensive use of aluminum alloys, since it has low cost, augmented toughness, resilience toward oxidation, elastic modulus lesser degree, and excellence wear [29]. Dokšanović, Džeba and Markulak [30] provided a detailed description regarding the usability of the aluminum alloys and indicated them a favorable choice with respect to their structures that is of significant value when used in chemically active environments and other remote locations [31–33]. They are further suitable when used in the treatment plants of water and sewage. One of the most common examples where the aluminum alloy is used is the construction of the rotating crane bridge (as shown in Fig. 3.12) developed in a sewage treatment plant.

Besides, after the Second World War, a variety of experimental bridges were developed in the United States. A road bridge designed in Canada is one such example which was explicitly developed in an area that has significant temperature-based fluctuations ranging from −40 to +40 °C. Common type of alloys, and their usage is given in Table 3.2 [35].

Fig. 3.12 Aluminum alloy in rotating crane bridge [34].

Table 3.2 Some of common alloys and their applications.

Common alloys	Applications
Babbitt metal	Used to manufacture bearing as it has a low weight of friction with steel
Bell metal	It is used to cast bells
Brass	Cheap cost jewelry is made from it. Along with manufacturing hose nozzles, the dies of stamps and connectors
Bronze	It is used to prepare metal money, weighted gears, and tools of electric hardware
Coin metal	It is used in the United States prominently for making metal money
Duralumin	It is used to manufacture aircraft, boats, cars and other heavy machinery owing to its hardness
Monel	It is used to prepare corrosion-free containers
Phosphor bronze	It is used to make springs and boat propellers
Solder metals	It is used in the process of mixing two metals
Sterling silver	It is used to make jewelry and art sculptures
Type metal	It is used to prepare the characters of typing for printings as well as for making ornamental goods, such as statuettes and candlesticks

Table 3.3 Efficacy and price of the metal and alloys.

Materials	Electrical conduction ($W\,m\,K^{-1}$)	Corrosion resistibility	Density ($g\,cm^{-3}$)	Price ($10,000 t^{-1})
Aluminum alloy (5052)	140–160	Average	2.66	0.21
Copper alloy (aluminum brass)	29–100	Good	8.4–8.9	0.8
Titanium alloy (TA2)	17	Excellent	4.5	2.8

Apart from it, some of the synthetic alloys such as titanium, titanium alloy, copper alloy, and aluminum alloy are major constituents of multistage flash (MSF) devices. Table 3.3 presents electrical conduction, oxidation resistibility, and the cost of these materials [36].

Besides the uses mentioned above of alloys, stainless steel is used in manufacturing wire and ribbon their other forms, i.e., screening, staple, belt, cable, weld, metalizing, catheter, and suture wire. Alloys of Gold and Silver are necessary elements for jewelry. White gold, an alloy of Gold, Silver, Palladium, and Nickel are essential to prepare a substitute for Platinum. Some of the alloys are also required in the petrochemical industry. Additionally, they are used as corrosion resisting compounds [37–39]. The alloys which have higher temperature are used in the aircraft industry, and its uses are related to wire welding, etc. [40].

3.4 Conclusion

Alloys are manufactured through the metallic components and thus possess metallic properties. They possess profound characteristics of electric conduction, resistibility for oxidation, and ductility. Moreover, they are highly flexible to deform or mold into required shapes. The two major categories of synthetic alloys are substitutional and interstitial alloys. Alloys that have stoichiometry and quartz-like organizations are known as intermetallic compounds. Moreover, the most commonly found alloy in the world is steel due to a higher level of firmness and flexibility. Several uses of alloys have been found in prominent industries and manufacturing as they have unique characteristics such as toughness, corrosion resilience, and a higher degree of solidity than pristine metals. Traditionally, four methods are involved in preparing alloys, such as; fusion methods, reduction method, electrodeposition method, and powder metallurgy methods. Alloys have been making a significant contribution toward humanity in terms of their extensive usage in manufacturing heavy materials, defense production, cosmetics, automobile, etc.

References

[1] K. Midander, Metal release from powder particles in synthetic biological media (Doctoral dissertation), Chalmers University of Technology, Sweden, 2006.

[2] I.M. Alarifi, H.M. Nguyen, A. Naderi Bakhtiyari, A. Asadi, Feasibility of ANFIS-PSO and ANFIS-GA models in predicting thermophysical properties of Al2O3-MWCNT/oil hybrid nanofluid, Materials 12 (21) (2019) 3628.

[3] A. Hajizadeh, N.A. Shah, S.I.A. Shah, I.L. Animasaun, M. Rahimi-Gorji, I.M. Alarifi, Free convection flow of nanofluids between two vertical plates with damped thermal flux, J. Mol. Liq. 289 (2019) 110964.

[4] A. Asadi, I.M. Alarifi, H.M. Nguyen, H. Moayedi, Feasibility of least-square support vector machine in predicting the effects of shear rate on the rheological properties and pumping power of MWCNT–MgO/oil hybrid nanofluid based on experimental data, J. Therm. Anal. Calorimet. (2020) 1–16.

[5] S. Nadeem, A. Alblawi, N. Muhammad, I.M. Alarifi, A. Issakhov, M.T. Mustafa, A computational model for suspensions of motile micro-organisms in the flow of ferrofluid, J. Mol. Liq. 298 (2020) 112033.
[6] D. M. Patel. (2018). Alloys and inter-metallic compounds. Available at http://www.vpscience.org/materials/Study%20Material%20on%20Alloys%20and%20Intermetallic%20Compounds%20By%20Dr.%20D.%20M.%20Patel.pdf.
[7] Goyal, S. (n.d.). List of important alloys and their uses. Available at https://www.jagranjosh.com/general-knowledge/list-of-important-alloys-and-their-uses-1482236932-1.
[8] M. Trojanowska-Tomczak, R. Steller, J. Ziaja, G. Szafran, P. Szymczyk, Preparation and properties of polymer composites filled with low melting metal alloys, Polym.-Plast. Technol. Eng. 53 (5) (2014) 481–487.
[9] Density. (n.d.). Available at: https://resources.saylor.org/wwwresources/archived/site/wp-content/uploads/2011/04/Density.pdf.
[10] LibreText. (2019). Substitutional alloys. Retrieved from https://chem.libretexts.org/Bookshelves/Inorganic_Chemistry/Map%3A_Inorganic_Chemistry_(Housecroft)/06%3A_Structures_and_energetics_of_metallic_and_ionic_solids/6.07%3A_Alloys_and_Intermetallic_Compounds/6.7A%3A_Substitutional_Alloys.
[11] LibreText, Intermetallic compounds, 2019, Retrieved from https://chem.libretexts.org/Bookshelves/Inorganic_Chemistry/Map%3A_Inorganic_Chemistry_(Housecroft)/06%3A_Structures_and_energetics_of_metallic_and_ionic_solids/6.07%3A_Alloys_and_Intermetallic_Compounds/6.7C%3A_Intermetallic_Compounds.
[12] J.J. Palop, L. Mucke, E.D. Roberson, Quantifying biomarkers of cognitive dysfunction and neuronal network hyperexcitability in mouse models of Alzheimer's disease: depletion of calcium-dependent proteins and inhibitory hippocampal remodeling, in: Alzheimer's Disease and Frontotemporal Dementia, Humana Press, Totowa, NJ, 2010, pp. 245–262.
[13] Alloys. (n.d.). Available at: https://courses.lumenlearning.com/boundless-chemistry/chapter/alloys/.
[14] Halin, D. S. (2018). Metal alloy. Available at https://www.researchgate.net/publication/326173232_metal_alloy. Doi. 10.13140/rg.2.2.20821.52961.
[15] What is an alloy? 2018. Available at: https://www.metalsupermarkets.com/what-is-an-alloy/.
[16] Morales, E. V. (Ed.),, Alloy Steel: Properties and Use, BoD–Books on Demand, 2011.
[17] Structure, Properties, and Applications of Various Alloys. (n.d.). Available at: https://www.globalspec.com/reference/46878/203279/preparation-of-alloys.
[18] G.S. Reddy, S.R. Sahu, R. Prakash, M. Jagannatham, Synthesis of cobalt-rich alloys with high saturation magnetization: a novel synthetic approach by hydrazine reduction method, Results Phys. 12 (2019) 652–661.
[19] Y.D. Gamburg, G. Zangari, Theory and Practice of Metal Electrodeposition, Springer Science & Business Media, 2011.
[20] D.S. Jayakrishnan, Electrodeposition: the versatile technique for nanomaterials, in: Corrosion Protection and Control Using Nanomaterials, Woodhead Publishing, 2012, pp. 86–125.
[21] T.G. Vo, S.D.S. Hidalgo, C.Y. Chiang, Controllable electrodeposition of binary metal films from deep eutectic solvent as an efficient and durable catalyst for the oxygen evolution reaction, Dalton Trans. 48 (39) (2019) 14748–14757.
[22] J.H. Weber, Y.E. Khalfalla, K.Y. Benyounis, Nickel-Based Superalloys: Alloying Methods and Thermomechanical Processing, 2016. https://www.sciencedirect.com/science/article/pii/B978012803581803383X?via%3Dihub.
[23] J.E. Newson, Powder metallurgy, Proc. Inst. Mech. Eng. C J. Mech. Eng. 154 (1) (1946) 208–215.

[24] H.G. Prashantha Kumar, M.A. Xavior, Processing of graphene/CNT-metal powder, Powder Technol. 45 (2018).
[25] lymie, R., Alloy Formation, Methods of Production, and Its Various Applications, 2017, Available at: https://steemit.com/science/@lymierikxz/alloy-formation-methods-of-production-and-its-various-applications.
[26] A.R. Marder, The metallurgy of zinc-coated steel, Prog. Mater. Sci. 45 (3) (2000) 191–271.
[27] A. Pola, M. Tocci, F.E. Goodwin, Review of microstructures and properties of zinc alloys, Metals 10 (2) (2020) 253.
[28] G. Katarivas Levy, J. Goldman, E. Aghion, Zincrospects of zinc as a structural material for biodegradable implants—a review paper, Metals 7 (10) (2017) 402.
[29] S. Kumar, S.K. Singh, J. Kumar, Q. Murtaza, Synthesis and characterization of Al-alloy by mechanical alloying, Mater. Today Proc. 5 (2) (2018) 3237–3242.
[30] T. Dokšanović, I. Džeba, D. Markulak, Applications of aluminium alloys in civil engineering, Tehnički vjesnik: znanstveno-stručni časopis tehničkih fakulteta Sveučilišta u Osijeku 24 (5) (2017) 1609–1618.
[31] K.G. Kumar, M. Rahimi-Gorji, M.G. Reddy, A.J. Chamkha, I.M. Alarifi, Enhancement of heat transfer in a convergent/divergent channel by using carbon nanotubes in the presence of a Darcy–Forchheimer medium, Microsyst. Technol. 26 (2) (2020) 323–332.
[32] I.M. Alarifi, A.B. Alkouh, V. Ali, H.M. Nguyen, A. Asadi, On the rheological properties of MWCNT-TiO2/oil hybrid nanofluid: an experimental investigation on the effects of shear rate, temperature, and solid concentration of nanoparticles, Powder Technol. 355 (2019) 157–162.
[33] S. Uddin, M. Mohamad, M. Rahimi-Gorji, R. Roslan, I.M. Alarifi, Fractional electro-magneto transport of blood modeled with magnetic particles in cylindrical tube without singular kernel, Microsyst. Technol. 26 (2) (2020) 405–414.
[34] F.M. Mazzolani, Structural applications of aluminium in civil engineering, Struct. Eng. Int. 16 (4) (2006) 280–285.
[35] Common Alloys. (n.d.). Spectro Analytical Labs Limited. Available at: https://www.spectro.in/Common-Alloys.html.
[36] C.P. Wang, H.Z. Wang, G.L. Ruan, S.H. Wang, Y.X. Xiao, L.D. Jiang, Applications and prospects of titanium and its alloys in seawater desalination industry, in: IOP Conference Series: Materials Science and Engineering (vol. 688, No. 3, p. 033036), IOP Publishing, 2019, November.
[37] B. Souayeh, M.G. Reddy, P. Sreenivasulu, T. Poornima, M. Rahimi-Gorji, I.M. Alarifi, Comparative analysis on non-linear radiative heat transfer on MHD casson nanofluid past a thin needle, J. Mol. Liq. 284 (2019) 163–174.
[38] A. Asadi, I.M. Alarifi, V. Ali, H.M. Nguyen, An experimental investigation on the effects of ultrasonication time on stability and thermal conductivity of MWCNT-water nanofluid: finding the optimum ultrasonication time, Ultrason. Sonochem. 58 (2019) 104639.
[39] I.M. Alarifi, A.G. Abokhalil, M. Osman, L.A. Lund, M.B. Ayed, H. Belmabrouk, I. Tlili, MHD flow and heat transfer over vertical stretching sheet with heat sink or source effect, Symmetry 11 (3) (2019) 297.
[40] A.H. Musfirah, A.G. Jaharah, Magnesium and aluminum alloys in automotive industry, J. Appl. Sci. Res. 8 (9) (2012) 4865–4875.

CHAPTER 4

Synthetic rubber

Abbreviations

CR	polychloroprene
EPDM	ethylene propylene diene monomer
GRG	general rubber goods
NBR	acrylonitrile butadiene rubber
SBR	styrene-butadiene rubber
SI	polysiloxane

4.1 Introduction

Synthetic rubbers can be interchangeably called as elastomers, which are the long-chain polymers. They possess some important chemical, physical, and mechanical characteristics. Synthetic rubbers are attributed to have the features of the chemical stability, higher resilience toward abrasion, and strength followed by superficial steadiness [1–3]. Most of the defined features are implanted from naturally found polymers while undergoing the process of cross-linking and with the help of additives. Some of the elastomers and their features are summarized below in Table 4.1 [4].

Hint statement
Synthetic rubbers are also combined with natural rubber and used as polymers. These alloyed materials are used for coating the papers, backing the carpet, and for shaping the objects like gloves.

One of the significant characteristics of elastomers is their ability to stretch, double increasing length, and to retain their actual length as soon as they are released. Natural rubber serves as the elastomer that is made of the units of isoprene, which are connected in cis-1,4 configurations which implants excellent characteristics, i.e., augmented elasticity and firmness to natural rubber. This, natural rubber is produced as latex (water emulsion) and is acquired from the tree of *Hevea brasiliens*. This tree has abundant growth in Brazil, Malaysia, and Indonesia. The person who revealed that

Synthetic Engineering Materials and Nanotechnology
https://doi.org/10.1016/B978-0-12-824001-4.00012-0

Table 4.1 Certain characteristics of some elastomers [4].

	Durometer hardness range	Tensile strength at room temp. psi	Elongation at room temp. %	Temp. range of service (°C)	Weather resistance
Natural rubber	20–100	1000–4000	100–700	−500–80	Fair
Styrene-butadiene rubber (SBR)	40–100	1000–3500	100–700	−55–110	Fair
Polybutadiene	30–100	1000–3000	100–700	−60–100	Fair
Polyisoprene	20–100	1000–4000	100–750	−55–80	Fair
Polychloroprene	20–90	1000–4000	100–700	−55–100	Very good
Polyurethane	62–95 A 40–80 D	1000–8000	100–700	−70–120	Excellent
Polyisobutylene	30–100	1000–3000	100–700	−55–100	Very good

it was possible to vulcanize the latex by heating it with sulfur or any other agent was Charles Goodyear [5].

The process of vulcanization of rubber undergoes chemical reactivity, which results in developing a strong connectivity between elastomers. Through vulcanization, the molecules of long-chain develop elasticity. Also, the crosslinking implant the molecules with the strength to bear the load support. Similarly, synthetic rubber is composed of elastomers that are able to be crosslinked, i.e., polyisoprene, ethylene-propylene-diene terpolymer, and polybutadiene [4]. Additionally, another major constituent of synthetic rubber is thermoplastic elastomers that are not able to undergo from crosslinking, and the alterations are made for their usability, i.e., automobile buffers, cables, and in the coating of cables. They are also recyclable. Although, they do not have such potential to completely acquire the characteristics of natural rubber as they lack the required capabilities to perform varied applications as thermostat rubber can do.

Notably, the major use of rubber is identified in the tire manufacturing industry. Apart from it, synthetic rubber is also a crucial component to produce garden hose, slipper or shoe, squeezed constituents, and plasticizers [4].

4.1.1 Natural rubber vs synthetic rubber

Natural constituents contribute to forming a fundamental structure of natural rubber as it is made from the elastic polymer of hydrocarbon. More specifically, latex or a milky colloidal suspension, which is found from the plant's sap, is the fundamental constituent of the elastomer of hydrocarbon polymer [6]. The composition of natural rubber is based on dense particles that are dispensed from the milky white fluid, which is termed as latex. Additionally, latex is obtained from the tree that is found in tropical and subtropical regions. Acquire natural rubber, the method of polymerization is employed, which includes isoprene (2-methyl-1,3-butadiene-$C_5H_8)_n$, which can be demonstrated as *cis*-1,4-polyisoprene (Fig. 4.1) [7].

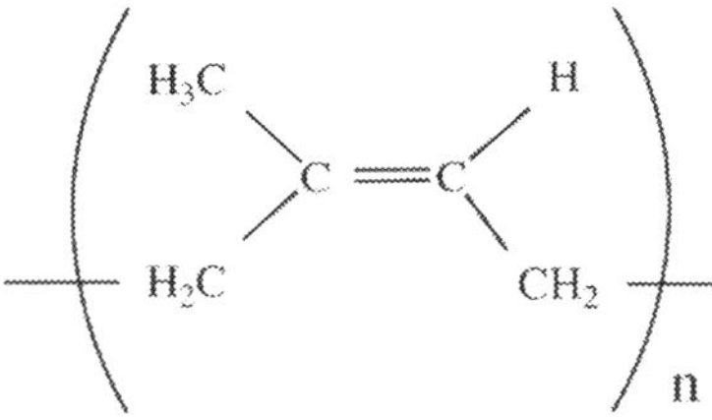

Fig. 4.1 The polymerization of natural synthetic rubber [7].

In contrast to this, synthetic rubbers are polymers that are termed as elastomers. Unlike natural rubber, for manufacturing synthetic rubber, the process of polymerization with monomers is employed. Some of those monomers are chloroprene (2-chloro-1,3-butadiene), 1,3-butadiene, isoprene (2-methyl-1,3-butadiene), 1,3-butadiene, and isobutylene (methyl propane), and isoprene (2-methyl-1,3-butadiene) where the trivial volume of isoprene is included due to cross-link [6]. Moreover, natural rubber has several roles, ranging from domestic use to its incorporation in industrial production. It further plays a greater role in making the finished goods and in the manufacture tires and tubes with overall consumption of 56% for the past few years [6]. While the rest of the proportion as vested with general rubber goods (GRG) that did comprise of tires and tubes.

In the year 1770, Joseph Priestley grasped the ability of rubber to remove the marks of a pencil. Further, in 1791, the first-ever commercial use of a water-resistant surface was introduced. Thus, vulcanizing natural rubber (the process of cross-linking the sulfur with polymers) was initiated in 1839. This development was a breakthrough with respect to the use of rubber for commercial production of tires. Consequent to a dearth in supply of natural rubber amid WW-II, the role of synthetic polymer grew in order to meet the growing demand [8].

Different types of synthetic rubber include; Neoprene, Buna, and butyl rubbers are manufactured to deliver particular characteristics, and they are favorable when used in different applications. Especially for producing tires Styrene-butadiene and butadiene rubber are used [9]. For instance, the process of synthesizing Neoprene is undergone through 3-butadiene and 2-chloro-1,3-butadiene. That has been shown in Fig. 4.2. [11].

Since both the natural and synthetic rubber possesses different characteristics based on their origin, composition, kind of monomers involved, and features, a detailed description of their differences is elucidated in Table 4.2.

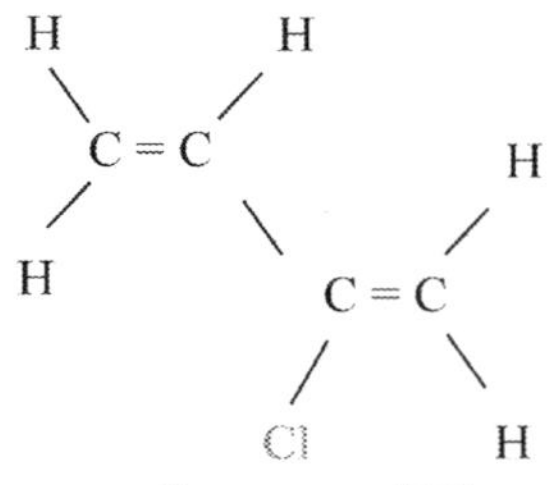

Fig. 4.2 The synthesizing structure of neoprene [10].

Table 4.2 Difference between natural and synthetic rubber [12].

	Natural rubber	Synthetic rubber
Description		
Origin	Naturally obtained from natural sources (from the tree of *Hevea brasiliensis*)	Synthetically manufactured polymers wherein the conditions are specified and controlled
Major constituent	Found in the saps of the plant	It is made from petroleum products while adopting the methods of polymerization or emulsion
Kinds of monomers	Monomers of *cis*-1-4 isoprene	Kinds of monomer differ based on different types of synthetic rubber
Features	The features are not convenient to alter or implant	Features can be implanted according to usability

4.1.2 Drawbacks of natural rubber

Natural rubber is a polymer of isoprene and possesses characteristics such as: good wear resistance, high elasticity, tensile strength, and elongation. The natural rubber, regardless of overlap density, gel, molar mass, and its distributive patterns, demonstrates characteristics that surpass the properties of synthetic rubber. However, some potential drawbacks are associated with the natural rubber [13] which include:

- Attracted by oxidizing.
- Attracted by water.
- Attracted by solvent.
- Attracted by nonpolar solvent.
- Little durability.
- Poor strength.

Further shortcomings of natural rubber include:

- Get damaged under the action of strong acid, ozone, oils, greases and fats.
- Not recommended for use in alcohols, esters, or among the aromatic solution.
- Low-temperature properties.
- Higher raw material prices.
- Does not apply to hot water being.
- Cannot be used in power transmission systems.

4.1.3 Theoretical context

4.1.3.1 Elastomers

Elastomers are compounds that can be deformed when they are given pressure [14]. They further possess properties such as; elasticity and higher molecular mass (Fig. 4.3). Since the categorization of rubbers is based on their origin, i.e., natural and synthetic, natural rubber is acquired from latex that is produced in plants. Whereas drenched hydrocarbons manufacture synthetic rubbers through chemical reactions. Moreover, various varieties of rubbers are available, where few of them are recognized to have machine characteristics in relation to natural rubber [16].

Hint statement

The major property of elastomers is their ability to get strained in a twice proportion as compared to the original extent. Also, they acquire their real shape once they are allowed to end up their dispersion.

4.1.3.2 Vulcanization

Synthetic rubbers are kind of polymers that are produced from dienes; thus, they obtain double bonds in their molecular chains. Vulcanization, in this regard, serves as a process that helps in altering the properties of the rubber through creating the polymer chains from cross-linking (Fig. 4.4). The overall process is executed through the presence of Sulfur [18]. This process is useful in transforming the degraded characteristics of polymers into a firm, adhesive, and adaptable rubber; since the level of vulcanization puts greater impacts on the characteristics. Moreover, the process of vulcanization with the use of the atoms of sulfur is mostly employed way to make the cross-links. Notably, the basis of this method was laid by Charles Goodyear [19, 20].

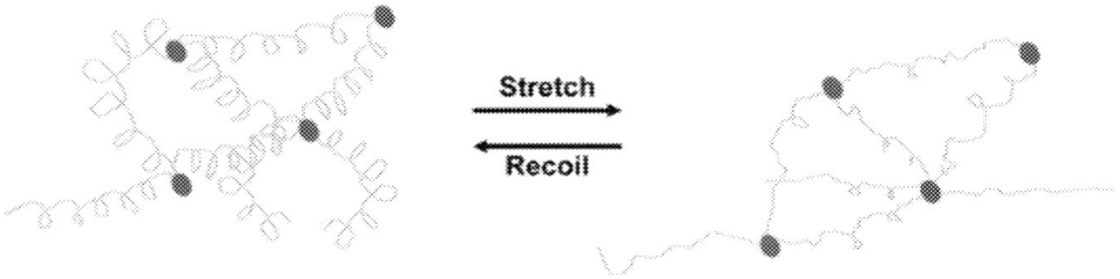

Fig. 4.3 The structure of elastomers [15].

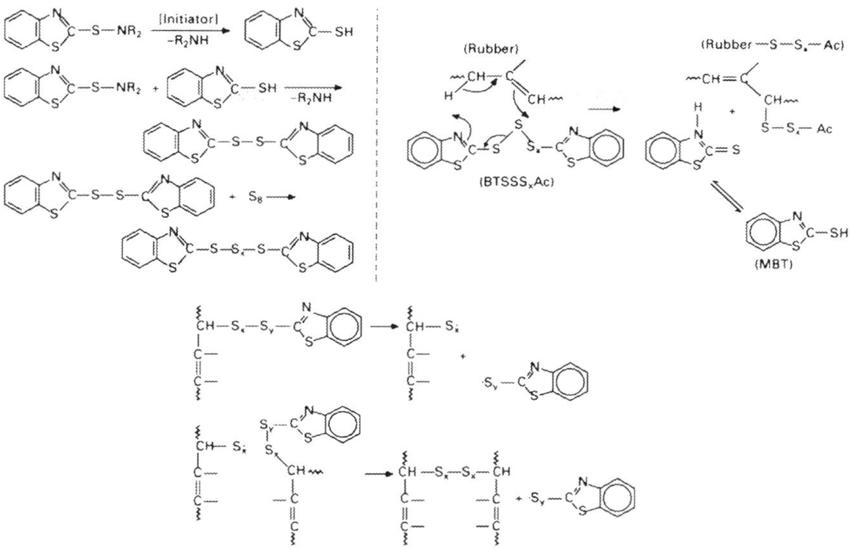

Fig. 4.4 Vulcanization of the polymer chains [17].

Hint statement
Vulcanization is a useful process for transforming the degraded features of polymers into a stable, adhesive, and flexible rubber.

4.1.4 Mechanical characteristics of rubbers

The exhibiting curve of stress deformation toward an elastomer shows that nearly all the curves embody the elastomers that deform. Thus, it can be inferred that elastomers demonstrate elasticity in conventional attitude.

Nevertheless, when the chains are extended, added elastic deformation takes place due to the elasticity of the links that originates a high modulus of elasticity (E). The sum of the cross-links is a determining factor for the elasticity of rubber or the volume of sulfur, which is supplemented to the material. If the addition of Sulfur is lower, it will produce a less hardened and elastic rubber. Whereas, if the concentration is added up, it will result in the restricted unfolding of the chains, which will make the rubber harden, firmer, and flimsier. The hardness in rubber is a symbol of its rigidness in relation to a modest level of stress. Commonly used method to bring hardness in rubber is the shore A, where it is measured with a tool that is called as durometer. This measurement is determined based on the diffusion of the frusto-conical tip, compared to the calibrated metal spring reaction [16].

Hardness is one of the most widely measured properties in rubber materials, as it is a practical way of determining the degree of vulcanization. The various rubber materials used in this study are NR, NBR, SBR, EPDM, and polyblends of NR with NBR and EPDM, in the raw state and in the reinforced state. From the mechanical strength measurements shown in Table 4.3, the reinforced rubbers have higher hardness values than the raw rubbers with the highest hardness observed in reinforced NBR, NR, and NBR at 79, 77, and 75, respectively [21].

Hint statement

Some of the major elastomer constituents of synthetic rubber that can be undergone with cross-linking are ethylene-propylene-diene terpolymer, polybutadiene, and polyisoprene.

Table 4.3 Mechanical strength of rubber and rubber compounds.

Sample	Reinforced	Hardness	Tensile (MPa)
Natural rubber	+	77.15	17.59
Styrene-butadiene rubber (SBR)	+	75.36	15.4
Nitrile butadiene rubber (NBR)	+	79.74	25.29
Ethylene propylene diene monomer (EPDM)	+	64.86	10.79
Nitrile rubber + nitrile butadiene rubber (NR + NBR)	+	70.22	20.19
Nitrile rubber + ethylene propylene diene monomer rubber	+	70.4	20.99
Natural rubber	–	43.24	3.04
Styrene-butadiene rubber	–	48.15	1.53
Ethylene propylene diene monomer (EPDM)	–	32.7	0.58
Nitrile butadiene rubber (NR + NBR)	–	51.75	2.02
Nitrile butadiene rubber (NR + NBR)	–	35.29	15.95
Ethylene propylene diene monomer (EPDM)	–	35.11	13.21

4.1.5 Chemical processing of natural and synthetic rubber

The process of synthesizing rubber comprises operations that involve material for value addition via chemical reactivity, irreversible alteration of physical features, and flow (Fig. 4.5). The steps of crucial importance, during chemical modification in tire production, include: breaking down the polymer through mastication on a mill, extrusion, calendaring, and vulcanization [22]. The inception of processing rubber, commences from the harvesting, at the time of collecting latex sap. Further, to make this material bulky, acid is included in latex. This fluid is reshaped into sheets through a mill that dispenses water to make it dry. Afterward, prevulcanization occurs where chemical reactions take place on sheets at low temperatures. The heating hardens it, and finally, the process is finalized [23].

Vulcanization involves chemical procedure wherein rubber is blended with activators, accelerator, and Sulfur at the temperature of 140–160 °C (Fig. 4.6). Cross-linking occurs among long molecules of rubber, to add up the tensile strength, tenderness, and resilience toward weather [24].

- Process of introducing crosslinking between polymeric chains

Cross-linking aims to make bonding in polymers. The methods of cross-linking the synthetic materials are categorized into cross-linking with polymerization and post-cross-linking [25].

- Vulcanization agent Rubber–Sulfur.

Vulcanizing the rubber with mostly used vulcanizing agent Sulfur initiates the cross-linking of rubber that is not saturated [26]. However, Sulfur, as a vulcanizing agent, does not proceed swiftly. When rubber and Hydrocarbons make chemical reactions, this process generally involves double bonds or C=C, along with it every cross-linking needs 40–55 atoms of Sulfur without accelerator. Therefore, vulcanization is not corrosion resilience

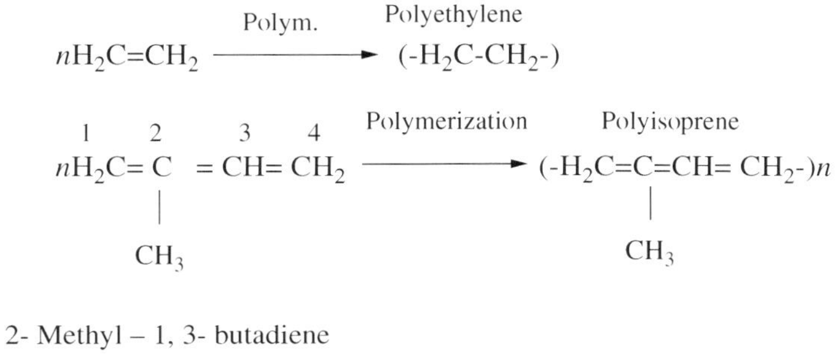

Fig. 4.5 Example of synthetic rubber.

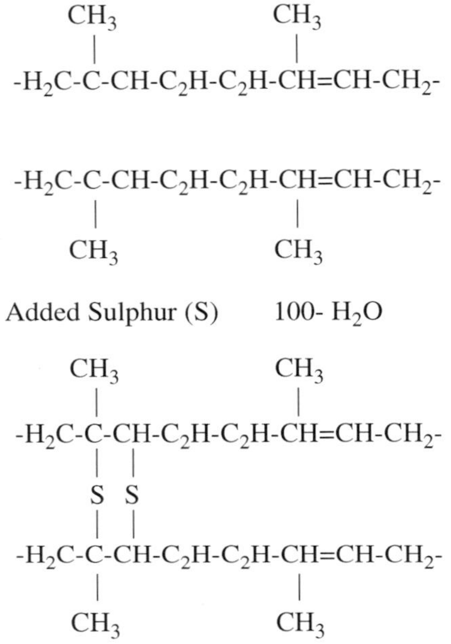

Fig. 4.6 Vulcanization of synthetic rubber [24].

and could not add a significant level of mechanical characteristics for rubber usability. The strategy to counter this shortcoming is the usage of the accelerator, which has now part of rubber compounding [27].

4.1.5.1 Example of synthetic rubber

Buna N rubber or Nitrile rubber: It is composed of acrylonitrile and butadiene—two copolymers. Since acrylonitrile has fluctuating organic fluid when blended with butadiene, a chemical reaction is thus stimulated to form Buna N rubber [28].

Example of Buna N.

$$\underset{\text{1, 3- butadiene}}{CH_2{=}CH\text{-}CH{=}CH_2} + \underset{\text{Acrylonitrile}}{nCH_2{=}CH\text{-}CN} \xrightarrow{\text{Polymerization}} \underset{\text{NBR/Buna-N}}{(\text{-}CH_2\text{-}CH{=}CH\text{-}CH_2\text{-}CH\text{-}CN\text{-})n}$$

Buna-S rubber: It is commonly found synthetic rubber and is also known as a random copolymer that is composed by emulsion polymerization of a blend that contains 1.3 butadiene and styrene with peroxide catalyst at 5 °C. Another name of Buna-S is Styrene-butadiene rubber (SBR), where Bu denotes butadiene and Na refers to sodium while styrene alludes S. This process undergoes vulcanization [29]. A brief description of the differences between Buna-N and Buna-S is provided in Table 4.4. The difference between degrees of resistibility of both types of synthetic rubber shown in Table 4.5.

Example of Buna S Rubber.

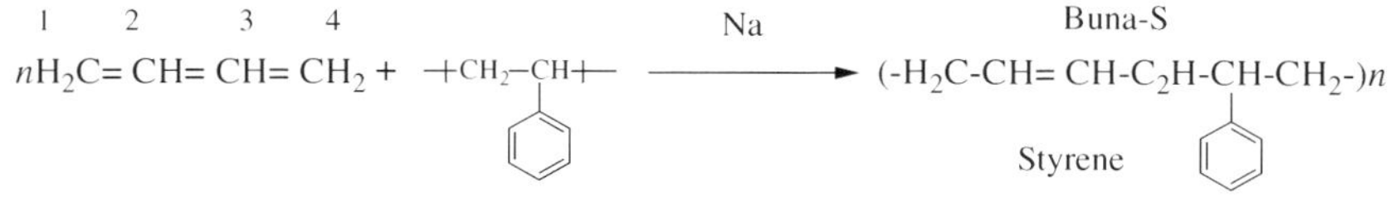

Table 4.4 The difference between Buna N and Buna-S rubber.

Buna-N	**Buna-S**
Nitrile butadiene rubber (NBR) Butadiene $CH_2{=}CH{-}CH{=}CH_2$	Styrene-butadiene rubber (SBR) Butadiene $CH_2{=}CH{-}CH{=}CH_2$
Monomers used	
Acrylonitrile $CH_2{=}CH{-}CH$	Styrene 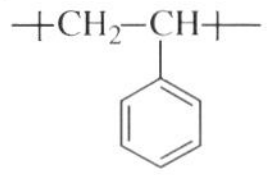
Repetitive unit	
• A mixture of rubber butadiene and acrylonitrile (15%–40%) undergoes emulsion polymerization • The initiator used is peroxide	• 75% styrene and 25% butadiene in the presence of $K_2S_2O_8$ undergoes emulsion polymerization
Properties	
• It's highly resistant to heat • It's resistant to oils and other organic solvents	• Buna-S is highly elastic and resistant to abrasion • It's easily attracted by oxygen and ozone • It's not resistant to oils and other organic solvents • It can be vulcanized by using Sulfur
Uses	
Used for making gaskets, conveyor belts, synthetic leather, etc.	Used for making automobiles, tires, shoes, floor mats, belts, etc.

Table 4.5 Difference between degrees of resistibility of both types of synthetic rubber.

Resistance types	Buna-N	Neoprene
Sunlight	Weak	Very good
Ozone	Weak	Good
Oxidization	Good	Good
Gas permeability	Good	Good
Mostly resistance to	Cases, oils and fuels Aliphatic hydrocarbon	Moderate chemicals and acids Oils, fats, greases, and solvers
Least resistance to	Ketones and aromatic oils Sunlight, weather, and flames	Esters and ketones Chlorinated aromatic, and nitro hydrocarbon

4.2 Methods and techniques for producing synthetic rubber

Although a natural source founds natural rubbers, certain methods are employed to enhance the physical or chemical process. Similarly, synthetic rubbers are processed through different methods, particularly to implant and add up the properties via physical or chemical means. The following are the methods and techniques to produce synthesized rubber.

4.2.1 Polymerization process

Mostly, the synthetic rubber is made from the same methods and techniques that are used to manufacture polymers [30–33]. Synthetic rubber (cis-1,4-polyisoprene) (Fig. 4.7) is manufactured by radical chain polymerization wherein the monomers (isoprene) are involved. Chemicals are detached during the pyrolyzing natural rubber, which leads to figure out the founding block. The types of polymerization are block or bulk polymerization, polymerization in solutions, bead or pearl polymerizations, and emulsion polymerization. The process is used to acquire high molecular weight.

Bead polymerization is processed by water with the dispersal of the unsolvable monomers to make tiny droplets of a specified volume. It is akin

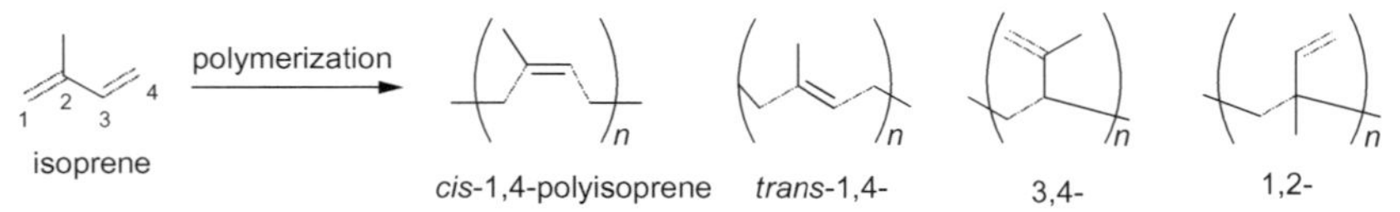

Fig. 4.7 Polymerization synthetic rubber of polyisoprene [34].

to bulk polymerization, whereas it is created through microreactors that facilitate the process of heat prevention. Amid the process, it stipulates the usage pf colloids to avert droplet merger [35, 36]. Lastly, emulsion polymerization is performed by the addition of a water unsolvable monomer to a fluid of an emulsifier. Thus, the process of emulsion polymerization is carried out to manufacture synthetic rubber [8].

4.2.2 Compounding

Rubber is mostly mixed with additives. Through the process of compounding, chemicals are added for vulcanization, i.e., sulfur. Additives comprise of fillers that perform either to augment the rubber's mechanical characteristics (reinforcing fillers) or to advance the rubber in order to diminish the cost (nonreinforcing fillers). The method of compounding is used to design the specific rubber for the purpose of mollifying available applications with relation to properties, cost, and process [37]. Fig. 4.8 shows a simple process of rubber compounding.

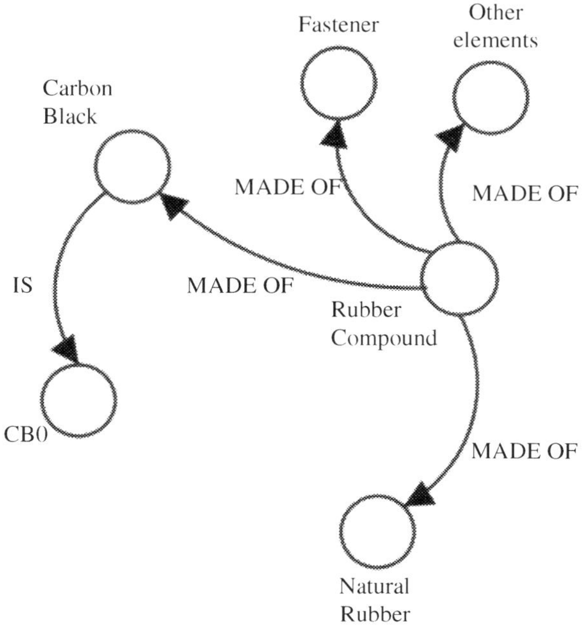

Fig. 4.8 The simple process of rubber compounding [38].

4.2.3 Mixing

The additives are required to mix well along with the vase rubber to get the appropriate dispersion of constituents. Untreated rubber is held with an increased level of thickness; thus, the mechanical working of the rubber can add up the heat to 150 °C (300 °F). When agents involved in vulcanization at the initial phase of mixing, vulcanization takes place [37].

However, in order to avert hasty vulcanizing, the process of two-stage mixing is adopted. Stage 1 is related to carbon black and further nonvulcanization additives that are associated with raw rubber. The appellation of master batch is employed to describe the first-stage mixture. On the other hand, stage two occurs upon the completion of the stage-1 process of mixing. Here, time is allowed for cooling. In this stage, mixing is processed by adding the vulcanizing agents (Fig. 4.9) [37].

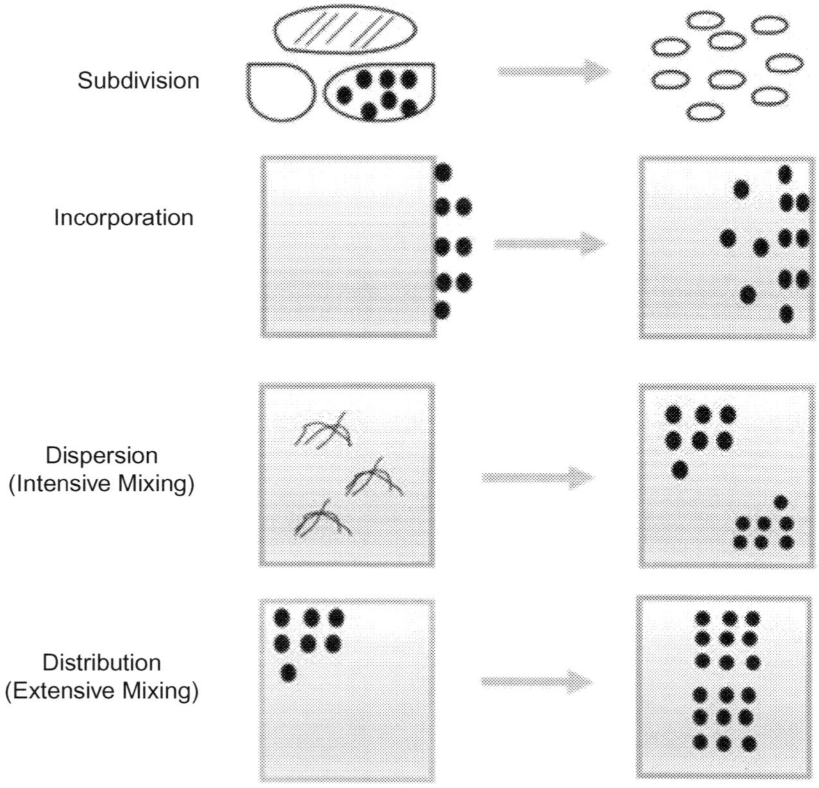

Fig. 4.9 Schematic view of mixing rubber phases.

4.2.4 Latex processing

In Latex processing, rubber is dispersed in the form of particles amid reaction with watery suspension. Generally, the types of rubber can be modified into latex. Varied techniques are employed to acquire multiple levels of diffusions, which may include colloid milling for coarse, ultrasonic dispersal of the particles, ball milling, and simple stirring or colloid milling for coarse dispersal. Latex processing is employed to acquire thin stuff: hot water containers, condoms, catheters, gloves, balloons, gloves, and bladders (Fig. 4.10).

Although, slurry or coarser particle dispersal is suitable for latex foam that is applicable in carpet backing of rubber. For blending emulsion or latex dispersal, below-mentioned additives are commonly used [37].

- Stabilizers were to uphold the particles of rubber for suspension stability. Some of the stabilizers are inorganic complexes such as soaps and phosphates, and sulfonates.
- *Sulfur*: It serves as a curing agent and is used to accelerate the process with xanthates, thiazoles, dithiocarbamates, and thiurams.
- *Other materials*: Solidifying materials along with cellulose derivatives, glue, and casein.
- Deterioration inhibitors added with amine-based antioxidants, substituted phenols, and waxes.
- PH adjusters and formaldehyde, ammonia, sodium, and potassium hydroxide.
- *Poisonous substances*: These substances usually include halogen derivatives of halogen, used to eliminate degenerative impurities by bacteria and fungi.

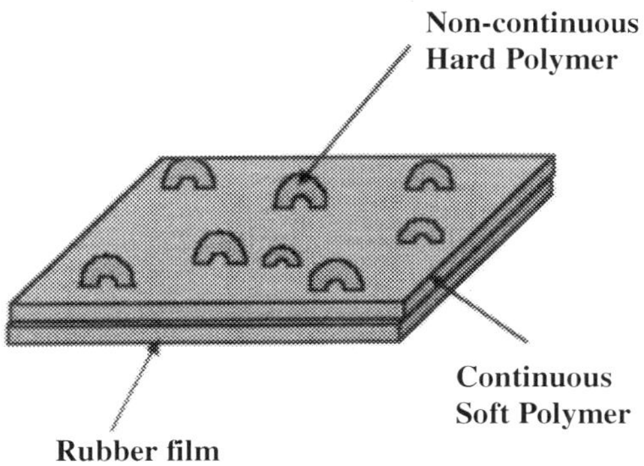

Fig. 4.10 Gloves were manufacturing from latex [39].

- Coagulant drugs and agents of gelling that contain calcium, aluminum salts, magnesium, formic and acetic acid, cyclohexylamine acetate, and ammonium acetate.

4.2.5 Milling machine process

The milling machine process is employed to mechanize the irregular surfaces through providing the workpiece alongside the rotating cutter that comprises of cutting edges. The machine of milling has a spindle that is motor-driven and functions to rotate around the cutter [30]. Fig. 4.11 exhibits rubber mill rolls that are used in the milling machine process.

4.2.6 Calendering

The process of calendering is held through heated rolls that are produced from the precompounded mass, which is paste-like. This mass is processed by roll nips, which results in the formation of a rubber sheet. Acrylonitrile Butadiene Styrene (ABS) and Poly Vinyl Chloride (PVC) are major components of producing calendered thermoplastics [40]. Fig. 4.12 illustrates the process of rubber calendering.

4.2.7 Mixing machine process

Mixing processes are general operations in industrial process engineering. The main goal of these processes is to make a heterogeneous physical system homogeneous by using manipulating operations. Generally, efficient mixing can be challenging to achieve, especially on an industrial scale [32]. Fig. 4.13 shows a rubber mixer.

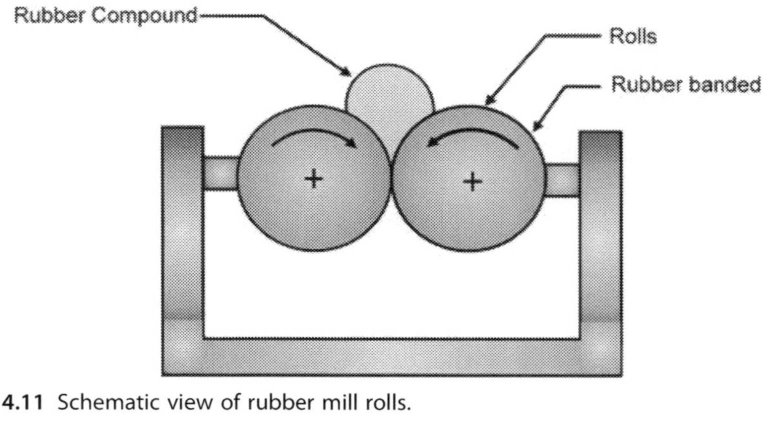

Fig. 4.11 Schematic view of rubber mill rolls.

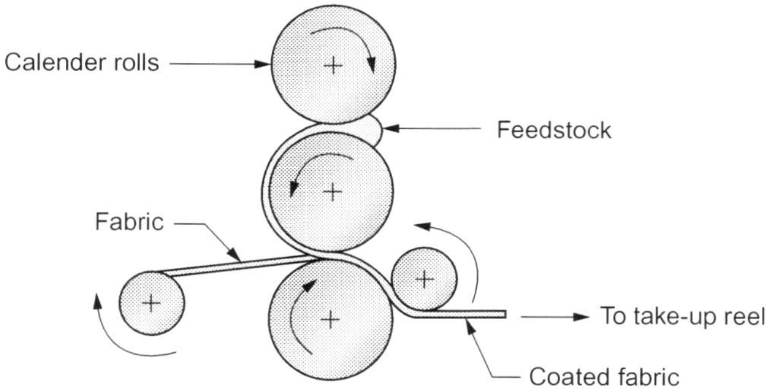

Fig. 4.12 Schematic view of rubber calendering and coating process.

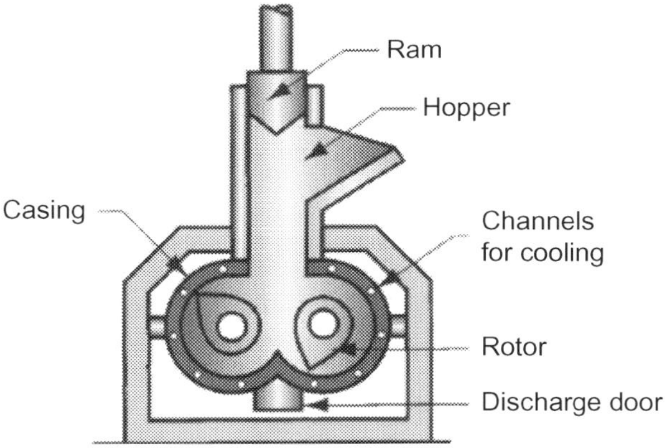

Fig. 4.13 Schematic view of rubber mixers.

4.3 Applications of synthetic rubber

One of the most prominent and leading applications of synthetic rubber is in manufacturing tires (Fig. 4.14) as they are the central components of automobiles, including heavy vehicles, i.e., airplanes and bicycles [40]. The major synthetic rubbers which are used to produce tires are butyl and butadiene rubbers. They are further used in preparing the carcass, tread, and sidewall in which characteristics of rubber are used to separate the required quality and efficiency in the final product of tire. Apart from producing tire, synthetic rubber is a major component of industrial things such as medical equipment, sports goods as well as in dying the paper and materials. Aside

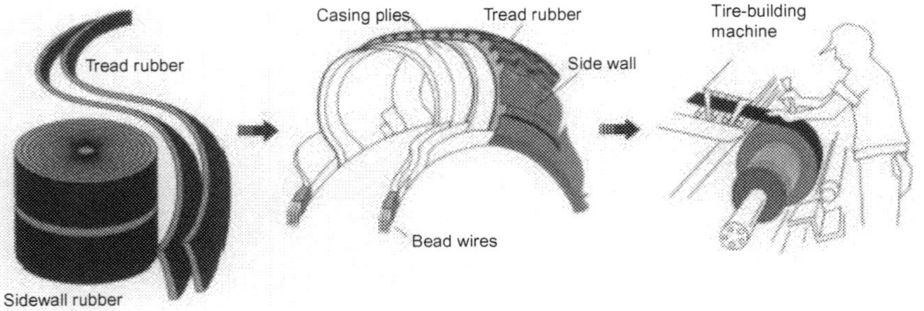

Fig. 4.14 Use of rubber in tire manufacturing [41].

from the extensive usage of synthetic rubber, it has also surpassed the role of natural rubber more broadly as they possess optimized performance, i.e., greater resilience and resistance toward chemicals [42, 43]. Despite these varied applications of synthetic rubber, there are projections for the hike in the demand of synthetic rubber in the future due to its extensive use for manufacturing the tires [44].

Besides the aforementioned applicability of synthetic rubber, they are also used with natural rubber in the form of polymers or compounds. Their combined materials are used in coating papers, backing the carpet, and combined materials are to provide perfect shape to the objects like gloves. Moreover, mattresses and sponges are also made from the extended rubber. Their applications are also apparent for making gloves, the thin structures of condoms, balloons, while in thicker coatings, they are used to make objects like engine mounts and other numerous items.

Another significant contribution of synthetic rubber is found in manufacturing the sealing trips. Due to the appropriate weather conditions, it is majorly used in EPDM (ethylene propylene diene monomer rubber) where the systems of sealing are made through an extrusion and ultimate developments, which are usually assembled as coupled, coated or dyed. More specifically, the Neoprene is a kind of polymer that plays a major role as chloroprene for resisting oil. It has the property of heat resistance more than 150 °C that is favorable for producing specific materials such as parts of machinery and pipelines that transport petroleum products [45–47]. Also, it is used as insulation in electrical cables. More specifically, some of the synthetic rubbers are described in the following, along with their indispensable applications [48]:

- Nitrile rubber is acquired from copolymerizing butadiene, along with acrylonitrile, an indispensable component for resisting fats, oils, and diluters.

- Butyl rubber stems by copolymers of isoprene or butadiene and isobutane, which are resistant to get dissolved, serve as a major component of producing tire's tubes.
- Buna rubber comprises of various monomers that work as a chemical agent, acquired by copolymerizing. It serves as the major component in creating tires for cars. It accounts for around 60% of the global manufacturing of synthetic rubber.
- Polychloroprene (CR) is used to produce gaskets, automobile fan belts, laptop sleeves, and hoses.
- Styrene-butadiene (SBR) are used for manufacturing bus tires, airplane tires, carousel conveyor, and footwear sole.
- Ethylene propylene diene monomer (EPDM) is the major component for producing solar panel heat gleaners, machine-driven vibrators, electric insulation, and convector.
- Acrylonitrile butadiene (NBR) is used in lab gloves, oil seals, synthetic leather, Vee belts, and toric joint.
- Polysiloxane (SI) coatings, sealants, molds (in dentistry, etc.).

Additionally, more uses of synthetic rubber are found in chewing gums, belts, and molding [41]. Door and window profiles, hoses, straps, floor-covering, dampeners, and floorings in the automotive industry are produced from synthetic rubbers [49, 50].

4.4 Conclusion

Synthetic rubber or elastomers are the polymers that are characterized by long molecular chains. While the natural rubber is made from the latex that is blended with other materials, it's to acquire the desired quality and usability. Through cross-lining, the strength of the rubber is augmented. On account of their higher degree of physical, chemical, and mechanical properties, such as: higher levels of durability and chemical resilience, they have surpassed the demand of products that are made from the natural rubber. The monomer that is the major component of natural rubber is cis-1-4 isoprene, whereas, for obtaining synthetic rubber, there is no specification for monomers. Synthetic rubber is prominently used in manufacturing automobile parts such as tires. Polymerization, latex processing, compounding, and mixing are the methods through which this synthetic form of rubber is made, and the required features are implanted in the synthetic rubbers. Moreover, it is projected that synthetic rubber will completely swap the demand for natural rubber in the future.

References

[1] S.O. Adesanya, A.S. Onanaye, O.G. Adeyemi, M. Rahimi-Gorji, I.M. Alarifi, Evaluation of heat irreversibility in couple stress falling liquid films along heated inclined substrate, J. Clean. Prod. 239 (2019) 117608.
[2] O.E. Al-Harbi, S.M. Almansour, Y.A.M. Alomair, N.B. Altulohi, M. Osman, I.M. Alarifi, Evaluate the mechanical and thermal behavioural deformation of auxetic composite nano-carbon fibers, in: ASME 2019 International Mechanical Engineering Congress and Exposition, American Society of Mechanical Engineers Digital Collection, 2019.
[3] A.G. Almotery, S.M. Alsaiari, A.Y. Alyoussef, F.M. Alsahli, M.O. Alharbi, M.A.A.T. El-Bagory, I.M. Alarifi, Fabrication and characterization of the recycling of composite palm materials, shell, leaves and branches in Saudi Arabia, in: ASME 2019 International Mechanical Engineering Congress and Exposition, American Society of Mechanical Engineers Digital Collection, 2019.
[4] A. Al-kawaz, Synthetic Rubber. Petrochemical Industry, 2017, Retrieved from http://www.uobabylon.edu.iq/eprints/publication_3_1057_164.pdf.
[5] R.N. Datta, Rubber Curing Systems, vol. 12, iSmithers Rapra Publishing, 2002.
[6] T. Sumathia, A. Srilakshmia, V.S. Kotakadib, D.V.R. Saigopala, Role of fungal enzymes in polymer degradation: a mini review, Res. J. Pharm. Biol. Chem. Sci. 2 (5) (2012) 1694.
[7] Surya, I., Maulina, S., & Ismail, H. (2018, January). Effects of alkanolamide and epoxidation in natural rubber and epoxidized natural rubbers compounds. In IOP Conference Series: Materials Science and Engineering (vol. 299, p. 012061).
[8] J. Cawley, C.J. Ruhm, The economics of risky health behaviors, in: Handbook of Health Economics, vol. 2, Elsevier, 2011, pp. 95–199.
[9] B. Hazeltine, Household technologies, in: Field Guide to Appropriate Technology, Academic Press, 2003, pp. 665–729.
[10] Neoprene: The First Synthetic Rubber. n.d. Available at: https://chlorine.americanchemistry.com/Science-Center/Chlorine-Compound-of-the-Month-Library/Neoprene-The-First-Synthetic-Rubber/.
[11] Ophardt, C. (2019). Natural and Synthetic Rubber. Available at: https://chem.libretexts.org/Bookshelves/Organic_Chemistry/Map%3A_Organic_Chemistry_(Smith)/Chapter_31%3A_Synthetic_Polymers/31.5%3A_Natural_and_Synthetic_Rubbers.
[12] Yashoda. (2016). Natural Rubber vs Synthetic Rubber. Available at: https://pediaa.com/difference-between-natural-rubber-and-synthetic-rubber/.
[13] Vaysse, L., Bonfils, F., Thaler, P., & Sainte-Beuve, J. (2009). Natural Rubber. Available at: https://www.researchgate.net/publication/47377711_Natural_Rubber.
[14] D.R. Askeland, P.P. Phulé, Ciencia e ingeniería de los materiales, Thomson, 2004.
[15] V.R. Sastri, Plastics in Medical Devices: Properties, Requirements, and Applications, William Andrew, 2013.
[16] J.M. Arguello, A. Santos, Hardness and compression resistance of natural rubber and synthetic rubber mixtures, Journal of Physics: Conference Series 687 (1) (2016, February) 012088. IOP Publishing.
[17] Y. Ikeda, Understanding network control by vulcanization for sulfur cross-linked natural rubber (NR), in: Chemistry, Manufacture and Applications of Natural Rubber, Woodhead Publishing, 2014, pp. 119–134.
[18] C.S.S.R. Kumar, A.M. Nijasure, Vulcanization of rubber, Resonance 4 (1997) 55–59.
[19] V.V. Rajan, W.K. Dierkes, R. Joseph, J.W.M. Noordermeer, Progress in Polymer Science 31 (9) (2006) 811. Retrieved from https://www.aceprodcon.com/how-rubber-is-made-for-industrial-uses/.

[20] M.A. Akiba, A.S. Hashim, Vulcanization and crosslinking in elastomers, Prog. Polym. Sci. 22 (3) (1997) 475–521.
[21] R. Pornprasit, P. Pornprasit, P. Boonma, J. Natwichai, Determination of the mechanical properties of rubber by FT-NIR, Journal of Spectroscopy 2016 (2016).
[22] R.L. Bebb, Chemistry of rubber processing and disposal, Environ. Health Perspect. 17 (1976) 95–102.
[23] Anon n.d. How Rubber Is Made. Available at: https://www.aceprodcon.com/how-rubber-is-made-for-industrial-uses/.
[24] A.B. Nair, R. Joseph, Eco-friendly bio-composites using natural rubber (NR) matrices and natural fiber reinforcements, in: Chemistry, Manufacture and Applications of Natural Rubber, Woodhead Publishing, 2014, pp. 249–283.
[25] T. Oyama, N. Shimada, A. Takahashi, Utilization of polyarylates having chemically introduced diazonaphthoquinone structure for reaction development patterning, J. Photopolym. Sci. Technol. 28 (2) (2015) 219–227.
[26] J. Wręczycki, D.M. Bieliński, R. Anyszka, Sulfur/organic copolymers as curing agents for rubber, Polymers 10 (8) (2018) 870.
[27] Anon n.d. Vulcanization & Accelerator. Available at: http://www.nocil.com/Downloadfile/DTechnicalNote-Vulcanization-Dec10.pdf.
[28] Anon, The Synthetic Rubber Production Process, 2018, Available at: https://www.aquasealrubber.co.uk/uncategorized/the-synthetic-rubber-production-process/#:~:text=Buna%20N%20rubber%20(also%20known,occurs%2C%20producing%20Buna%20N%20rubber.
[29] Buna-S Rubber, 2010. Available at: http://synthetic-rubbers.blogspot.com/2010/12/buna-s-rubber.html#:~:text=Buna%2DS%20Rubber%3A&text=The%20rubber%20obtained%20is%20also,rubber%20in%20its%20physical%20properties.
[30] Anon (n.d. Milling Operations. Available at: http://uhv.cheme.cmu.edu/procedures/machining/ch8.pdf.
[31] S.A. Ashter, Thermoforming of Single and Multilayer Laminates: Plastic Films Technologies, Testing, and Applications, William Andrew, 2013. Figure 4.6 shows rubber calendering and coating process.
[32] V. Mosorov, Applications of tomography in reaction engineering (mixing process), in: Industrial Tomography, Woodhead Publishing, 2015, pp. 509–528.
[33] Anon (n.d.). Rubber Chemical Compound. Available at: https://www.britannica.com/science/rubber-chemical-compound/Synthetic-rubber-production.
[34] Anon n.d. New Catalysts for Regio- and Stereospecific Polymerization. Available at: http://www2.riken.jp/lab/organometallic/engl/research_1_e.html.
[35] L. Luo, N.A. Shah, I.M. Alarifi, D. Vieru, Two-layer flows of generalized immiscible second grade fluids in a rectangular channel, Mathematical Methods in the Applied Sciences 43 (3) (2020) 1337–1348.
[36] B. Mahanthesh, I.L. Animasaun, M. Rahimi-Gorji, I.M. Alarifi, Quadratic convective transport of dusty Casson and dusty Carreau fluids past a stretched surface with nonlinear thermal radiation, convective condition and non-uniform heat source/sink, Physica A: Statistical Mechanics and Its Applications 535 (2019) 122471.
[37] Health and Safety Executive. (n.d.). Introduction to rubber processing and safety issues. Retrieved from https://www.hse.gov.uk/rubber/introduction-to-rubber-processing.pdf.
[38] S. Bandini, F. Sartori, CKS-net: a conceptual and computational framework for the management of complex knowledge structures, in: IICAI, 2005, December, pp. 1742–1761.
[39] E. Yip, P. Cacioli, The manufacture of gloves from natural rubber latex, J. Allergy Clin. Immunol. 110 (2) (2002) S3–S14.

[40] S.A. Ashter, Thermoforming of Single and Multilayer Laminates: Plastic Films Technologies, Testing, and Applications, William Andrew, 2013. Figure 4.11 shows rubber calendering and coating process.
[41] Tire. (n.d.). Available at: http://www.madehow.com/Volume-1/Tire.html.
[42] R. Asmatulu, K.S. Erukala, M. Shinde, I.M. Alarifi, M.R. Gorji, Investigating the effects of surface treatments on adhesion properties of protective coatings on carbon fiber-reinforced composite laminates, Surf. Coat. Technol. 380 (2019) 125006.
[43] K.G. Kumar, M.N. Khan, M. Osman, A.R. Alharbi, M. Rahimi-Gorji, I.M. Alarifi, Slip flow over a non-Newtonian fluid through a Darcy–Forchheimer medium: numerical approach, Modern Physics Letters B 33 (35) (2019) 1950448.
[44] D. Misurelli, C. Raymond, Synthetic Rubber, U.S. International Trade Commission, Washington, DC, 1997. Retrieved from: https://www.usitc.gov/publications/docs/pubs/industry_trade_summaries/pub3014.pdf.
[45] K.G. Kumar, B.S. Avinash, M. Rahimi-Gorji, I.M. Alarifi, Optical and electrical properties of Ti1-XSnXO2 nanoparticles, J. Mol. Liq. 293 (2019) 111556.
[46] A. Asadi, I.M. Alarifi, V. Ali, H.M. Nguyen, An experimental investigation on the effects of ultrasonication time on stability and thermal conductivity of MWCNT-water nanofluid: finding the optimum ultrasonication time, Ultrason. Sonochem. 58 (2019) 104639.
[47] M. Bayat, I.M. Alarifi, A.A. Khalili, T.M. El-Bagory, H.M. Nguyen, A. Asadi, Thermo-mechanical contact problems and elastic behaviour of single and double sides functionally graded brake disks with temperature-dependent material properties, Sci. Rep. 9 (1) (2019) 1–16.
[48] Uses of Synthetic Rubber. (n.d.). Retrieved from: http://sciencehomeworkhelp.weebly.com/uses-of-synthetic-rubber.html.
[49] H.S. Fogler, Elements of Chemical Reaction Engineering, third ed., Prentice-Hall India Publications, 2002.
[50] R.H. Perry, D.W. Green, Perry's Chemical Engineering Hand Book, seventh ed., 1997.

CHAPTER 5

Synthetic foam

Abbreviations

AFF aqueous film foam
AR alcohol resistant
CTE coefficient of thermal expansion
EAM energy absorbing material
FFFP film-foaming fluoroprotein foam
HGMs hollow glass microspheres
HOVs human-operated vehicles
kN kiloNewton
PI polyimide
PP polypropylene
PU polyurethane
PVC polyvinyl chloride
ROVs remotely operated vehicles
SEM scanning electron microscope
TEC Trelleborg Emerson & Cuming
TOB Trelleborg Offshore, Boston

5.1 Introduction

The word syntactic is derived from the Greek word "syntaxis," which refers to a science that deals with blending and order. Thus, the mixture of air foams that comprises of a mechanism that produces compounds and has void is termed as synthetic foam. The word foam further reflects the cellular structure of composites. The advent of epoxies has been laid the foundation for making synthetic foam as an organized material [1]. The process of enhancing the qualities of already available materials and introducing new goods and expertise are the result of the increased necessity and pursuit of humans [2].

Synthetic foams are composites that are developed through the hollow spheres implanted with the polymeric medium. In the past three decades, simple two-phase polymer matrix foams were used to manufacture with a hollow glass of polymer sphere. The viability of this material was as a buoyance compound that was used in maritime and undersea engineering [3, 4]. The profound viability and characteristics of synthetic foams multiplied its

Synthetic Engineering Materials and Nanotechnology
https://doi.org/10.1016/B978-0-12-824001-4.00009-0

development and usability in the sectors of aeronautics, space, and maritime [5]. The use of synthetic foams was also introduced in sandwich-structured compounds.

Significant numbers of studies, theoretical [6], and tentative [7, 8] have been published that provided the properties of its core sandwich compounds. Also, different published studies examined the impacts of dissimilarity in microballoon-sized fractions based on their physical and mechanical characteristics [9, 10]. Notably, when it is adding up to the volume of microballoon fraction, it dwindles the density and firmness of synthetic foam [11].

Hint statement

Synthetic foams are produced by adding hollow thin-walled tiny components along with the matrix compounds that are called as microspheres or microballoons.

Nevertheless, while maintaining the volume fraction steady, variation in the characteristics of synthetic foams will be more useful. In order to understand the approach, an example of utilizing microballoons with dissimilar wall-thickness can be taken into consideration as dissimilarity in wall thickness can lead to further augmentation in the mechanical strength of synthetic foams [12]. Perhaps, when it comes to assessing the degree of absorbing water, synthetic foams have a lesser degree owing to their closed-cell structure in comparison with matrix materials and open-cell foams [11, 13]. Synthetic foams are manufactured by the addition of hollow thin-walled tiny components along with the matrix compounds that are termed as microspheres or microballoons [14]. While, the mechanical characteristics of synthetic foams can be augmented by determining a suitable mixture of matrix materials and hollow microspheres [15]. Microspheres with varied wall-thickening and diameter in volume fraction, are also used in matrix materials to enhance features of the synthetic foams [16]. Due to the reliance on real-life applicability, synthetic foams are undergone with several techniques of loading, such as compressive, tensile, and shear loading [17].

Additionally, it has been investigated that upon compression, the stress–strain curve of synthetic foams that are made up of glass are categorized in three unique regions, i.e., the linear elastic, plateau, and densification regions [18]. More specifically, the liner region refers to the elastic deforming of synthetic foam. The plateau region, on the other hand, having a trivial alteration in stress, is parallel to the unproductive disaster of the glass projectile owing to the devastation of microspheres. On the contrary, the densification region corresponds to the filling up the cavity with the help of fragments and the crushing of the matrix process. The region is further associated with an increased level of stress that leads to trivial changes in strain [18].

The major disadvantage of synthetic foams that are constituted from glass-based microspheres is their arduousness and fragility as it is vulnerable to severe damages when they encounter huge strain [16]. Fig. 5.1 displays the scanning electron microscope (SEM) images of synthetic foams [17].

Yousaf et al. (2020) produced synthetic foams and combined the microspheres in the matrix at the nominal concentration ranges from 0% to 40%. Two kinds of microspheres (551 and 920 grades) (Fig. 5.1) long with varying diameter and wall thickness were blended into matrix materials to have knowledge about their degree of responsiveness under the condition of compression. The prescribed mass of all samples of materials was

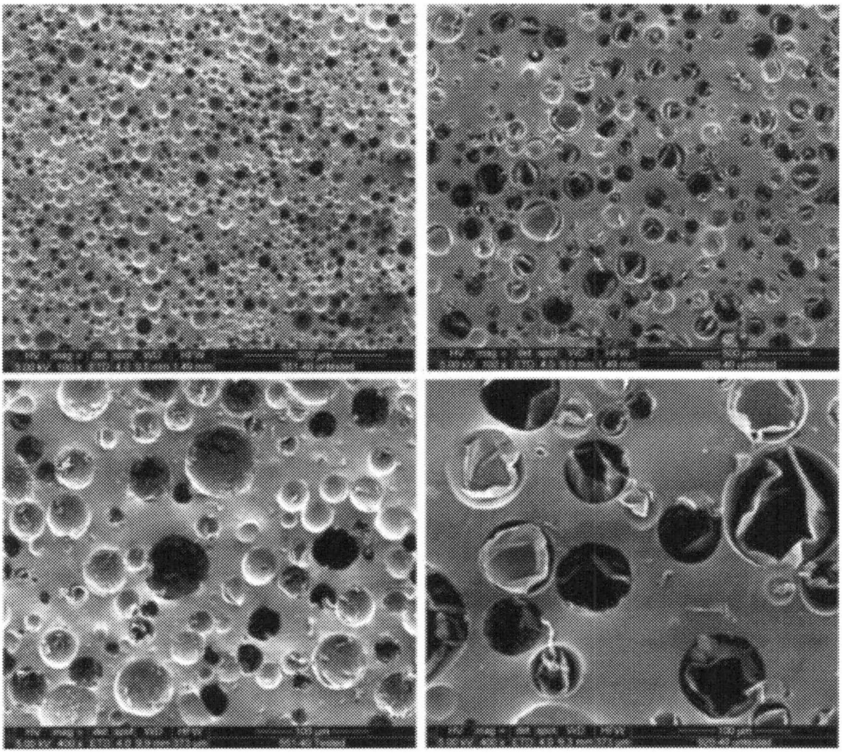

Fig. 5.1 SEM images of syntactic foams 551–40% (above) and 920–40% (below) at different magnifications [17].

accumulated through gauging the density. The microspheres were composed of enclosing a fluorocarbon and acrylic copolymer [17].

Rendering to the study of Lansing [19], mechanical characteristics of foams which are made up of gelatinized starch and prepolymer polyurethane include; energy absorbency, compact pressure, modulus, and hypothetical moderating of static shock. On the contrary, physical features have been figured out based on their thickness, the volume of the cell, and cell count. More importantly, Lansing [19] analyzed the data for assessing the mechanical properties on an Instron universal testing frame model 5567 that was installed with a 5 kN load cell and # T489–74 density platens valued for 100 kN [19].

Lansing [19] also gauged the density of foam through Ohaus Scout Pro SPE2001 analytical balance and Ohaus scale density determination kit along with extracted water [19]. The framework of the kit has been built on the Archimedes principle of displacement, as provided in Fig. 5.2 [19]. One of the distinctive features of the kit is its ability to gauge the compactness of taster that usually drifts in fluid, i.e., acetone [19].

5.1.1 Types of synthetic foams

Two fundamental types of polymers are two-phase synthetic foam and three-phase synthetic foam (Fig. 5.3). Two-phase synthetic foam comprises of the binder phase and hollow particles. However, an inadequate concentration of binder is found to fulfill the site of interstitial among particles. On the other hand, three-phase goes through the binder, microballoons, and

Fig. 5.2 Density determination kit setup [19].

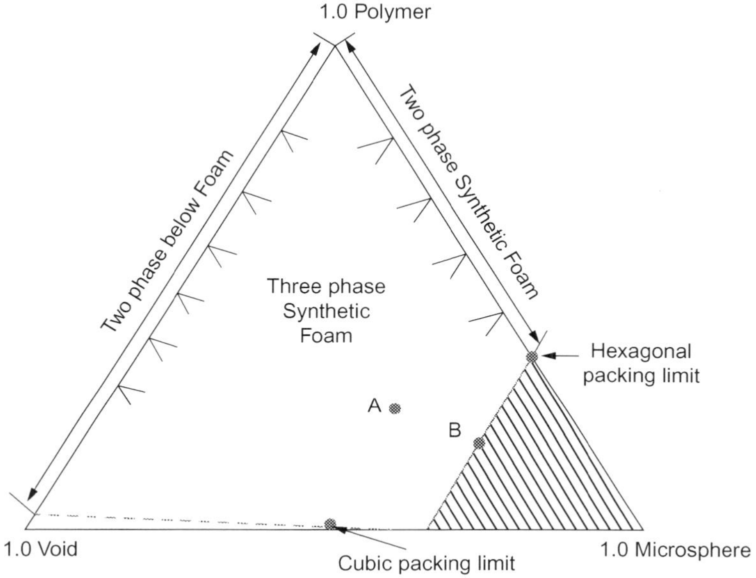

Fig. 5.3 Kinds of synthetic foam.

interstitial void. The material of two-phase is held with the adequate matrix material and hollow particles that can fulfill the vacant of interstices. Generally, most of the synthetic foams are two-phase, i.e., and epoxy matrix possesses microballoons of glass. While three-phase foams are of greater importance as they have lesser density. Thus, the materials of three-phase foaming have more specified characteristics [20].

The process of acquiring low apparent densities require air in the binder. When, the volume of filler surpasses the specified quantity (67 vol ~ o), the volume of binder beings to reduce in comparison with a concentration that is found among the microspheres. If the binder cannot encompass the entire concentration of microspheres, it results in disruption of the homogeneity of the entire mechanism. Consequently, it can originate cavities, which can be evidenced by open pores that are visible on the exterior. Therefore, there is a negative degenerative effect on the characteristics of the materials. Synthetic foams and the characteristics and density of their constituent materials are not hinged on the ratio of the binder. However, there is a major role of the volume, polydispersity, bulk and apparent density, sphericity, and tenderness, and uniformity of the microspheres. Therefore, the fluid concentration of the mixture is related to the volume of the microspheres and bulk or apparent density (Fig. 5.4) [21].

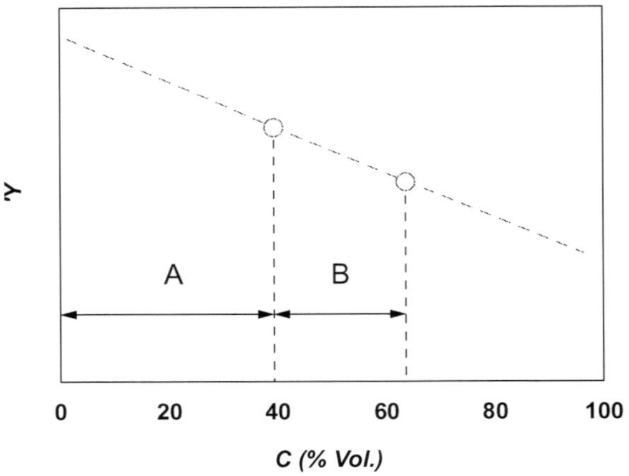

Fig. 5.4 The relationship between the apparent density ('Y) of synthetic foam and the filler concentration (C) the casting composite (A), the molding compositions (B).

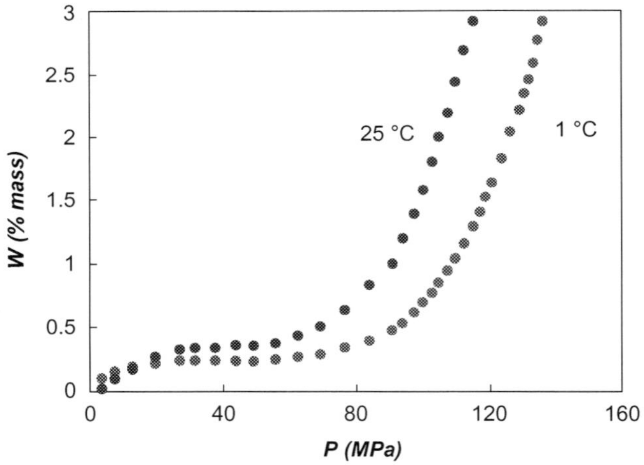

Fig. 5.5 Two different temperature of synthetic foam for water absorption (W) vs pressure (MPa).

An indispensable criterion of synthetic foam is water absorptions owing to the wider applicability in the marine industry. A surge is found in water absorption when there is growing hydrostatic coercion upon examining the hydrostatic coercion destruction (Fig. 5.5). The proportion of accumulating water absorption can be done by applying Eq. (5.1) [22].

$$\rho = \frac{(m_1 - m_2)}{m_1} \times 100 \tag{5.1}$$

where m_1 denotes the initial weight, while m_2 shows the final weight [22].

The earlier technology to obtain foams was composed of a binder (as phenol-formaldehyde) and phenolic microspheres. Further technology to obtain foams includes template carbonization that comprised of hollow microspheres, high porosity melamine, along with cellular mechanism in the form of binder. Nowadays, syntactic carbon foams are produced by microspheres and binders (pitches and polymers), when the fundamental compounds are positioned according to their efficacy and usability [23]. Table 5.1 presents the properties of carbonized synthetic foams.

Hint statement

The mechanical foaming method is eco-friendly and cost-effective.

Apart from carbonized synthetic foams, synthetic foamed plastics serve as another important material and is interchangeably termed as expanded plastics or polymeric foams. It has an apparent density that is reduced considerably, with the help of several cells that are dispensed from the mass. However, the modulus and compressive strength describe widely accepted criteria for attributing the mechanical characteristics of rigid plastics [24]. Table 5.2 displays some of the properties of synthetic foamed plastics.

Some of the significant advantages of synthetic foam are that it has a lower extend toward moisture absorption, and it is characterized to have higher compressive strength in contrast with open-cell structure foams [25]. In the study by Khanna and Gopalan [26], during reinforcing, the rigid polyurethane foams, graphite powders, glass beads, and e-glass chopped fibers were taken. Since, there was no role of chopped glass fiber to reduce the modulus and compressive strength of the foam. Moreover, the graphite

Table 5.1 Properties of carbonized synthetic foams (phenolic binder and phenolic microspheres).

	Filler binder ratio (~mass)			
Parameter	**(90/10)**	**(60/40)**	**(40/60)**	**(30/70)**
Apparent density (kg/m^3) prior to carbonization after carbonization	130 150	210 220	300 300	410 390
Ultimate compression strength at 900 °C (MPa)	1	4	8	17
Compression modulus at 900 °C (MPa)	42	120	294	715

Table 5.2 Properties of synhetic foamed plastics.

Parameter	Binder type: Crosslinked polystyrene	Binder type: Organosilicon resin
Apparent density (kg/m^3)	510	400
Ultimate compression strength (MPa)	35	52
Operating temperatures (°C)	~30 + 175	~50 + 430

powder and glass beads played a significant role in adding up the quality of compressive modulus. Whereas, the compressive strength of the foam remained unaffected. The process of reinforcing the use of *E*-glass fibers contributed to add up the deformation response and tensile of foam. However, reinforced foams caused deterioration in the compressive strength of foams [26]. Through deformation, solid compounds are reshaped. The preliminary transformation is not complex, and it transforms dies or tools synthetically to produce the required level of resilience. The procedure of deformation usually involves in manufacturing materials in a progressive manner from simple to intricate structure [27]. Fig. 5.6 distinguishes between compressive strength and deformation of two synthetic foamed polymers.

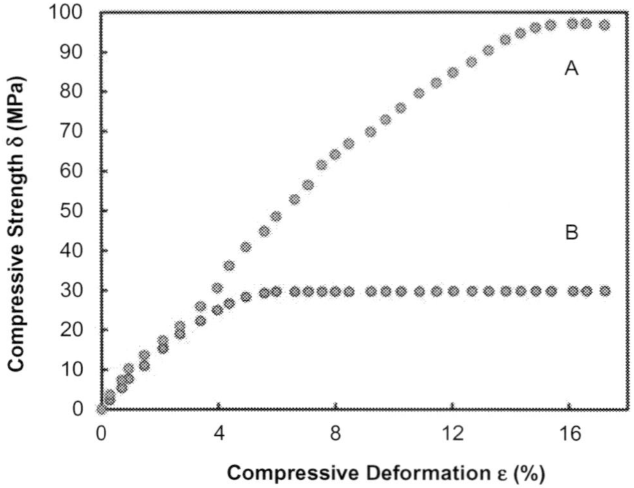

Fig. 5.6 Comparison between compressive strength and deformation, (A) epoxy synthetic foam with phenolic microspheres and (B) epoxy resin.

5.2 Methods and techniques

Synthetic foam is produced through a polymer while employing mechanical, physical, and chemical foaming methods. However, the techniques to the progression of synthetic foam involve foam extrusion and foam injection molding through which supercritical fluid, i.e., carbon dioxide or nitrogen, light-weighted and highly stabilized synthetic foams are obtained [28, 29].

Hint statement

Chemical foaming methods, due to their thermal biodegradability, are suitable for the polymers that have melting viscidness.

5.2.1 Foaming process

The process of foaming polymers can be divided into three phases: cell formation, cell growth, and cell stabilization. Fig. 5.7 exhibits a graphical representation of the foaming process [30].

5.2.1.1 Cell formation

During the process of foaming, a blowing agent is included in a molten polymer. Afterward, a huge volume of gas is obtained by consecutive chemical

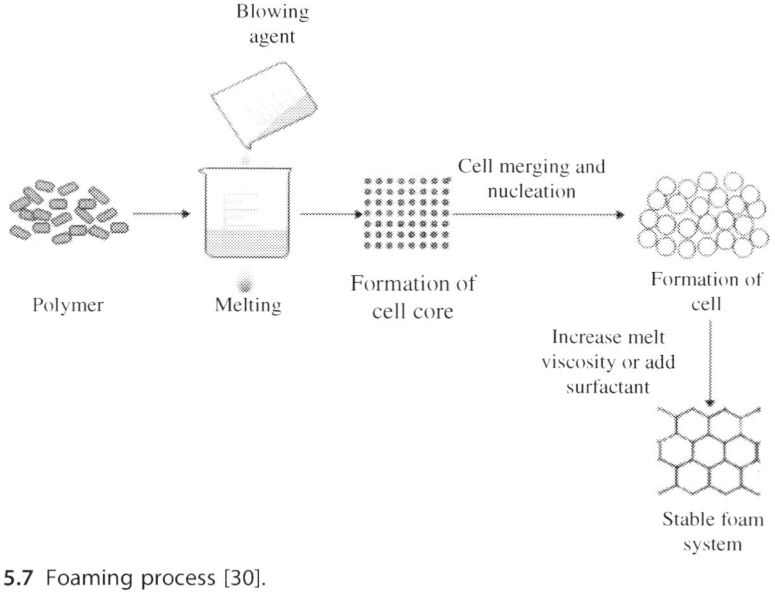

Fig. 5.7 Foaming process [30].

reactions, which results in forming a polymer solution. Once the volume of gas is increased, the suspension transforms into a supersaturated solution. Then, gas is discharged from the solution. Consequently, the discharged gas shapes the cell nucleus by nucleation (Fig. 5.7) [31, 32].

5.2.1.2 Cell growth

Once the cells are formed, the pressure made by gas into the cell is inversely relational to the range of the cell. Following this, if the cell is smaller, it will have to bear more pressure into the cell. If two cells are of different sizes and nearby to each other, the transition of gas would be from the smaller cell to the larger cell. Subsequently, the two cells get merged. Consequential to the nucleation, the number of cells is added along with the extension in the diameter of the cells, leading toward the growth of cells (Fig. 5.7) [33, 34].

5.2.1.3 Cell stabilization

When the process of construction and development of cells takes place, the size of the foam is expanded, making the cell walls diluter. Consequently, the foaming mechanism gets unsteady. However, in such cases, the cells maintain their stability through cooling or by increasing the surfactants (Fig. 5.7) [35, 36].

Hint statement

Synthetic foams have lower compactness, heat resistance, sound resistance, a higher degree of firmness, and are more resilience toward corrosion.

5.3 Methods to produce synthetic foam

Foaming methods ae divided into several methods. The following are some of the major methods of its classification, i.e., mechanical foaming, physical foaming, and chemical foaming [30].

5.3.1 Mechanical foaming

In the process of mechanical foaming, the air is mixed with polymer resins through mechanical mixing. This results in providing resin the shape of the foam (Fig. 5.8) [30]. The mechanical process of mechanical foaming provides certain advantages, such as:

- It has no supplementary foaming agent and comprises simple processes and tools that are easier to use.

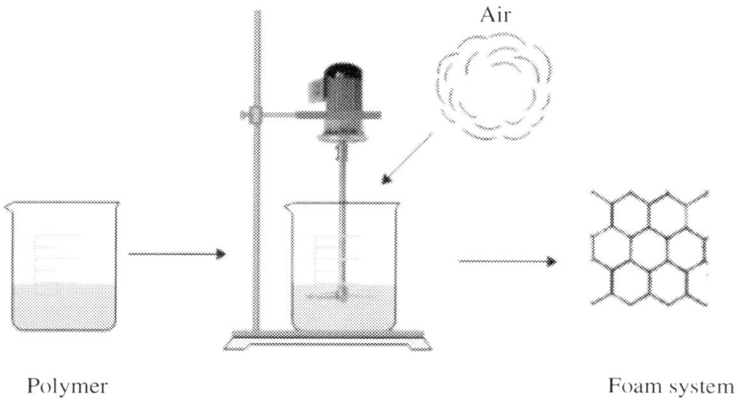

Fig. 5.8 Mechanical foaming method [30].

- It is environmentally friendly and has nonhazardous elements.
- It is cost-effective and has higher efficacy [30].

In contrast to its various advantages, the high demand for the equipment is regarded as its significant demerit [30].

5.3.2 Physical foaming

The process of physical foaming is associated with a low boiling point of the fluid and a polymer. They are mixed and consequently processed to be foamed by asserting pressure and heat (Fig. 5.9) [37]. The process of physical foaming has certain benefits, some of which include:

- It is cost-effective as the physical agents, i.e., carbon dioxide and nitrogen, that are involved in this process are available at relatively lower prices.
- The method does not bring about any environmental degradation and does not dispense harmful aftereffects on the environment. Thus, it is a mostly adopted method.
- The process of physical foaming completes without any residue and produces a less intense impact on the characteristics of foaming polymers [30].

Besides, the above stated useful applications of physical foaming method, potential disadvantages of the method are:

- It is not technically convenient. Rather, it stipulates advance technical specialties.
- The equipment that is prerequisite in this method involves a special injection molding machine and auxiliary tools [30].

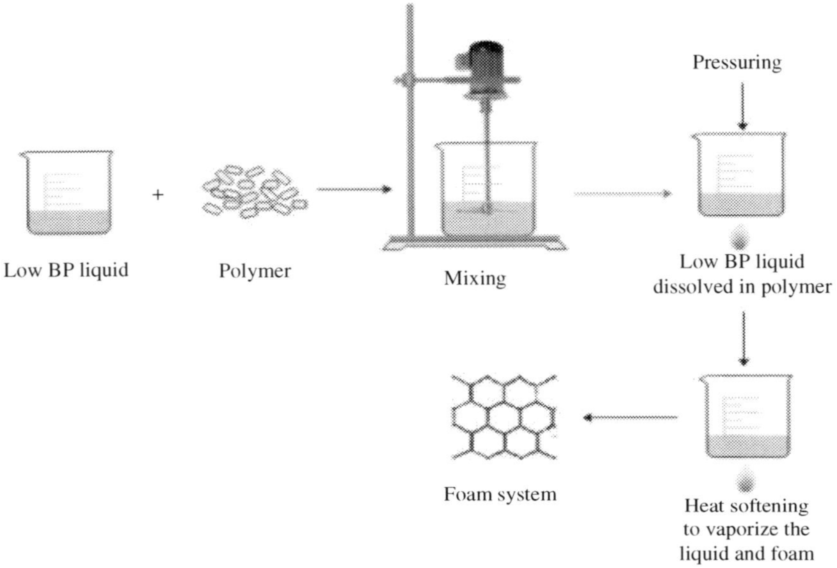

Fig. 5.9 Physical foaming method [37].

5.3.3 Chemical foaming

The inclusion of two different methods holds the process of chemical foaming. In the first method, a blowing agent is mixed with a molten polymer in order to release gas through decomposing. Consequently, the polymer is obtained under high compression and heat (Fig. 5.10) [30]. The second method, on the other hand, involves chemical reactions that take place between two polymers to yield inactive gases, resulting in the production of foam (Fig. 5.6) [30].

Certain advantages in relation to adopting chemical foaming methods are summarized as follows:

- Chemical foaming methods are thermally decomposable at a certain temperature, resulting in the release of one or more gases. Therefore, the method is suitable for polymer resins that are possessed with melting viscidity at a certain range of heat.
- Unlike, physical foaming method, it can be executed with a regular injection molding machine.

Other than this, some common disadvantages of the chemical foaming method are:

- It needs higher accuracy in mold manufacturing.
- The mold is not available at a reasonable cost.
- It further entails the second clamping pressure tool that is required amid the process of developing increased pressure [30].

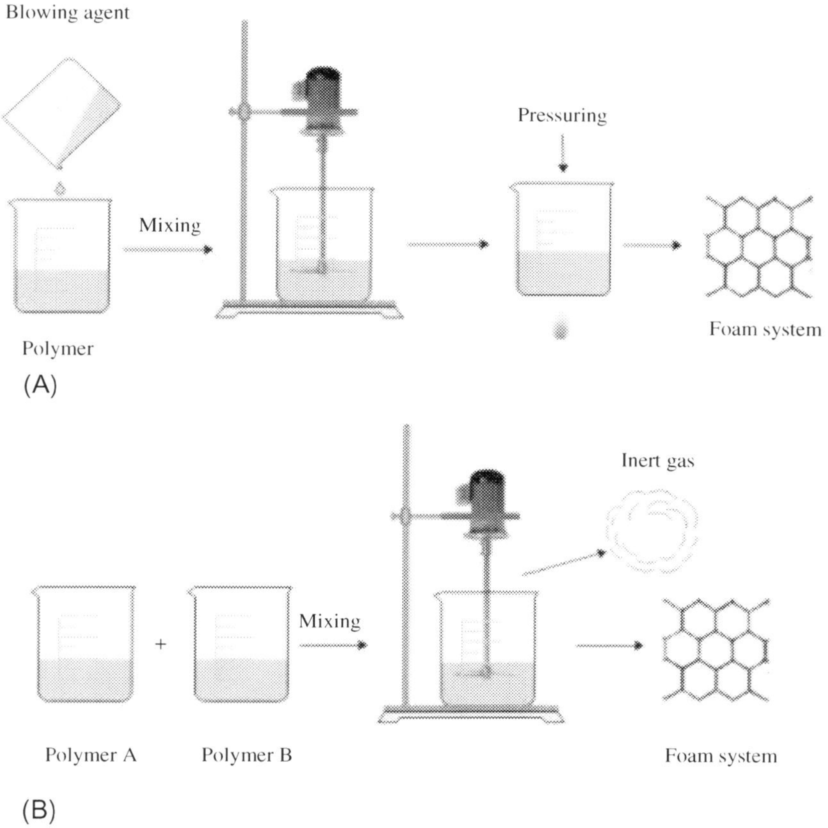

Fig. 5.10 Chemical foaming method (A) blowing agent and foaming system (B) [30].

5.4 Techniques of processing synthetic foam

5.4.1 Extrusion molding of foam

In order to provide a better understanding of foam extrusion molding, the co-rotational twin-screw extruder is taken as an example (Fig. 5.11) [38]. Further, a supercritical fluid (carbon dioxide or nitrogen) was introduced as a blowing agent. Since there are various heating points in the extruder, it was treated with heat at a certain temperature where the rate of discharge and the speed of the screw was predetermined in line with the conditions of procedure. Afterward, a polymer composite or a polymer blend was included in the extruder hopper to extrude according to the procedural conditions. The supercritical fluid was exposed to the site of the barrel by a syringe. The fluid then mixed into the molten polymer fluid while

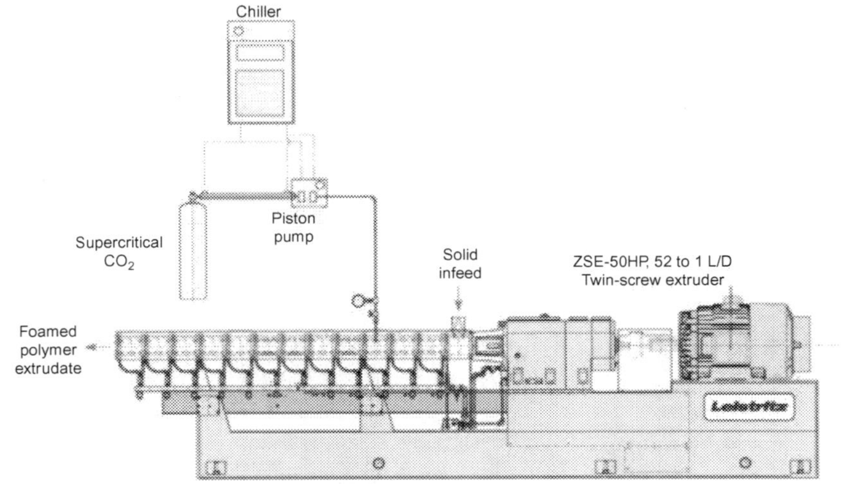

Fig. 5.11 Schematic diagram of the extrusion process through twin-screw extruder [38].

diminishing the melt thickness of it. When the temperature of foaming has dwindled, the compactness of the cell is increased, which augments the firmness of melt. Consequently, foams, based on microcells, are produced [39, 40].

5.4.2 Injection molding of foam

An injection model machine is taken as an example to elucidate the foam injecting molding technique. The injection machine comprises of various injection temperatures, beginning from the hopper to the nozzle (Fig. 5.12) [41]. The parameters of injection molding are predetermined, which include, injection temperature, the pressure of injection, its speed, and the temperature of the mold, time to hold, and release temperature. Notably, these parameters are determined based on their procedural conditions. Primarily, a polymer composite or a blend is transformed into plastic by the procedural condition. Further, a supercritical fluid was added into the barrel, following certain measurements, resulting in forming a polymer met to obtain a standardized polymer or gas solution [42–46]. The mold cavity was then filled with the solution, which led to a swift decline in pressure. This might have produced standardized cell nucleation and cell development. However, the tangible presence and properties of inserted foams have a reliance on the raw compounds, mold pattern, and procedural conditions [46, 47].

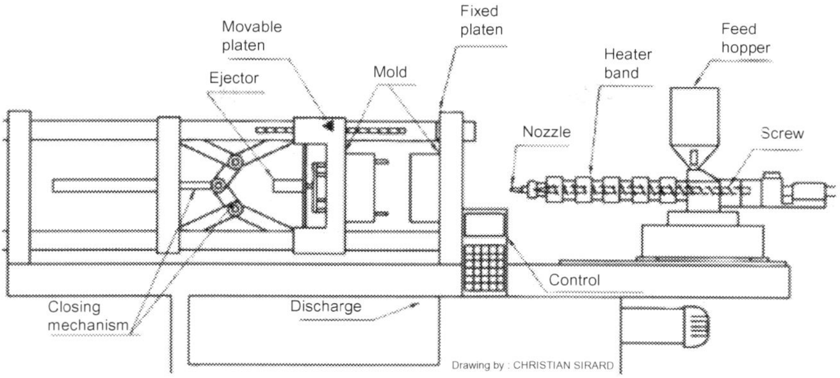

Fig. 5.12 Schematic diagram of the injection molding machine [41].

5.5 Applications of synthetic foam

Synthetic foams are composites that are extensively used in myriad industries, such as; aerospace, automobile, sporting goods, and undersea industries owing to their trivial volume, outstanding phonic features, and floating capability [48]. Since synthetic foams are produced from glass microspheres, they are applicable to be used in lightweight functions and thus occupied significance in the submarine industry [8].

Some of the major characteristics of synthetic foams include; low compactness, heat resistance, sound resistance, maximum firmness, more resilience toward corrosion, etc. [49–51]. These characteristics make them highly productive when used in the public and industrial sectors. Presently, polymer foam has become one of the major components of polymer engineering [52]. For instance, Polyurethane (PU) Soft Foams are used in producing articles of furniture, composite materials and textiles, car headrests, and sporting stuff [53]. It is also a major constituent to manufacture freezers and their containers [54]. Additionally, Polypropylene (PP) Foam has scope in the manufacturing of ammunition, automotive parts, aircraft engineering, and thermal insulating materials [55]. Phenolic foam is a compound that is a prerequisite of the manufacturing industries of automobiles, aircraft, electric goods, and architecture [56]. However, the polyamide foams are necessary elements to make aircraft and also used in marine engineering [57]. Some of the chief polymer foams and their usage are summarized as shown in Table 5.3:

More specifically, apart from the above polymer foams, a plethora of manufacturing and engineering industries are overwhelmed with the multipurpose applications of synthetic foams that are discussed in Table 5.4.

Table 5.3 Polymer foams and their applications.

Polymer foams	Major applications
Polyurethane (PU) soft foam	Used in manufacturing articles of furniture, composite materials, textiles, shoes, and hats. It is also used in car headrests and sports goods [53], and in producing refrigerators, freezers, and their containers [54]
Polypropylene (PP) foam	They are primarily used in the arms industry, automobile, aerospace, in developing materials for thermal insulation, along with the formation of objects that are shockproof and irresistible [58–60]. Their usage is also beneficial in packaging and construction materials [55]
Phenolic foam	This type of foam is commonly used in the architecture, automotive, and space industry, manufacturing of iron, steel, and other electric compounds [56]. Also, it serves as the major component in developing heat-insulating and corrosion resilient infrastructure and superfast automobiles [30]
Polyimide (PI) foam	It possesses a higher degree of thermal insulating and steadiness and works as an efficient fire resistance compound. This type of foam has major applicability in producing goods related to aerospace, aircraft, and marine engineering [57]

Table 5.4 Application of synthetic foams in different industries.

No.	Description	Applications
1	Electric Industry	• Electrical covering compounds [1]
2	Aerospace	• Ablative heat-protective compounds, void plasters, abradable seals [1] • Aerospace's red parts are usable in commercial airplanes [1]
3	Defense	• They are further used in developing radar compulsive material, conductive coverings, and impact material [1] • Missile supports [1] • Tooling slabs (mold design, light weight mixtures; thermoforming tooling) [1]
4	Energy Absorbing Material (EAM)	• To develop shockproof materials to mitigate shockwave to save humans and substructure [1] • Syntactic foam shields are made for mitigating blasts along with metal blast safety panels [48]

Table 5.4 Application of synthetic foams in different industries—cont'd

No.	Description	Applications
5	Marine	• Use of syntactic foam slabs in the onward and in post-free-flood areas of submarines by Trelleborg Emerson & Cuming (TEC) and Trelleborg Offshore, Boston (TOB; Boston, MA) [48] • To produce wheels and covers of submarines, remotely operated vehicles (ROVs), and human-operated vehicles (HOVs), which are used in deep-sea exploration [48] • To reinforce hollow parts within the airplanes [48] • To make airplane radomes, rub strips, and covers for screws and guide vanes, and to envelop penetrating electronic apparatus [48]
6	Spacecraft Structures and Components	• Used to make space shuttle. Spray-on syntactic foams containing glass HGMs, chopped glass fiber, cork, and other composites that are used as insulation for the outside petroleum tank and to manufacture solid rocket boosters on the space shuttle. Low CTE of syntactic foams has been beneficial to produce space glasses that are crucial to stabilizing amid variation in heat [48]
7	Consumer products	
	Sports Equipment	• Used to manufacture Adidas Fevernova soccer balls that were used in the 2006 World Cup [48]
	Furniture	• Saran microballoons made of polyvinylidene chloride are combined with epoxy or polyester to produce a synthetic wood. Saran microspheres are also a significant component to make synthetic marble. They manufacture bathroom fixtures, bathtubs, etc. [61]
	Food Containers	• To obtain eccolite products that are used in food and beverage containers [48]

Continued

Table 5.4 Application of synthetic foams in different industries—cont'd

No.	Description	Applications
8	Composite Tooling and Vacuum-Forming Plug Assists	• It is used to make composite compounds tooling boards and plug assists (a mechanical device) due to their thermal steadiness, low volume, low thermal conductivity, and convenient machining [48]

5.6 Conclusion

Production of epoxies is considered as the milestone of preparing synthetic foams. The hollow spheres made compounds are the synthetic foams that are distinguished due to their profound mechanical, physical, and chemical properties. Moreover, the methods to obtain synthetic foam includes; mechanical foaming, physical foaming, chemical foaming. Whereas, foam extrusion molding and foam injection molding are the widely-practiced techniques to produce synthetic foam. To conclude, synthetic foams are used in the production of numerous products that are of greater importance. Besides, myriad applications of synthetic foam have been described from the automotive and aerospace industries to defense, i.e., maritime and subsea industries. Thus, it is worth mentioning that owing to their all-inclusive properties and usability, and synthetic foams have been used as indispensable and inevitable compounds in manufacturing and engineering industries to fulfill the growing and distinct needs of humans.

References

[1] Lair, E., O'Sullivan, B., & Carlisle, K. (2019). The Evolution of Syntactic Foam Technology to Meet Increasingly Diverse Subsea Applications and Demands. Retrieved from: https://wwz.ifremer.fr/rd_technologiques/content/download/136095/file/Session%202%20Lair.pdf.
[2] J. Rakowska, Best practices for selection and application of firefighting foam, in: MATEC Web of Conferences, vol. 247, EDP Sciences, 2018, p. 00014.
[3] G.M. Gladysz, K.K. Chawla, Syntactic and composite foams: proceedings of an engineering conferences international (ECI) conference, J. Mater. Sci. 41 (13) (2006) 3959–3960.
[4] D. Choqueuse, P. Davies, R. Martin, Ageing of composites in underwater applications, Ageing Compos. (2008) 467–498.
[5] E. Woldesenbet, S. Peter, Volume fraction effect on high strain rate properties of syntactic foam composites, J. Mater. Sci. 44 (6) (2009) 1528–1539.

[6] L. Bardella, F. Genna, Elastic design of syntactic foamed sandwiches obtained by filling of three-dimensional sandwich-fabric panels, Int. J. Solids Struct. 38 (2) (2001) 307–333.
[7] N. Gupta, S. Sankaran, On the characterisation of syntactic foam core sandwich composites for compressive properties, J. Reinf. Plast. Compos. 18 (14) (1999) 1347–1357.
[8] A. Corigliano, E. Rizzi, E. Papa, Experimental characterization and numerical simulations of a syntactic-foam/glass-fibre composite sandwich, Compos. Sci. Technol. 60 (11) (2000) 2169–2180.
[9] J.R.M. d'Almeida, An analysis of the effect of the diameters of glass microspheres on the mechanical behavior of glass-microsphere/epoxy-matrix composites, Compos. Sci. Technol. 59 (14) (1999) 2087–2091.
[10] H.S. Kim, M.A. Khamis, Fracture and impact behaviours of hollow micro-sphere/epoxy resin composites, Compos. A: Appl. Sci. Manuf. 32 (9) (2001) 1311–1317.
[11] N. Gupta, E. Woldesenbet, Microballoon wall thickness effects on properties of syntactic foams, J. Cell. Plast. 40 (6) (2004) 461–480.
[12] Z.M. Ariff, S.S. Samsudin, N.A.A. Mutalib, Compressive behaviour of syntactic foam filled with epoxy hollow spheres having different wall thickness, 14th International Conference "Polymeric Materials, 2010, pp. 1–10.
[13] M. Puterman, M. Narkis, S. Kenig, Syntactic foams I. Preparation, structure and properties, J. Cell. Plast. 16 (4) (1980) 223–229.
[14] H.S. Kim, P. Plubrai, Manufacturing and failure mechanisms of syntactic foam under compression, Compos. A: Appl. Sci. Manuf. 35 (9) (2004) 1009–1015.
[15] N. Gupta, W. Ricci, Comparison of compressive properties of layered syntactic foams having gradient in microballoon volume fraction and wall thickness, Mater. Sci. Eng. A 427 (1–2) (2006) 331–342.
[16] N. Gupta, R. Ye, M. Porfiri, Comparison of tensile and compressive characteristics of vinyl ester/glass microballoon syntactic foams, Compos. Part B 41 (3) (2010) 236–245.
[17] Z. Yousaf, M. Smith, P. Potluri, W. Parnell, Compression properties of polymeric syntactic foam composites under cyclic loading, Compos. B: Eng. 186 (2020) 107764.
[18] R. Huang, P. Li, T. Liu, J. Xu, The 3D failure process in polymeric syntactic foams with different cenosphere volume fractions, J. Appl. Polym. Sci. 136 (19) (2019) 47491.
[19] B.J. Lansing, Mechanical and Physical Characterization of Foams Made of Gelatinized Starch and Pre-Polymer Polyurethane, Rochester Institute of Technology, 2016.
[20] G.M. Gladysz, B. Perry, G. McEachen, J. Lula, Three-phase syntactic foams: structure-property relationships, J. Mater. Sci. 41 (13) (2006) 4085–4092.
[21] F.A. Shutov, Syntactic polymer foams, in: Chromatography/Foams/Copolymers, Springer, Berlin, Heidelberg, 1986, pp. 63–123.
[22] D. Shastri, H.S. Kim, A new consolidation process for expanded perlite particles, Construct. Build Mater. 60 (2014) 1–7.
[23] E. Galimov, E. Sharafutdinova, N. Galimova, Technologies for producing syntactic carbon foams with specified operational properties, in: MATEC Web of Conferences, vol. 298, EDP Sciences, 2019, p. 00023.
[24] Foamed Plastics, Van Nostrand's Scientific Encyclopedia, 2006, https://doi.org/10.1002/0471743984.vse3339.pub2.
[25] C. Hiel, D. Dittman, O. Ishai, Composite sandwich construction with syntactic foam core—a practical assessment of post-impact damage and residual strength, Composites 24 (5) (1993) 447–450.
[26] S.K. Khanna, S. Gopalan, Reinforced polyurethane flexible foams, WIT Trans. State-of-the-Art Sci. Eng. (2005) 20.
[27] National Research Council, Unit Manufacturing Processes, National Research Council, 1995. Available at: https://www.nap.edu/read/4827/chapter/10.

[28] D. Zhou, Y. Xiong, H. Yuan, G. Luo, J. Zhang, Q. Shen, L. Zhang, Synthesis and compressive behaviors of PMMA microporous foam with multi-layer cell structure, Compos. Part B 165 (2019) 272–278.
[29] W. Gong, H. Fu, C. Zhang, D. Ban, X. Yin, Y. He, X. Pei, Study on foaming quality and impact property of foamed polypropylene composites, Polymers 10 (12) (2018) 1375.
[30] F.L. Jin, M. Zhao, M. Park, S.J. Park, Recent trends of foaming in polymer processing: a review, Polymers 11 (6) (2019) 953.
[31] S. Pérez-Tamarit, E. Solórzano, A. Hilger, I. Manke, M.A. Rodríguez-Pérez, Multi-scale tomographic analysis of polymeric foams: a detailed study of the cellular structure, Eur. Polym. J. 109 (2018) 169–178.
[32] V. Kakumanu, S.S. Sundarram, Dual pore network polymer foams for biomedical applications via combined solid-state foaming and additive manufacturing, Mater. Lett. 213 (2018) 366–369.
[33] R. Kuska, S. Milovanovic, S. Frerich, J. Ivanovic, Thermal analysis of polylactic acid under high CO2 pressure applied in supercritical impregnation and foaming process design, J. Supercrit. Fluids 144 (2019) 71–80.
[34] S. Wang, A. Ameli, V. Shaayegan, Y. Kazemi, Y. Huang, H.E. Naguib, C.B. Park, Modelling of rod-like fillers' rotation and translation near two growing cells in conductive polymer composite foam processing, Polymers 10 (3) (2018) 261.
[35] M. Barmouz, A.H. Behravesh, The role of foaming process on shape memory behavior of polylactic acid-thermoplastic polyurethane-nano cellulose bio-nanocomposites, J. Mech. Behav. Biomed. Mater. 91 (2019) 266–277.
[36] J. Heimann, A.M. Matz, B.S. Matz, N. Jost, Processing of open-pore silicon foams using graphite composite as space holder, Sci. Technol. Mater. 30 (1) (2018) 23–26.
[37] S. Chen, W. Zhu, Y. Cheng, Multi-objective optimization of acoustic performances of polyurethane foam composites, Polymers 10 (7) (2018) 788.
[38] B. Chen, L. Zhu, F. Zhang, Y. Qiu, Process development and scale-up: twin-screw extrusion, in: Developing Solid Oral Dosage Forms, Academic Press, 2017, pp. 821–868.
[39] T. Ellingham, L. Duddleston, L.S. Turng, Sub-critical gas-assisted processing using CO2 foaming to enhance the exfoliation of graphene in polypropylene + graphene nanocomposites, Polymer 117 (2017) 132–139.
[40] G. Wang, G. Zhao, L. Zhang, Y. Mu, C.B. Park, Lightweight and tough nanocellular PP/PTFE nanocomposite foams with defect-free surfaces obtained using in situ nano-fibrillation and nanocellular injection molding, Chem. Eng. J. 350 (2018) 1–11.
[41] S. Jocelyn, Identification et réduction du risque pour les interventions de maintenance et de production sur des presses à injection de plastique en entreprises (Doctoral dissertation), École Polytechnique de Montréal, 2012.
[42] I.M. Alarifi, V. Movva, M. Rahimi-Gorji, R. Asmatulu, Performance analysis of impact-damaged laminate composite structures for quality assurance, J. Braz. Soc. Mech. Sci. Eng. 41 (8) (2019) 345.
[43] M. Akermi, N. Jaballah, I.M. Alarifi, M. Rahimi-Gorji, R.B. Chaabane, H.B. Ouada, M. Majdoub, Synthesis and characterization of a novel hydride polymer P-DSBT/ZnO nano-composite for optoelectronic applications, J. Mol. Liq. 287 (2019) 110963.
[44] I.M. Alarifi, A.B. Alkouh, V. Ali, H.M. Nguyen, A. Asadi, On the rheological properties of MWCNT-TiO2/oil hybrid nanofluid: an experimental investigation on the effects of shear rate, temperature, and solid concentration of nanoparticles, Powder Technol. 355 (2019) 157–162.
[45] EL-Bagory, T. M., Alarifi, I. M., & Younan, M. Y., Prediction of mechanical properties for curved dumbbell-shaped specimen at different orientation angles of ring hoop tension test, Adv. Eng. Mater. 21 (9) (2019) 1900191.

[46] R.C.N. Barbosa, R.D.S.G. Campilho, F.J.G. Silva, Injection mold design for a plastic component with blowing agent, Procedia Manufact. 17 (2018) 774–782.
[47] L. Wang, Y. Hikima, S. Ishihara, M. Ohshima, Fabrication of lightweight microcellular foams in injection-molded polypropylene using the synergy of long-chain branches and crystal nucleating agents, Polymer 128 (2017) 119–127.
[48] N. Gupta, S.E. Zeltmann, V.C. Shunmugasamy, D. Pinisetty, Applications of polymer matrix syntactic foams, JOM 66 (2) (2014) 245–254.
[49] M.A. Alamir, I.M. Alarifi, W.A. Khan, W.S. Khan, R. Asmatulu, Electrospun nanofibers: preparation, characterization and atmospheric fog capturing capabilities, Fibers Polym. 20 (10) (2019) 2090–2098.
[50] I. Tlili, M. Osman, E.M. Barhoumi, I. Alarifi, A.G. Abo-Khalil, R.P. Praveen, K. Sayed, Performance enhancement of a humidification–dehumidification desalination system, J. Therm. Anal. Calorim. (2019) 1–11.
[51] A. Hajizadeh, N.A. Shah, S.I.A. Shah, I.L. Animasaun, M. Rahimi-Gorji, I.M. Alarifi, Free convection flow of nanofluids between two vertical plates with damped thermal flux, J. Mol. Liq. 289 (2019) 110964.
[52] J. Pinto, D. Morselli, V. Bernardo, B. Notario, D. Fragouli, M.A. Rodriguez-Perez, A. Athanassiou, Nanoporous PMMA foams with templated pore size obtained by localized in situ synthesis of nanoparticles and CO2 foaming, Polymer 124 (2017) 176–185.
[53] I. Gerges, M. Tamplenizza, F. Martello, C. Recordati, C. Martelli, L. Ottobrini, C. Lenardi, Exploring the potential of polyurethane-based soft foam as cell-free scaffold for soft tissue regeneration, Acta Biomater. 73 (2018) 141–153.
[54] J. Paciorek-Sadowska, M. Borowicz, B. Czupryński, M. Isbrandt, Effect of evening primrose oil-based polyol on the properties of rigid polyurethane–polyisocyanurate foams for thermal insulation, Polymers 10 (12) (2018) 1334.
[55] J. Hou, G. Zhao, G. Wang, L. Zhang, G. Dong, B. Li, Ultra-high expansion linear polypropylene foams prepared in a semi-molten state under supercritical CO2, J. Supercrit. Fluids 145 (2019) 140–150.
[56] J. Li, A. Zhang, S. Zhang, Q. Gao, W. Zhang, J. Li, Larch tannin-based rigid phenolic foam with high compressive strength, low friability, and low thermal conductivity reinforced by cork powder, Compos. Part B 156 (2019) 368–377.
[57] J. Yang, Y. Ye, X. Li, X. Lü, R. Chen, Flexible, conductive, and highly pressure-sensitive graphene-polyimide foam for pressure sensor application, Compos. Sci. Technol. 164 (2018) 187–194.
[58] V.S. Swarna, I.M. Alarifi, W.A. Khan, R. Asmatulu, Enhancing fire and mechanical strengths of epoxy nanocomposites for metal/metal bonding of aircraft aluminum alloys, Polym. Compos. 40 (9) (2019) 3691–3702.
[59] I. Tlili, M. Osman, I. Alarifi, H. Belmabrouk, A. Shafee, Z. Li, Performance enhancement of a multi-effect desalination plant: a thermodynamic investigation, Phys. A Stat. Mech. Appl. 535 (2019) 122535.
[60] I.M. Alarifi, Investigation the conductivity of carbon fiber composites focusing on measurement techniques under dynamic and static loads, J. Mater. Res. Technol. 8 (5) (2019) 4863–4893.
[61] T.F. Anderson, H.A. Walters, C.W. Glesner, Castable, sprayable, low density foams and composites for furniture, marble, marine, aerospace, boats and related applications, J. Cell. Plast. 6 (4) (1970) 171–178.

CHAPTER 6

Synthetic biosources

Abbreviations

ADCA	aminodeacetylcephalosporanic acid
ASF	Anderson-Schulz-Flory
ATP	adenosine triphosphate
BNC	bacterial nanocellulose
Cat-DO	catechol 1,2-dioxygenase
CBF	circulatory fluidized bed
CNCs	cellulose nanocrystals
Co	carbon monoxide
CO_2	carbon dioxide
DHS-D	dehydroshikimate dehydratase
DNA	deoxyribonucleic acid
DSM	Dutch Multinational Corporation
ER	enoate reductase
FTS	Fischer-Tropsch synthesis
G1P	glucose-1-phosphate
IPM	inert porous media
MW	meter water equivalent
OPXBIO	OPX biotechnologies
PCA-D	protocatechuate decarboxylase
PSA	pressure swing adsorption
SRB	sulfate-reducing bacteria
UDP	uridine diphosphate glucose

6.1 Introduction

The field of synthetic biology rose to its zenith in 2000 as a result of the historical decline in the purchasing value of deoxyribonucleic acid (DNA) sequencing, which was consequent to an unprecedented modification in the conduct of molecular science. In spite of applying a single gene method (through removing or implanting), the principles that are used in engineering were adopted in biology, and complicated multiple gene concepts, i.e., pathways and complete genomes were created [1]. This further included other developments such as the genome transplantation [2], which assisted intricate plans in order to manufacture complete microorganisms' genomes. They were further relocated to the cells of bacteria [3]. Along with it, the

Synthetic Engineering Materials and Nanotechnology
https://doi.org/10.1016/B978-0-12-824001-4.00013-2

CRISPR/Cas9-mediated gene excision was developed to perform functions that were not able to be implemented earlier before. Although this progression was practicable, it was comparatively convenient [4, 5]. Presently, synthetic biology acquired a unique feature as compared to its erstwhile presence during the 19th and 20th centuries that envisages distinct implant characteristics in the contemporary structure in order to manufacture viable products [6]. Some of the plans are to manufacture the organisms that create therapeutic or industrial molecules capable of making biofuels in a productive manner through the use of hydro-energy and solar rays. Furthermore, their functions may include operating and trace the activities for naturally found organisms while providing a scientific examination of these activities [7–9].

Most of the microorganisms are composed of one cell and are termed as unicellular. They are not visible from the naked eye, and due to their very small size, artificial magnification tool is used to observe their presence. Among microorganisms, differences are found in terms of organization, size, metabolism, and habitation.

Bacteria are one of the microorganisms which are abundantly found everywhere on the earth. Also, they live inside the human body. However, not all bacteria are harmful, but some of them are useful. Protists are the unicellular microorganisms that do not fall in the category of fungi, plants, or animals. The examples of protists are algae and protozoa. They form the fundamental structure of various food chains while making available the nutrients to other organisms. However, viruses are the acellular microorganisms since they do not comprise cells. But their fundamental constituents are genetics materials such as Ribonucleic acid or Deoxyribonucleic acid and proteins [10] (Fig. 6.1).

Hint statement

sMutual plants intent to create the molecules that can remove land pollution and are able to trace and give alert about the molecular markers that are useful for environment and medicines.

Synthetic biology is acquiring boom on manufacturing and industrial grounds, which is evident from its amplified usability in various sectors such as medicine, agriculture, and petroleum sectors. Above all the synthetic biology is the potential user to cope up with the overwhelming environmental hazards, i.e., climate change and running out of fresh water. However, it has manifold risks related to ecology [11].

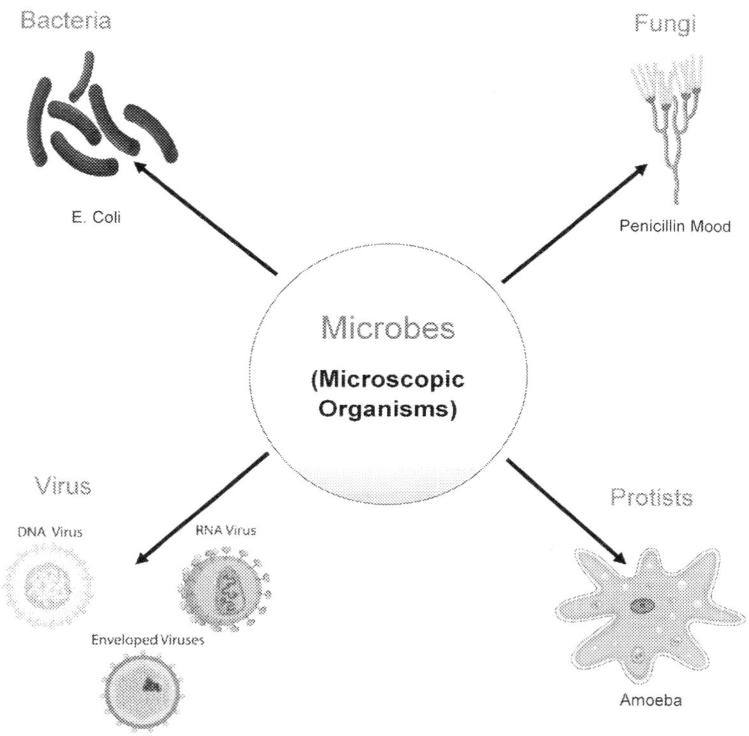

Fig. 6.1 The biology education chart for microorganism diagram types.

The advent of the 21st century is marked with the rise in the use of fuels, water resources, and food that is consequent to the rapid demographic bane and faster inclusion of technological assistance. Humans are now more inclined to hinge on renewable goods in the form of biomass to yield the anticipated compounds. The prerequisite to maintain the progress is to manufacture the goods through the compounds that can be replenished and have characteristics to remain favorable for the climate. They must not possess features of the combustible fuel or compounds that cannot be replenished and have risks related to health and well-being. Also, the compounds must be feasible in terms of economic value.

Nanocellulose is the notion of reflecting the cellulose-made compounds that have distinguished chemical and physical characteristics in terms of their functionality and their source of mining, such as: rigid bacterial nanocellulose (BNC), cellulose nanocrystals (CNCs), and nanofibrillated cellulose (NFC) [12].

The term bioresources denote the naturally found compounds that can be replenished and are recyclable. They are present in organic matters and raw components of the humans and animals' waste materials.

Hint statement

Biosources are the elements that are found in nature but they are recyclable and are sustainable. They are found from the organic disposal and raw compounds that are resultant from the functions of humans and animal.

Moreover, manifold industries, factories, agricultural-led activities, forest management, maritime, and district management sectors are the biggest sources to create bioresources. However, the feedstocks of the bioresources are acquired through manufacturing industries; however, mills of palm oil are one of its examples.

The agricultural plants are considered as a source of manufacturing the bioproducts and can perform the function of transporting energy and are termed as specialty goods and platform chemicals. The geographical location of Malaysia and other countries are situated in tropical regions, a robust potential, associated with them in terms of forest cover, agricultural dominance, metropolitan mechanism, and maritime sectors. In addition to it, a maintainable life cycle of carbon (Fig. 6.2) is potentially associated with the progress in these sectors. In the carbon lifecycle, the solar energy

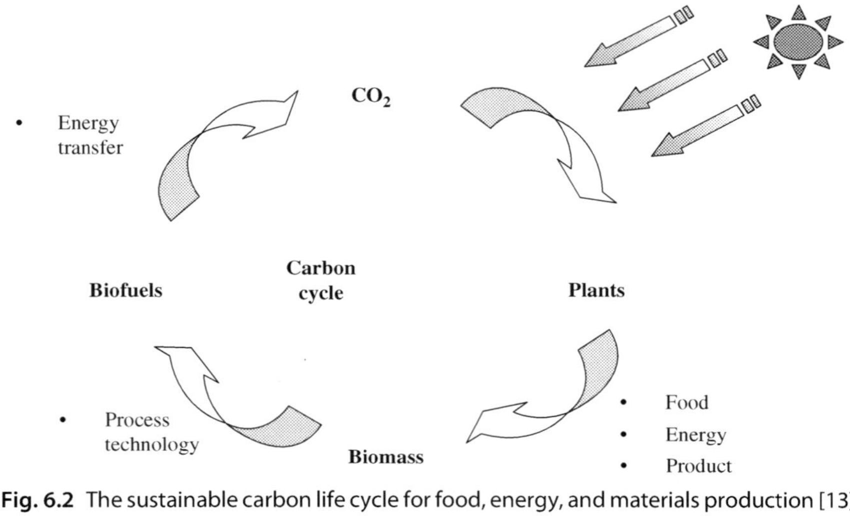

Fig. 6.2 The sustainable carbon life cycle for food, energy, and materials production [13].

circulates to make a carbon cycle while transforming it into energy, food, and other required compounds, according to the demands of local industries as the features of feedstocks, goods and usability are distinguished from country to country [13].

6.1.1 Classification of bioresources

The usage of bioresources is multidimensional; for instance, they do not only produce one compound but also other associated products are manufactured through it [14]. Different bioresources have been categorized that are stated in Table 6.1.

Technological developments in synthetic biology have made it possible to engineer the cell's genetic circuits for the purpose of programming new

Table 6.1 Classification of the different bioresources.

Types of bioresources	Characteristics
Primary bioresources	**o**They can be acquired from agricultural land, timberland, and aquatic environments **o**They have their specific uses **o**They are responsible for producing food items and other significant products and energy-generating materials **o**Wood, Algae, bamboo are some of the examples of bioresources [14]
Secondary bioresources	**o**The process of obtaining secondary bioresources involves primary processing and processing through the industrial means, in the form of bioproducts and residues. Additionally, secondary bioresources are also acquired as a result of maintaining the green regions, i.e., sports grounds, dikes, and parks **o**They have a lesser degree of quality degeneration as they are harvested through the virgin materials [15] **o**Likewise, secondary bioresources, virgin materials are the basic components of tertiary bioresources
Tertiary bioresources	**o**However, unlike secondary bioresources, they are combined with the unspecified portions, and the residues take place in a trivial ratio at the place of generation **o**Primary processing, reaping, and post-reaping activities are associated with the processing of large-scale production of tertiary bioresource **o**Besides, its production is possible on a small scale [15]

Continued

Table 6.1 Classification of the different bioresources—cont'd

Types of bioresources	Characteristics
Quaternary bioresources	oThey are unique due to their time frame, which can be divided into short, mid, and long categories, and it begins with the consumption of products oOn the level of short term, quaternary bioresources are acquired from the consumption of food and feed as, i.e., waste materials of humans and animals. These bioresources are produced within a short time period oQuaternary bioresources with mid-term delays are produced from days to months after consumption. Examples of such materials are the packaging stuff and newspapers as both of the goods are for one-time use only oHowever, the long-term time frame is associated with the duration of years or decennary. For instance, construction materials that are based on wood that is placed at the residential places may take more years to get degenerated into wasteful timber. Thus, the prolonged time is the responsible factor to deteriorate the potency of efficient usage of quaternary bioresources [15]

behavioral tendencies, the function of controlling logic, and the dynamic expression of genes in the cells. The technological advancement produces a collection of active sensors that are able to distinguish between the states of cells, to perform the function of deliverance and to create a regularized dosage of therapeutic biomolecules [16]. The origin of biological life is linked to expanding the gene-based natural network to know about the environment and to recognize the beneficial outcomes. Thus, many of the biological scientists envision to employ this strategy to utilize the difficult structuring of cells in order to functionalize therapeutic cells. The main purpose, three significant stages are involved:

- Program the sensors to recognize and respond to certain molecules that are enabled to distinguish the cells states (either it is stabilized or vulnerable) [16].
- Recognize a particular molecule and cell state through creating a regularized dosage for therapeutic molecules.
- Enable the cells to perform on delivery specific tasks (Fig. 6.3) [16].

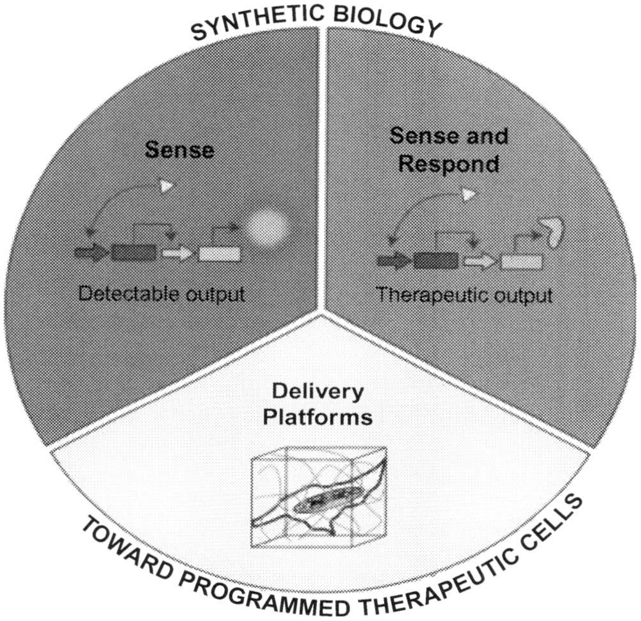

Fig. 6.3 Three steps that are involved in the engineering program therapeutic cells [16].

6.1.2 Biomass and its classification

When the biomass is transformed, it is known as a feedstock. There are various kinds of feedstocks available such as grains, agricultural waste, grasses, sugar crops, and algae. The type of feedstock to be used depends on the part of the plant. For instance, biodiesel is obtained from oilseeds, i.e., soybean. Natural oils, hydrocarbons, methane, biodiesel, and ethanol are the different types of biomass [17] (Fig. 6.4).

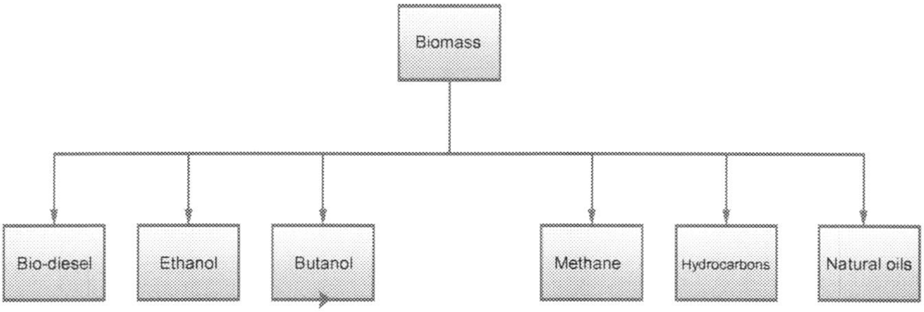

Fig. 6.4 Classification of biomass.

Biomass is attributed to have a lower energy density, which reflects its increasing need to meet the requirement of energy. The density of biomass has a greater impact on its combustion [18]. Biomass density can be calculated from V_{OB}/ha by first estimating the biomass of the inventoried volume and then "expanding" this value to take into account the biomass of the other aboveground components as follows:

$$\text{Above ground biomass density } (t/ha) = (V_{OB})\,(W_D)\,(B_{EF}) \tag{6.1}$$

$$B_{EF} = \text{Exp}^{[(3.213-0.506)\,\text{Ln}\,(BV)]} \text{ for } B_V < 190\,t/ha \tag{6.2}$$

where W_D = volume-weighted average wood density (1 of oven-dry biomass per m^3 green volume) and B_{EF} = biomass expansion factor (ratio of aboveground oven-dry biomass of trees to oven-dry biomass of inventoried volume).

Example 6.1

Broadleaf forest with a $V_{OB} = 300\,m^3/ha$ and weighted average wood density; $W_D = 0.65\,t/m^3$?

Solution:

Calculate biomass of V_{OB}: $= 300\,m^3/ha \text{ x } 0.65\,t/m^3 = 194\,t/ha$.

Calculate the B_{EF} $B_V > 190\,t/ha$, therefore $B_{EF} = 1.74$.

Calculate aboveground biomass density (Eq. 6.1): $= 1.74 \times 300 \times 0.65$.

Above ground biomass density (t/ha) = 338 t/ha.

Example 6.2

Broadleaf forest with a $V_{OB} = 150\,m^3/ha$, and weighted average wood density, $W_D = 0.55\,t/m^3$?

Solution:

Calculate biomass of V_{OB}: $= 150\,m^3/ha \times 0.55\,t/m^3 = 82.5\,t/ha$.

Calculate the B_{EF} (Eq. 6.2): $B_V < 190\,t/ha$, therefore $B_{EF} = 2.66$.

Calculate aboveground biomass density (Eq. 6.1): $= 2.66 \times 150 \times 0.55$

$$B_{EF} = 220\,t/ha$$

In the production of ethanol and other chemicals, glucose can be used while adopting specified microbial organizations [19]. Monosaccharide is the simplest form of glucose. It produces the polymer of monosaccharides that results in the complex form of sugar that is termed as a polysaccharide.

The standardized shorthand to represent sugar is the Fischer projection [20]. However, the structure of sugar that are organized while using a flat polygon, in order to show the ring is known as Haworth structures [21]. Fig. 6.5 shows the Fischer and Haworth projection formulae for Glucose.

Next includes carbohydrates that are composed of oxygen, hydrogen, and carbon. The empirical organization of carbohydrates is exhibited as CH_2O_n. Carbohydrates are organic compounds that are composed in the form of ketones or aldehydes along with various hydroxyl groups in carbon chains. Since the simplest form of carbohydrates is the monosaccharide, which can be found in the form of polyhydroxy aldehyde.

On the basis of structure, carbohydrates can be exhibited in three forms, such as Haworth structure, Hemiacetal structure, an open-chain structure. The open-chain structure comprises a long-straight-chain of carbohydrates. On the other hand, Hemiacetal structure is the foremost carbon of glucose that gets condensed along with –OH group of fifth carbon to form a ring-like structure. However, the Haworth structure in the form of carbohydrates that comprises pyranose ring structure [22], as shown in Fig. 6.6.

Caspersen strip is made of waxy subring, which is waterproof. The strip forces H_2O to move into the Symplast pathway across PM filters out toxics (Fig. 6.7).

- It moves into xylem by AT.
- H_2O moves into xylem down.

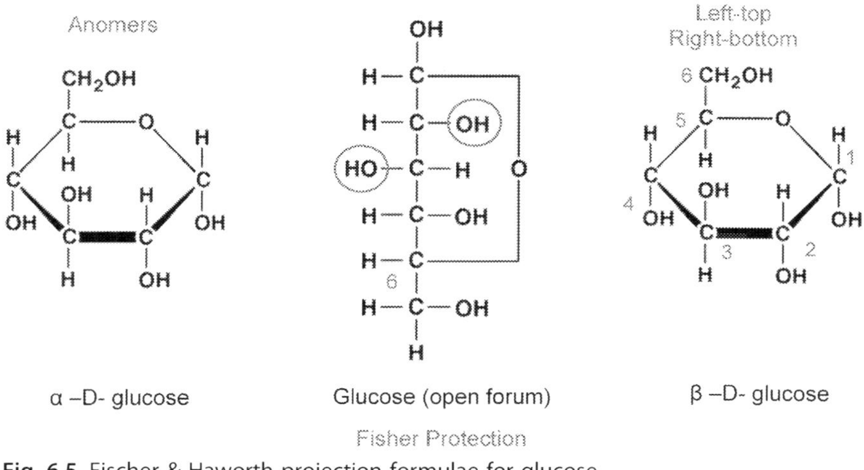

Fig. 6.5 Fischer & Haworth projection formulae for glucose.

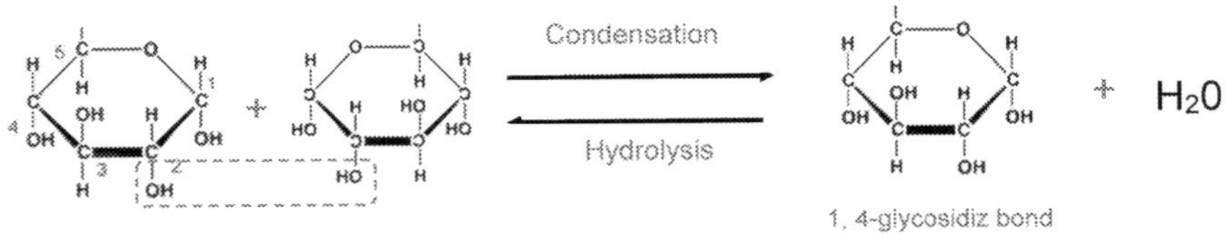

Fig. 6.6 The basics of carbohydrates bonds.

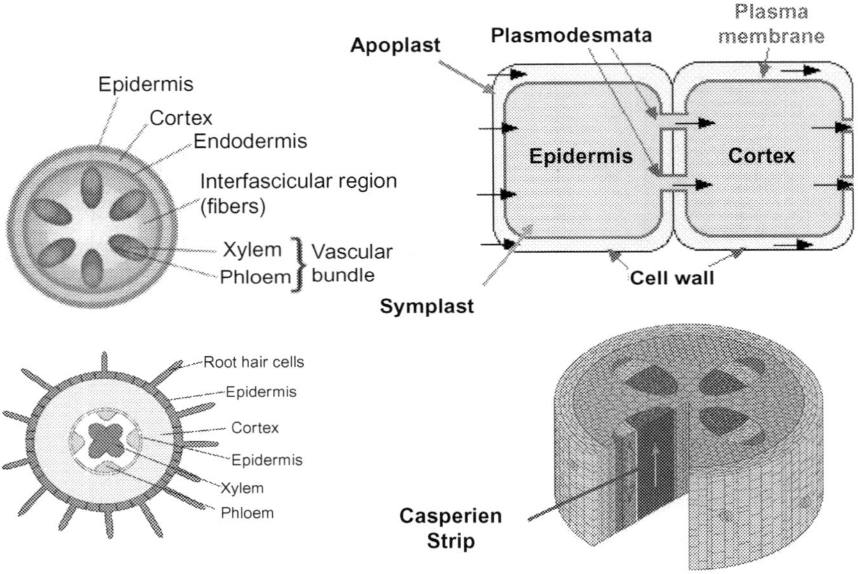

Fig. 6.7 Water transport in plants.

- H_2O moves cross endodermis.
- H_2O moves into xylem.

H_2O moves across root cells in pathways:

- The apoplast is a pathway that is called by tension.
- Symplast is a pathway through cytoplasm diffusion and osmosis.

Apart from gluconeogenesis, glycogen makes available a supplementary means of glucose. Since glycogen is composed of numerous glucose, and it works as a battery backup for the living body while providing an immediate means of glucose when required. Also, it accounts a site for preserving additional concentration of glucose when the level of glucose increases more than its normal range in blood. Metabolically, a significant element of a molecule is the branching of glycogen. As the breakdown of glycogen instigates from the brinks of molecules, the branches move toward the edges of molecules. The sites where glycogen is most abundantly found are skeletal muscle and liver. The process of breaking the glycogen comprises dispensing glucose-1-phosphate (G1P), reorganizing the rest of the glycogen (that is essential to maintain the process of breaking glycogen). Finally, it involves the transformation of G1P into G6P to keep continued the metabolism. Notably, G6P can be split into glycolysis, which is further transformed in the form of glucose through gluconeogenesis. Further, it is corroded in the pentose phosphate pathway [23]. The products and reactants that are concentrated as a result of glycogen breakdown are favorable as the concentration leads to the hydrolysis of glycogen for G1P. Whereas, for glycogen synthesis, the cell adopts another mechanism that is dissimilar to the glycogen breakdown. The process of synthesizing begins from G1P that is transformed into an activated intermediate, Uridine diphosphate (UDP)-glucose. Additionally, it increases the concentration of glucose in the amassing chain of glycogen. Consequent to this addition, the molecules of glycogen have to be reorganized for metabolism [23] (Figs. 6.8 and 6.9).

6.1.3 Glycogen metabolism

In order to maintain glucose level, glycogen metabolism has a significant role. It comprises two indispensable processes. Firstly, it involves synthesizing glucose while undergoing glycogenesis and gluconeogenesis. Secondly, it breaks down the glucose, which is termed as glycogenolysis. Reciprocally, controlling the glycogen and its break down contains several allosteric

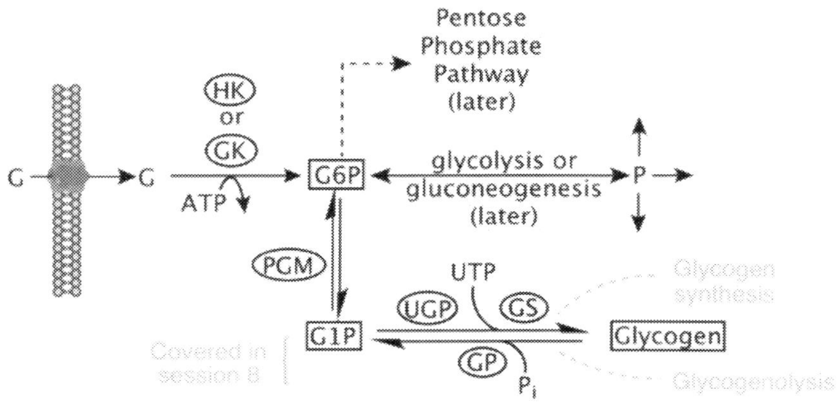

Fig. 6.8 Carbohydrate biosynthesis glycogen synthetic.

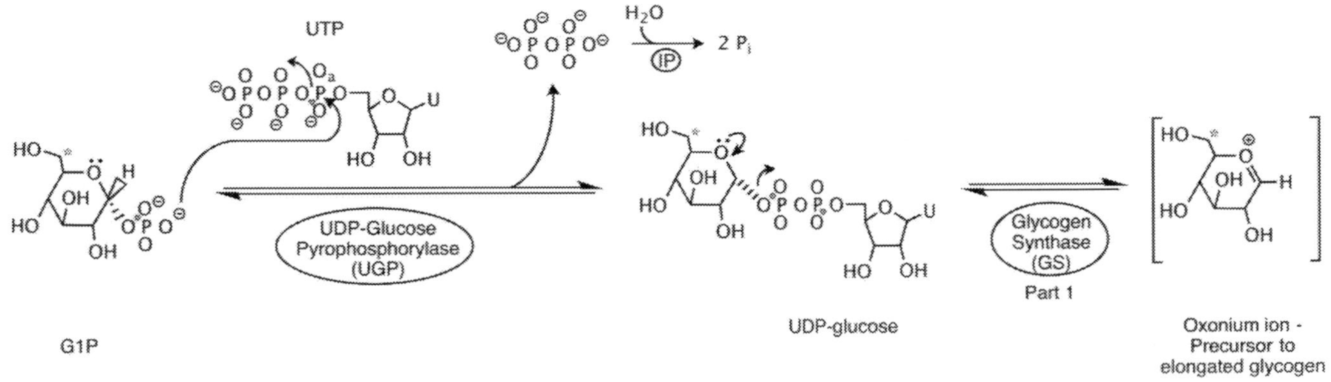

Fig. 6.9 The detailed pathway of glycogen synthetic.

elements and glucagon, epinephrine, and insulin, which facilitates the even supply of glucose, particularly, for producing an energy molecule, namely Adenosine triphosphate (ATP). Amid an increasing level of glucose in the blood, the excess glucose is converted into glycogen. Mainly liver cells constitute the significant mean of preserved glycogen among the animals [24] (Fig. 6.10).

6.2 Methods of synthetic biosources

The concerns regarding the degeneration of the environment are related to the usage of fossil fuels and other petroleum products and chemical substances. Regardless of the hazardous effects on the environment, the values of these commodities have not been lessened. Along with it, the demand for synthetic fuels and chemicals has gained the rise [12]. Another factor of the hike in demand is the fast depletion of energy resources that is resultant to the unprecedented rise in the population of the world and the growing industrialization of world economies. The energy that is produced from biomass is featured as a maintainable substitute to produce energy that has acquired greater importance in all sectors from governmental to the nongovernmental or commercial entities [25].

Moreover, the biofuels of first- and second-generation serve as the best source to tackle an imminent energy crisis. However, various concerns have been raised recently on the usage of these biofuels. But a significant development of the third-generation biofuels has provided the best alternative

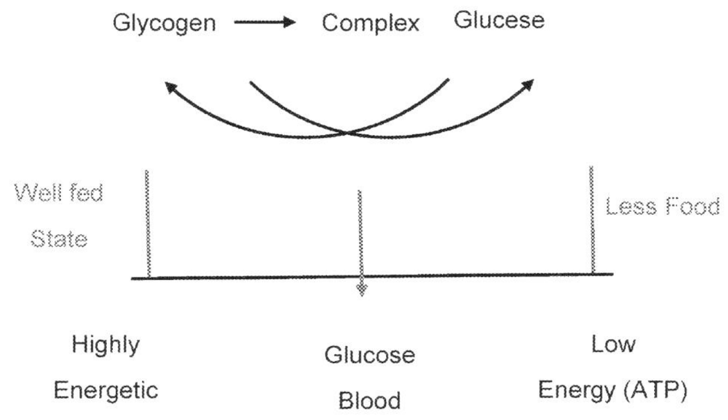

Fig. 6.10 Glycogen synthetic of glycogen metabolism.

solution. Third generation biofuels are produced through microbiological processing of industrial, agricultural, and domestic wastage [26].

Hint statement

Synthetic biofuels are being reckoned as the solution to the concentration of carbon dioxide. As they can efficiently replace the combustible fossil fuels that play a major role in environmental degradation. Similarly, they are best resources that can fulfill world's energy demand.

Imperatively, to counter the depletion of energy resources, there is a drastic need for sustainable products and energy conservative. For instance, the plant biomass is associated with the mere source that can produce sustainable products through organic compounds [27]. Similarly, to produce the liquid form of fuels, biofuels and biomass are the leading sources. They are regarded as more environmentally efficient as they have been reported to have less intensity for greenhouse gas emissions [28]. For growth and production of biomass at the ideal standards, inputs of nutrients for biofuel production are air, CO_2, light, and H_2O, and transporting energy and foods are called as outputs. The technologies that are prerequisites for a carbohydrate economy include; development of the feedstock of biomass, the transformation of biomass in the form of fuel, and availing the fuels. Fig. 6.11 exhibits a schematic illustration of the growth and production of biomass wherein the inputs are nutrients, H_2O, air, light, CO_2 for producing biofuel, and the outputs include fuel for transportation and food. Moreover, the three major technologies for producing energy means from biomass are biomass growth, fuel utilization, and fuel production (Fig. 6.11) [29].

Syngas, biosyngas, and Fischer-Tropsch derivatives are some of the examples of synthetic biofuels as synthetic gas or syngas is composed of the blend of CO_2, CO, and H_2 [30]. Syngas and biosyngas are useful substitutes of fuels, i.e., hydrogen that is obtained from compressing carbon monoxide (CO) and dioxide. Another way to obtain long and short-chain hydrocarbons is Fischer-Tropsch synthesis (FTS) [31]. More specifically, the Fischer-Tropsch process is transforming coal in the form of synthetic liquefied fuel [32]. The core functions of synthetic fuels and biofuels, i.e., biohydrogen, ethanol, biodiesel, and synthetic oil are combusting the engines [29–32]. The following are the methods that are employed to produce synthetic fuels.

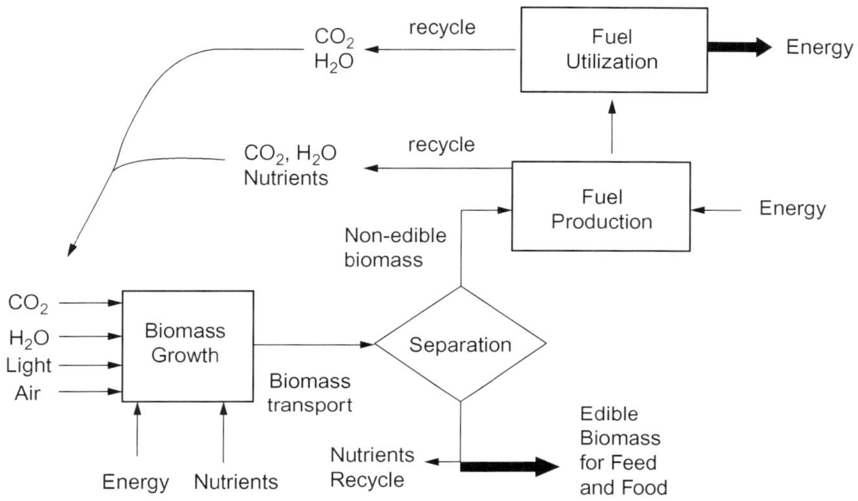

Fig. 6.11 Schematic representation of the growth and development of biomass [29].

> Hint statement
>
> Method that is employed to convert coal into a liquefied form is known as Fischer-Tropsch process.

6.2.1 Production of biofuels via Fischer Tropsch (FT) synthesis: Biomass-to-liquids

The process of transforming the feedstock into synthetic gas through using Fischer Tropsch is characterized by Fischer-Tropsch reaction where carbon monoxide and hydrogen along with hydrocarbons are involved [33].

$$CO + 2H_2 \rightarrow \text{“} - CH_2 - \text{”} + H_2O \tag{6.3}$$

where “$-CH_2-$” depicts a compound that is composed of paraffinic hydrocarbons, having the chains of varied length. Usually, the process is conducted in the temperature range of 150–300 °C to avert the production of high methane byproducts. Higher temperature brings about the additional rate of conversion and bolsters the production of the anticipated alkanes that have long chains [33]. Atmospherically, the range of temperature falls within one to various tens. The hydrogenation reaction of FT is stimulated through Co and Fe catalysts. However, the volume and dispersal of the products of hydrocarbon reactions are usually processed through Anderson–Schulz–Flory (ASF) chain polymerization kinetics model [34].

Fig. 6.12 exhibits the schematic procedure of Fischer Tropsch synthesis. A watery phase of bio-oil is transformed in highly critical settings, which produce syngas and are further spread in a turbine for producing electricity. It is advanced through dry restructuring reactors and water-gas shifting. It also includes the system of two pressure swing adsorption (PSA). In this way, there is an increase in the molar flow rate of CO and H_2 in syngas, which results in attaining the required stoichiometric ratio (H_2/CO) at the conduit of the Fischer–Tropsch reactor. Also, it entails a loop to reoccur the circulation of the unreacted syngas. Consequently, it augments the conversion of CO in the form of biofuels [34].

6.2.2 Fluidized bed gasification

Fluidized bed gasification is the widely used method of transforming biomass. Among the various available technologies, the air-blown circulatory fluidized bed (CBF) is mostly employed. The fluidized bed gasification process is attributed to close-couple combustion, having trivial or without intermediate cleaning of gas. Moreover, the end product of the process of transforming the biomass is electricity and heat. During the process, the gas is engineered through a fluidized bed gasifier that is usually functionalized at the temperature of 900 °C. Additionally, it comprises CO, H_2, H_2O, and CO_2 and significant concentration of hydrocarbon, i.e., Benzene, tars, CH_4, and C_2H_2. Subsequently, the gas requires more processing in the form of the catalytic reformer, through which the hydrocarbons are transformed into CO, H_2, H_2O, and CO_2 [36] as shown in Fig. 6.13 [37].

Mostly, syngas transformation into liquid fuels stipulates raw gas in trivial quantities and or without inert gases, the processing of gasification and reforming must be adopted with pristine oxygen in spite of air, while the use of steam is as moderator [36].

The fluidized bed reactor is termed as the most significant element of fluidization. In the reactor, fuel acts as a fluid in the presence of passive bed material. The behavior is created by compressing gas in fluidization from combustion and gasification in fluidized beds, as shown in Fig. 6.13 [37]. Those fluidization forms are a vapor, O_2, and air. The bed material that is mostly used is the Silica Sand, along with more bulk solids, particularly those who can show catalytic action [38].

6.2.3 Entrained flow gasification

To produce noncatalytic syngas (H_2 and CO) involving biomass, it entails intense temperature, usually at 1300 °C. In this instance, and entrained flow

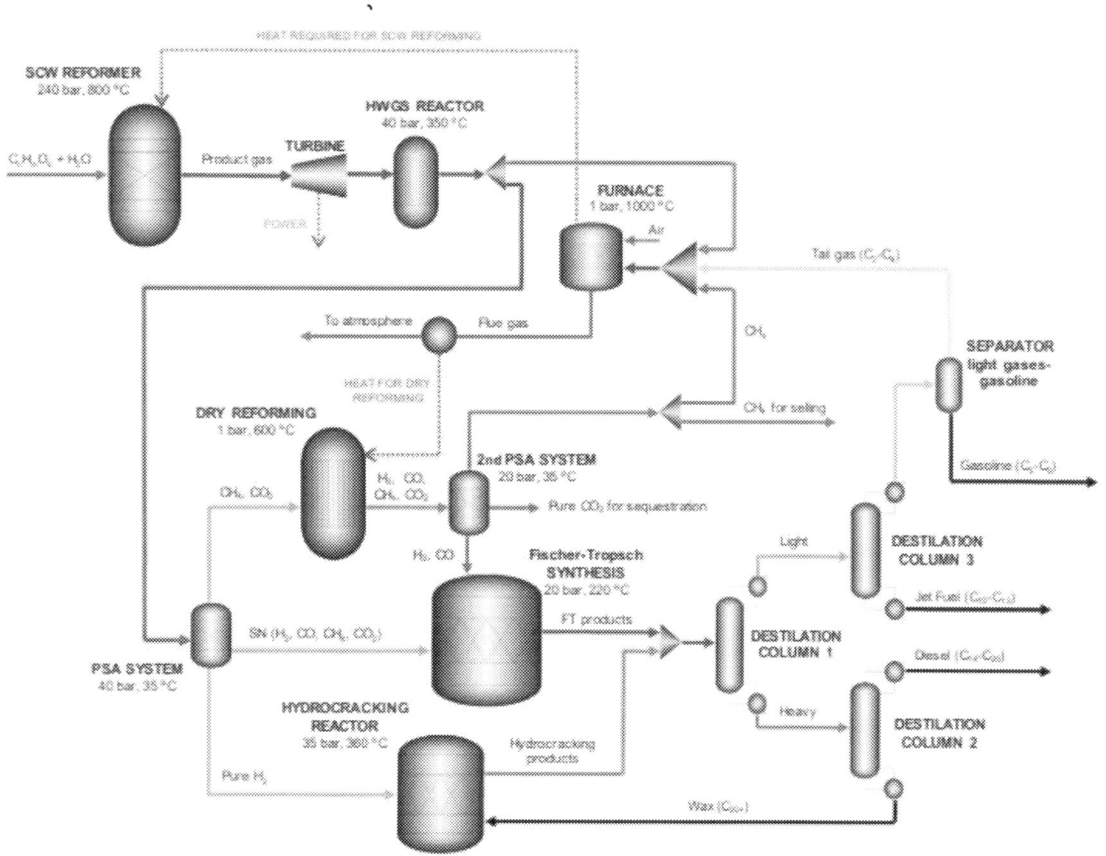

Fig. 6.12 Schematic diagram of the process of Fischer Tropsch synthesis [35].

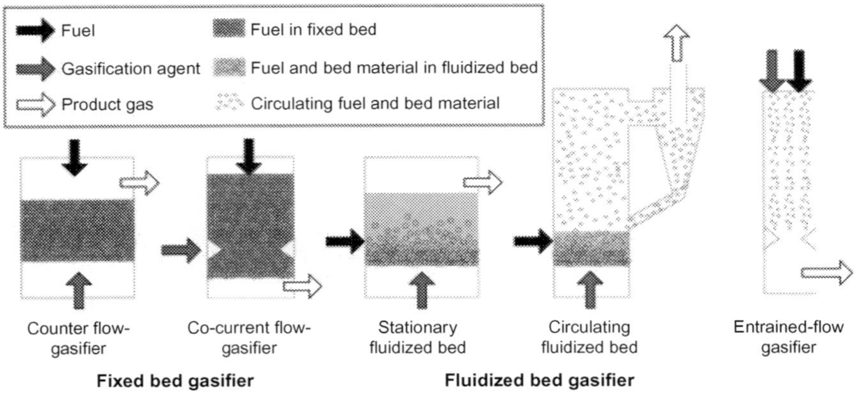

Fig. 6.13 Fluidized bed gasification [37].

gasification is reputed as the most often adopted reactor [39]. While realizing the fact that biomass is composed of mineral constituents (rash), and the entrained flow gasifier with a sluggish tendency deems it as the most favorable technology [40]. More specifically, these reactors require tiny elements of fuels for adequate conversion. It entails excessive refining of non-liquid fuels that are not energy conservative and usually dispense the constituents that are unable to be fed via customarily adopted pneumatic mechanisms [40]. Therefore, the emphasis of research and development is central to the strategies that could technically permit the feeding of fuels and could adjust the money matters of the entire chain. Consequently, the most prominent prior methods to adopt treatment are torrefaction and pyrolysis, as they both are suitable enough to facilitate cost-effective and productive outcomes in the form of the transformation syngas from biomass [36].

The entrained-flow gasifier has a feeding section that comprises oil, fluids, and an oil injection mechanism. It also contains a section that has a higher temperature for proceeding bio-oil gasification ($D=0.05$ cm and $L=50$ cm) and a chilling section. This feeding section facilitates the insertion of the natural spirit or bio-oil. The blend of oxygen with air may be inserted via an inlet-conduit for the purpose of atomization of the bio-oil or natural spirit. The whole phase of the composition of the gas is undergone with dry-gas. The samples of gas are collected at several places in the line of outlets. The samples are examined via gas chromatography to find the trivial compounds of CxHy, and CH_4, H_2, N_2, O_2, and CO (Fig. 6.14) [41].

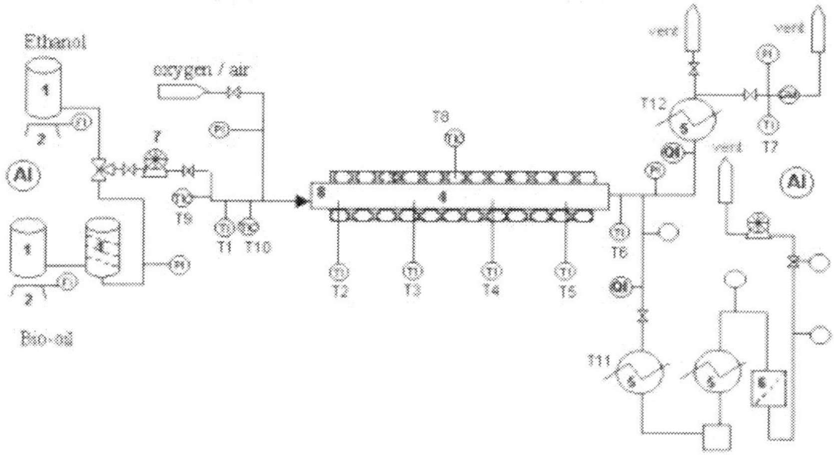

Fig. 6.14 Illustration of laboratory-based entrained flow gasifier [41].

6.2.4 Polygeneration

The aforementioned methods are related to the manufacturing of synthetic fuels with a greater concentration of H_2 and CO, which is required for the purpose of obtaining amplified efficacy of production like methanol and Fischer–Tropsch diesel. However, as a substitute method of it, Polygeneration is adopted. In this method, H_2 and CO are utilized in varied manners, perhaps producing electricity is one of them. Historically, the first experience of conversion of the waste into gas that composed of hydrocarbons was evidenced in Germany at the waste gasification plant of the Schwarze Pumpe [42]. The gas was processed to clean and to enable it for the usage as feedstock to manufacture methanol. In contrast, the rest of the quantity of gas (methane) utilized as fuel for a comprehensive chain of 75 MW_e to generate electricity [42]. For an integrated production of electricity and methanol, a process of Polygeneration is exhibited in Fig. 6.15. Fossil fuels, i.e., coal, are fed into a gasifier where a reaction takes place amid oxygen, which leads to the production of syngas. Some proportion of it is provided to a chemical synthesis plant to obtain methanol [44]. Methanol can be deposited or sold for using it at peak-time power generation [45] as shown in Fig. 6.15.

6.2.5 Biorefinery

The biorefinery is another notion that is associated with the integrated biofuels. This method occupied more acceptance in the society of the United

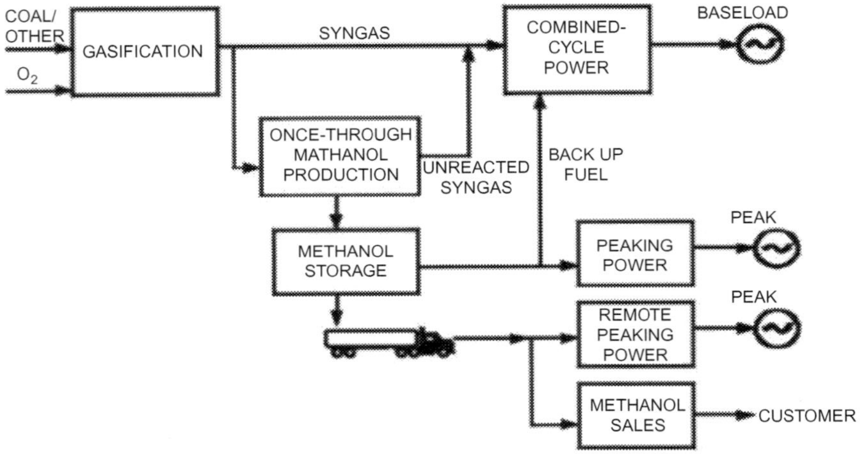

Fig. 6.15 A polygeneration energy system for methanol and electricity [43].

States of America, wherein, the traditional method of fermenting is amalgamated into thermo-chemical conversion along with syngas in the capacity of the intermediate product. This method is acknowledged for producing alcohol and ethanol via various kinds of biomass. A well-known and developed refinery of ethanol plants is anticipated to prepare alcohols that have lower than 1 \$/gallon [36]. In contrast with conventional means of refining oil, biorefinery transforms the biomass, and produces value-added chemicals, energy, and food while integrating tools and processes of biomass conversion [46] (Fig. 6.16).

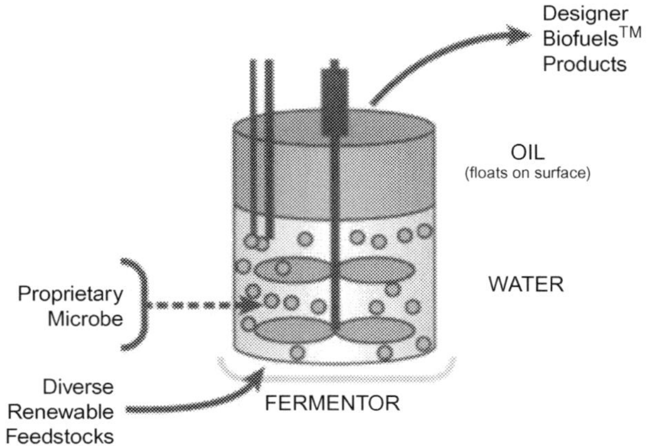

Fig. 6.16 The biorefinery technique associated with the integrated biofuels.

6.3 Technique to produce synthetic biofuels

6.3.1 Filtration combustion experiments

These experiments possess varied portions of syntshetic biogas blends, accompanied by air. They are processed while utilizing the schematic layout. Moreover, a reactive supply mechanism is involved in it, with an anticipated line of extraction, an inert porous media (IPM) reactor, and acquiring data product gas assess temperature evaluating the structure of a system. The process that aimed to reform the biogas was executed in a cylindrical shape conduit that was made up of Quartz and was 290 mm long, with 2 mm thickness of wall and diameter inside was 39 mm. It was processed under atmospheric compression. The reactor was enveloped with solid alumina spheres, which constituted a porous matrix of porosity of 40% [47]. It reflected an invalid proportion within the cylinder that was equipped with solid spheres and further accumulated via quotients between the unfettered bulk of pores and the entire volume of the solid matrix [48].

Guerrero et al. (2020) conducted experiments of filtration combustion with different proportions of biogas along with air (Fig. 6.17) [47]. The arrangement consisted of an IPM reactor, a supply system of reactants, an acquiring data system in order to make an analysis of product gas composition and temperature along with a model extraction line. Restructuring biogas was processed in a quartz-made cylindrical-shaped tube that has a length of 290 mm, whereas the thickness of the wall accounted for 2 mm, along with a diameter of 39 mm. The process was initiated amid air pressure. The solid alumina spheres (Al_2O_3, 5.5 mm diameter) was used to protect the reactor that formed a porous matrix having a porosity of ~40%. It shows the void segmentation in the cylinder that was occupied by solid spheres [48]. The insulation of the reactor was done with Fiberfrax insulation blankets that accounted for 11- and 3-mm thickness internally and externally. The supply of fuel for the mixtures of biogas was produced via natural gas local distribution. The molar concentration was, on average, 96% CH_4/4% C_2H_6. In comparison, the concentration of CO_2 was 99.9%. Dry air was obtained through a compressor that was reciprocating. The whole process was undergone at room temperature of 20 ± 1 °C. Moreover, for quantitative control, three thermal mass flow controllers were used that were blended prior to its presence inside the reactor. Consequently, a homogenous fuel-air mixture was obtained [47].

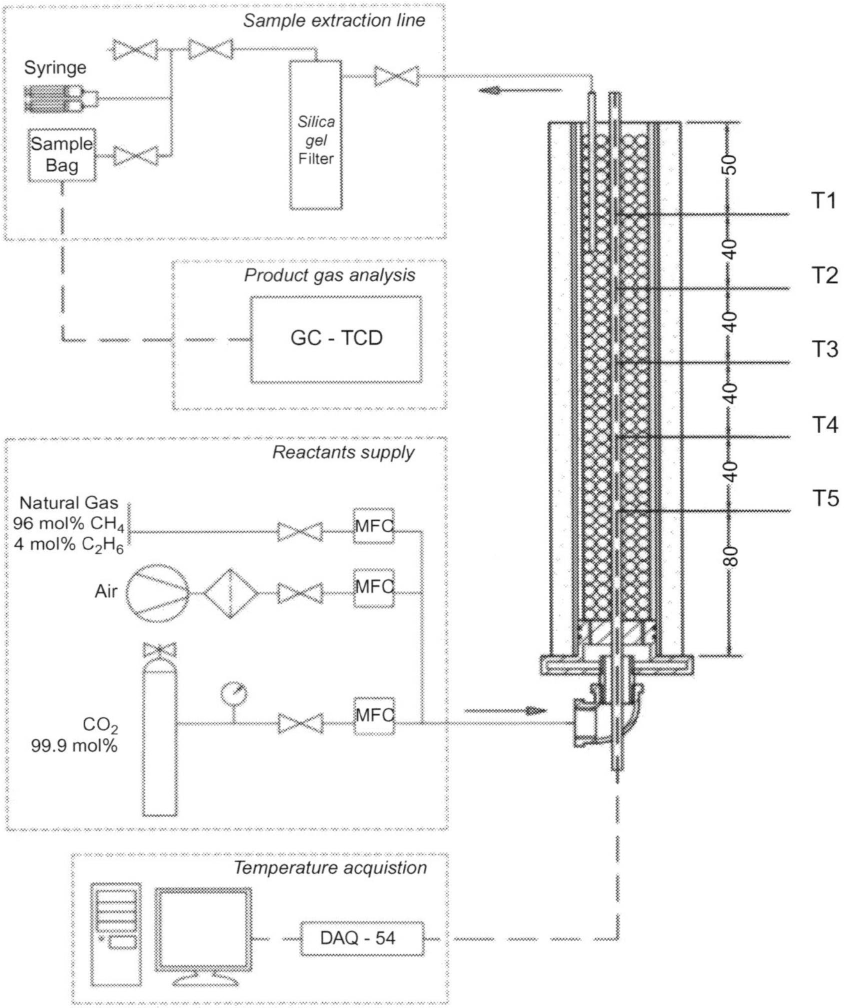

Fig. 6.17 Filtration combustion experiments [47].

6.4 Applications of synthetic biosource

Since technological development has made it possible to create efficient products that are composed of the wasteful natural materials that are consequent to anthropogenic and animal activities. Synthetic biology is significant as it contributes to producing economically beneficial and sustainable bioproducts, transforming the green chemical from the agricultural residues, making petrochemical substitutes, reshaping the ordinarily used chemical with adding efficacy, and optimizing the pharmaceutical industry. Below are some of the major applications of synthetic biology.

6.4.1 Delivering economic, renewable BioAcrylic

A significant petrochemical is "acrylic" which is a major component of the industry and consumer goods. The constituents of acrylic not only enhance the durability of paints but also make it free from odor. Moreover, the firmness and durability of adhesives are also amplified, and the quality of cleaning agents is also improved. In the meantime, the industry of Acrylic has a worth of $8 billion in the world economy. OPX Biotechnologies (OPXBIO) is transforming the biobased and replenishable acrylic to make it analogs to the effectiveness of petro-acrylic. Consequently, these efforts not only have made it cost-effective but also resulted in reducing its tendency of hazardous gaseous emission. Moreover, plummeting oil reliance and stabilization of the oil price market can be anticipated [49]. The production of lactic acid (2-hydroxypropanoic acid) is possible biotechnologically through fermenting carbohydrates. Further, it is dried catalytically to form acrylic acid. The process of Lactic acid is done through fermentation, which has a lower production cost, and it is not resource-intensive. Consequently, it has no environmental hazards and works as an appropriate alternative for the feedstock in order to produce bioacrylic acid (Fig. 6.18).

6.4.2 Making "green chemicals" from agricultural waste

The most effective and widely used chemical agents are the surfactants as they assist in even a combination of chemicals that are not more inclined to remain in steady condition, likewise oil and water. Presently, the major constituents of most of the surfactants are petro-chemicals and seeds of oil

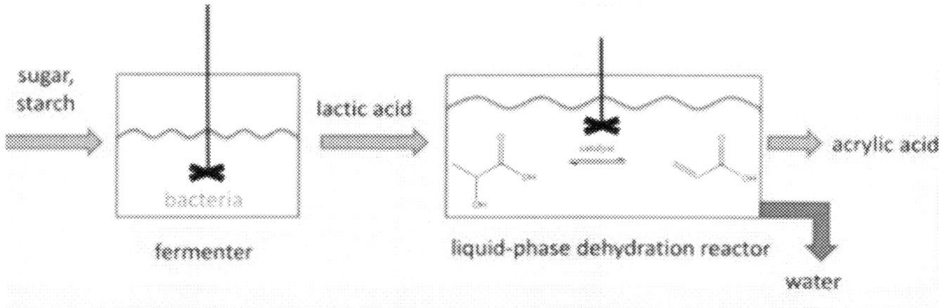

Fig. 6.18 Acrylic acid production from renewable energy resources [50].

(coconut or palm). Yearly, large-scale production of surfactants through petro-chemicals is responsible for the greenhouse gas emission that is parallel to gasoline inflammation of 3.6 billion gallons. Although the production from seeds is environmentally friendly, it has certain limitations as the use of seeds is related to the forest cover, and its excessive use can bring deforestation. Thus, to counter this shortcoming, microorganisms have been developed that are enable to transform the agricultural residues into productive surfactants [51–54]. They reshape the soybean hulls into useful surfactants. Notably, the hull is made up of a case from wood that preserves the soybean, and it is not digestible by humans and other living organisms [49]. The interaction of biosurfactants occurs with metal surfactants, which results in an orientation of the lipophobic head and the lipophilic tail in the outside environment. These developments also prevent oxidation. Consequently, the antimicrobial effect diminishes the biomass of sulfate-reducing bacteria (SRB). Moreover, it also hinders the formation of biofilms as these biofilms have a tendency to corrosion in the reservoirs. For example, *Bacillus licheniformis* and *Pseudomonas aeruginosa* are desaturated to possess significant antimicrobial effects on various strains of SRB (Fig. 6.19).

6.4.3 Developing a suite of biobased products and services

The first-ever use of synthetic biology was initiated by Dutch Multinational Corporation (DSM), a life sciences and materials sciences company that

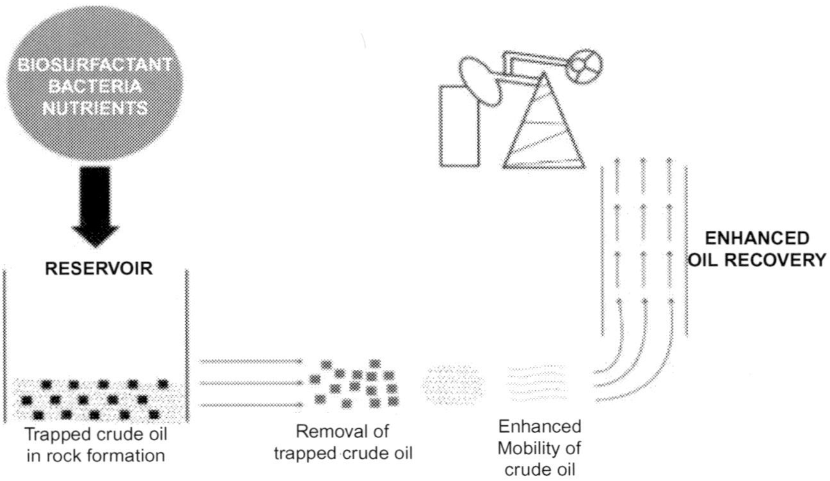

Fig. 6.19 Schematic diagram of biosurfactants in microbially enhanced oil recovery (MEOR) [55].

established the trend of enhancing the saleable manufacture of Cephalexin, which is a synthetic antibiotic. Commencing the microbial strain that could produce penicillin, the company has begun and improved the two enzyme-encoding genes with a single step straight fermentation process of adipoyl-7-ADCA (aminodeacetylcephalosporanic acid) that was consequent to the conversion of Cephalexin through two enzymatic phased. This process replaced the older one that comprised 13 phases. Therefore, it resulted in a reduction in cost and energy conservation [49].

6.4.4 Engineering low-cost sugars for petroleum substitute

Sugar components by nonfood biomass are utilized as significant components to produce diverse kinds of biofuels and sustainable chemical compounds that are money-intensive and do not remain economically stable due to the volatility in the price of feedstocks of petroleum. The progressed market of biofuel is projected to reach $ 21 billion gallons by 2022. In the customarily adopted methods of sugar fermentation, for the sake of manufacturing biofuels and sustainable chemicals, starch from corn or wheat or sucrose from sugar are used. However, biomass manufactured by Arivida, has supported the price fluctuations of unfinished goods, and they are energy conservative to manufacture biofuels and sustainable chemicals [49].

6.4.5 Creating economic advantage for a commonly used chemical

A highly valued chemical intermediate, adipic acid is a major constituent for making nylons, which has its usability associated with large-scale industries such as automobile parts and construction businesses. The accumulated worth of this industry has reached to $5.2 billion. However, the process that is involved in the production of adipic acid contributes 4 tons of carbon dioxide (CO_2) to the atmosphere per ton production of it. Meanwhile, this concentration can be diminished by using bio based processes by more than 20%. For instance, Verdezyne is using a cost-effective and eco-friendly fermentation processing for adipic acid, which will significantly reduce all the associated environmental and economic shortcomings [49]. Adipic acid is produced via fermentation that stipulates heterologous expression related to pathways and has an ability to transform intracellular metabolites into adipic acid. Further, it includes more pathways that are distinguished from each other on the basis of their theoretical yield. The cis, cis-muconic acid (ccMA) pathway (Fig. 6.20) instigates with 3-dehydroshikimate, which is regarded as a midway metabolite in the synthesis pathway of aromatic amino

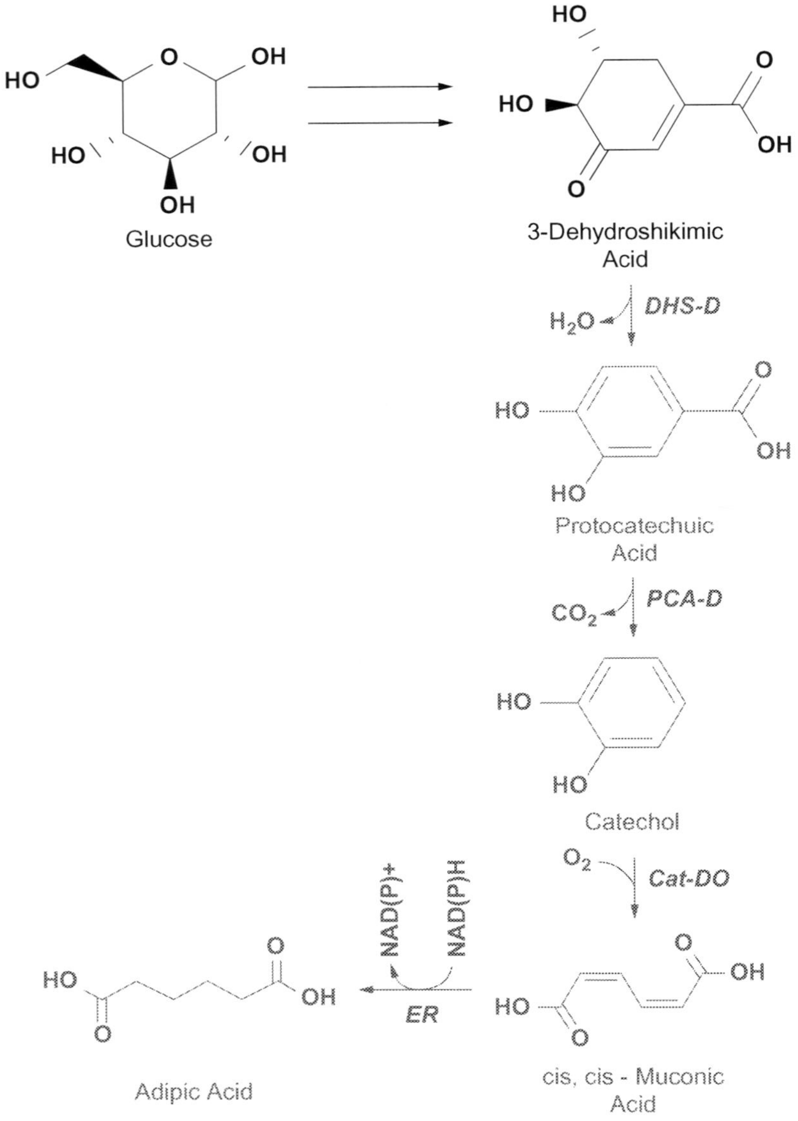

Fig. 6.20 The pathway for synthesizing adipic acid from glucose in *S. cerevisiae* [56].

acid [56]. Niu et al. [57] implemented this pathway in *E. coli* and achieved a final ccMA titer of 36.8 g/L, which was then hydrogenated to adipic acid using a platinum catalyst [57]. Specifically, the ccMA pathway was begun into a *S. cerevisiae* strain that empowered it to yield 140 mg/L of ccMA via glucose [58].

6.4.6 Increasing efficiency in bioprocessing of pharmaceuticals

The development of manufacturing Januvia as a cure for type II diabetes was conferred with the Presidential Green Chemistry Challenge Award in 2006. Subsequently, while using synthetic biology and other related technologies, transaminase was developed that was fully apt to enable a new bio catalytic route, which is currently under the process of commercialization [59–61]. Moreover, beginning with designing new enzymes variants, novel enzymes have been developed that are able to assist in detecting activities [62–64]. Further 1000-fold improvement was brought into the quality of these enzymes that were aimed to develop an enantioselective, hyper-active, sustainable, and pragmatic enzyme that could be labeled as unprecedented [5].

6.5 Conclusion

Synthetic biology is the branch of science that has made it possible to produce compounds that are not productive and cost-effective while providing no hazardous impact on climate. The genesis of synthetic biology is associated with initiating development in molecular science through which new methods of DNA sequencing were adopted that included pathways and complete genomes and its transplantation. More revolutions in molecular science were marked with creating microorganisms that had multipronged advantages. One significant development in synthetic biology was the initiation of synthetic BioSources that could meet the need for energy resources. As the shortage of energy resources and hazards of climate have made the realization of manufacturing synthetic biofuels that are equally cost and quality efficient. Some of the methods that led to the production of biofuels synthetically were Fischer Tropsch, Fluidized Bed Gasification, Entrained Flow Gasification, and Polygeneration. Apart from it, the usability of synthetic BioSource was not confined to it, but they are now prerequisites for various large-scale engineering and pharmaceutical engineering.

References

[1] J.I. Glass, N. Assad-Garcia, N. Alperovich, S. Yooseph, M.R. Lewis, M. Maruf, J.C. Venter, Essential genes of a minimal bacterium, Proc. Natl. Acad. Sci. 103 (2) (2006) 425–430.

[2] C. Lartigue, J.I. Glass, N. Alperovich, R. Pieper, P.P. Parmar, C.A. Hutchison, J.C. Venter, Genome transplantation in bacteria: changing one species to another, Science *317* (5838) (2007) 632–638.

[3] D.G. Gibson, J.I. Glass, C. Lartigue, V.N. Noskov, R.Y. Chuang, M.A. Algire, C. Merryman, Creation of a bacterial cell controlled by a chemically synthesized genome, Science *329* (5987) (2010) 52–56.
[4] J.A. Doudna, E. Charpentier, The new frontier of genome engineering with CRISPR-Cas9, Science 346 (6213) (2014) 1258096.
[5] M. Jinek, K. Chylinski, I. Fonfara, M. Hauer, J.A. Doudna, E. Charpentier, A programmable dual-RNA-guided DNA endonuclease in adaptive bacterial immunity, Science *337* (6096) (2012) 816–821.
[6] A.A. Cheng, T.K. Lu, Synthetic biology: an emerging engineering discipline, Annu. Rev. Biomed. Eng. 14 (2012) 155–178.
[7] J. Nielsen, J.D. Keasling, Engineering cellular metabolism, Cell 164 (6) (2016) 1185–1197.
[8] S. Jagadevan, A. Banerjee, C. Banerjee, C. Guria, R. Tiwari, M. Baweja, P. Shukla, Recent developments in synthetic biology and metabolic engineering in microalgae towards biofuel production, Biotechnol. Biofuels 11 (1) (2018) 185.
[9] J. Davies, Using synthetic biology to explore principles of development, Development 144 (7) (2017) 1146–1158.
[10] Types of microorganisms. Available at: https://courses.lumenlearning.com/microbiology/chapter/types-of-microorganisms/..
[11] Anon 2016. Synthetic Biology and Biodiversity. European Commission Report. Issue no. 15. Available at: https://ec.europa.eu/environment/integration/research/newsalert/pdf/synthetic_biology_biodiversity_FB15_en.pdf.
[12] B. Thomas, M.C. Raj, J. Joy, A. Moores, G.L. Drisko, C. Sanchez, Nanocellulose, a versatile green platform: from biosources to materials and their applications, Chem. Rev. 118 (24) (2018) 11575–11625.
[13] F.N. Ani, Utilization of bioresources as fuels and energy generation, in: Electric Renewable Energy Systems, Academic Press, 2016, pp. 140–155.
[14] D. Kantaram Jadhav, Bio-resourses & its utilization in health sector of India, Int. J. Environ. Sci. Nat. Resour. 1 (4) (2017) 109–111.
[15] I. Körner, Civilization biorefineries: efficient utilization of residue-based bioresources, in: Industrial Biorefineries & White Biotechnology, Elsevier, 2015, pp. 295–340.
[16] I.C. MacDonald, T.L. Deans, Tools and applications in synthetic biology, Adv. Drug Deliv. Rev. 105 (2016) 20–34.
[17] J.S. Tumuluru, C.T. Wright, R.D. Boardman, N.A. Yancey, S. Sokhansanj, A Review on Biomass Classification and Composition, Co-Firing Issues and Pretreatment Methods, 2011 Louisville, Kentucky, August 7–10, 2011, American Society of Agricultural and Biological Engineers, 2011, p. 1.
[18] P. Zlateva, R. Dimitrov, Possibilities of using algae for renewable energy production in the Black sea region, in: 2019 11th Electrical Engineering Faculty Conference (BulEF), IEEE, 2019, September, pp. 1–4.
[19] D.H. Vynios, D.A. Papaioannou, G. Filos, G. Karigiannis, T. Tziala, G. Lagios, Enzymatic production of glucose from waste paper, Bioresources 4 (2) (2009) 509–521.
[20] Mioy T. H. (2020). Sugars: Fischer & Haworth Projections. Available at https://www.mioy.org/uploads/1/3/9/1/13912136/sugars-fischer-haworth.pdf.
[21] Drawing Sugar Structures: Fischer Projections, Haworth Structures and Chair Conformers (n.d.). Available at: http://www.chtf.stuba.sk/~szolcsanyi/education/files/Chemia%20heterocyklickych%20zlucenin/Prednaska%206/Odporucane%20studijne%20materialy/Drawing%20sugar%20structures.pdf..
[22] Aryal, S. (2018). Carbohydrates—Definition, Structure, Types, Examples, Functions. Available at: https://microbenotes.com/carbohydrates-structure-properties-classification-and-functions/.

[23] Anon (n.d.). Glycogen Metabolism Notes. Available at: https://oregonstate.edu/instruct/bb450/summer09/lecture/glycogennotes.html.
[24] A. Bhat, K. Deepak, Carbohydrate metabolism. (Doctoral dissertation), University of Jammu, Jammu, India, 2017. Available at: file:///D:/Office%20Work/RBRA-DO66/Chapter%206%20comments/Glycogenesis.pdf.
[25] N. Gaurav, S. Sivasankari, G.S. Kiran, A. Ninawe, J. Selvin, Utilization of bioresources for sustainable biofuels: a review, Renew. Sustain. Energy Rev. 73 (2017) 205–214.
[26] A. Robles-Medina, P.A. González-Moreno, L. Esteban-Cerdán, E. Molina-Grima, Biocatalysis: towards ever greener biodiesel production, Biotechnol. Adv. 27 (4) (2009) 398–408.
[27] C.J. Cleveland, Encyclopedia of Energy, Elsevier, 2004.
[28] L.R. Lynd, J.H. Cushman, R.J. Nichols, C.E. Wyman, Fuel ethanol from cellulosic biomass, Science 251 (4999) (1991) 1318–1323.
[29] G.W. Huber, S. Iborra, A. Corma, Synthesis of transportation fuels from biomass: chemistry, catalysts, and engineering, Chem. Rev. 106 (9) (2006) 4044–4098.
[30] R. Picazo-Espinosa, J. González-López, M. Manzanera, Bioresources for third-generation biofuels, Biofuel's Eng. Process Technol. 6 (2011) 115–133.
[31] S. Srinivas, R.K. Malik, S.M. Mahajani, Fischer-Tropsch synthesis using bio-syngas and CO2, Energy Sustain. Dev. 11 (4) (2007) 66–71.
[32] Markov, S. A. (2012). Biofuels and Synthetic Fuels. Available at: https://www.researchgate.net/publication/281596949_Biofuels_and_Synthetic_Fuels/link/55ef8f9908aedecb68fdb943/download.
[33] A. Lappas, E. Heracleous, Production of biofuels via Fischer–Tropsch synthesis: biomass-to-liquids, in: Handbook of Biofuels Production, Woodhead Publishing, 2016, pp. 549–593.
[34] C.H. Bartholomew, Recent technological developments in Fischer-Tropsch catalysis, Catal. Lett. 7 (1–4) (1990) 303–315.
[35] F.J. Campanario, F.G. Ortiz, Fischer-Tropsch biofuels production from syngas obtained by supercritical water reforming of the bio-oil aqueous phase, Energ. Conver. Manage. 150 (2017) 599–613.
[36] A. Van der Drift, H. Boerrigter, Synthesis Gas From Biomass for Fuels and Chemicals, 31, ECN Biomass, Coal and Environmental Research, 2006.
[37] R. Rauch, A. Kiennemann, A. Sauciuc, Fischer-Tropsch synthesis to biofuels (BtL process), in: The Role of Catalysis for the Sustainable Production of Bio-fuels and Biochemicals, Elsevier, 2013, pp. 397–443.
[38] M. Siedlecki, W. De Jong, A.H. Verkooijen, Fluidized bed gasification as a mature and reliable technology for the production of bio-syngas and applied in the production of liquid transportation fuels—a review, Energies 4 (3) (2011) 389–434.
[39] A. Collot, suMatching gasifiers to coals, IEA Clean Coal Centre Reports 63 (2002).
[40] A. Van der Drift, H. Boerrigter, B. Coda, M.K. Cieplik, K. Hemmes, Entrained Flow Gasification of Biomass Ash Behaviour, Feeding Issues, System Analyses, ECN, Petten, 2004.
[41] R.H. Venderbosch, L. Van de Beld, W. Prins, Entrained flow gasification of bio-oil for synthesis gas, in: 12th European Conference and Technology Exhibition on Biomass for Energy, Industry and Climate Protection, 2002, pp. 17–21.
[42] B. Sander, G. Daradimos, H. Hirschfelder, Operating results of the BGL gasifier at the Schwarze Pumpe, in: Gasification Technologies Conference, 2003.
[43] Assessment, A. D. O. E, Commercial-Scale Demonstration of the Liquid Phase Methanol (LPMEOH™) Process, 2003. DOE/NETL-2004/1199.
[44] M.V. Mančić, D.S. Živković, M.L. Đorđević, M.N. Rajić, Optimization of a polygeneration system for energy demands of a livestock farm, Therm. Sci. 20 (Suppl. 5) (2016) 1285–1300.

[45] L. Zheng, N. Weidou, Z. Hongtao, M. Linwei, Polygeneration energy system based on coal gasification, Energy Sustain. Dev. 7 (4) (2003) 57–62.
[46] T.E. Amidon, S. Liu, Water-based woody biorefinery, Biotechnol. Adv. 27 (5) (2009) 542–550.
[47] F. Guerrero, L. Espinoza, N. Ripoll, P. Lisbona, I. Arauzo, M. Toledo, Syngas production from the reforming of typical biogas compositions in an inert porous media reactor, Front. Chem. 8 (2020) 145.
[48] D. Trimis, F. Durst, Combustion in a porous medium-advances and applications, Combust. Sci. Technol. 121 (1–6) (1996) 153–168.
[49] Anon, Current Uses of Synthetic Biology for Renewable Chemicals, Pharmaceuticals, and Biofuels, 2013. Available at: Synthetic-Biology-and-Everyday-Products-2012.pdf.
[50] Albert, J. (n.d.). Sustainable Production of Acrylic Acid. Available at: https://www.crt.tf.fau.eu/forschung/arbeitsgruppen/komplexe-katalysatorsysteme-und-kontinuierliche-verfahren/biomasse-und-nachhaltige-erzeugung-von-plattformchemikalien/nachhaltige-erzeugung-von-acrylsaure/.
[51] I.M. Alarifi, W.S. Khan, R. Asmatulu, Synthesis of electrospun polyacrylonitrile-derived carbon fibers and comparison of properties with bulk form, PLoS One 13 (8) (2018), e0201345.
[52] I.M. Alarifi, R.A. Alharbi, M.N. Khan, W.S. Khan, A. Usta, R. Asmatulu, Water treatment using electrospun PVC/PVP nanofibers as filter medium, Int. J. Mater. Sci. Res. 1 (2) (2018) 43–49.
[53] T.M. El-Bagory, M.Y. Younan, I.M. Alarifi, Failure analysis of ring hoop tension test (RHTT) specimen under different loading conditions, in: Pressure Vessels and Piping Conference (vol. 51623, p. V03AT03A024), American Society of Mechanical Engineers, 2018, July.
[54] I.M. Alarifi, Fabrication and characterization of electrospun polyacrylonitrile carbonized fibers as strain gauges in composites for structural health monitoring applications (Doctoral dissertation), Wichita State University, 2017.
[55] E.O. Fenibo, G.N. Ijoma, R. Selvarajan, C.B. Chikere, Microbial surfactants: the next generation multifunctional biomolecules for applications in the petroleum industry and its associated environmental remediation, Microorganisms 7 (11) (2019) 581.
[56] K. Raj, S. Partow, K. Correia, A.N. Khusnutdinova, A.F. Yakunin, R. Mahadevan, Biocatalytic production of adipic acid from glucose using engineered Saccharomyces cerevisiae, Metab. Eng. Commun. 6 (2018) 28–32.
[57] W. Niu, K.M. Draths, J.W. Frost, Benzene-free synthesis of adipic acid, Biotechnol. Prog. 18 (2) (2002) 201–211.
[58] K.A. Curran, J.M. Leavitt, A.S. Karim, H.S. Alper, Metabolic engineering of muconic acid production in Saccharomyces cerevisiae, Metab. Eng. 15 (2013) 55–66.
[59] G. Chinni, I.M. Alarifi, M. Rahimi-Gorji, R. Asmatulu, Investigating the effects of process parameters on microalgae growth, lipid extraction, and stable nanoemulsion productions, J. Mol. Liq. 291 (2019) 111308.
[60] B. Souayeh, M.G. Reddy, P. Sreenivasulu, T. Poornima, M. Rahimi-Gorji, I.M. Alarifi, Comparative analysis on non-linear radiative heat transfer on MHD Casson nanofluid past a thin needle, J. Mol. Liq. 284 (2019) 163–174.
[61] S. Kasaragadda, I.M. Alarifi, M. Rahimi-Gorji, R. Asmatulu, Investigating the effects of surface superhydrophobicity on moisture ingression of nanofiber-reinforced biocomposite structures, Microsyst. Technol. 26 (2) (2020) 447–459.
[62] K.G. Kumar, M. Rahimi-Gorji, M.G. Reddy, A.J. Chamkha, I.M. Alarifi, Enhancement of heat transfer in a convergent/divergent channel by using carbon nanotubes in the presence of a Darcy–Forchheimer medium, Microsyst. Technol. 26 (2) (2020) 323–332.

[63] S. Uddin, M. Mohamad, M. Rahimi-Gorji, R. Roslan, I.M. Alarifi, Fractional electro-magneto transport of blood modeled with magnetic particles in cylindrical tube without singular kernel, Microsyst. Technol. 26 (2) (2020) 405–414.
[64] I.M. Alarifi, H.M. Nguyen, A. Naderi Bakhtiyari, A. Asadi, Feasibility of ANFIS-PSO and ANFIS-GA models in predicting thermophysical properties of Al2O3-MWCNT/oil hybrid nanofluid, Materials 12 (21) (2019) 3628.

CHAPTER 7
Synthetic oil

Abbreviations

API	American Petroleum Institute
CDS	carbon dioxide splitting
CeO_2	cerium oxide
CH_4	methane
Cr_2O_3	chromium (III) oxide
dERC	direct electrochemical reduction of carbon dioxide
DME	dimethyl ether
EE	electric energy
ERC	electrochemical reduction of carbon dioxide
GTL	gas-to-liquid
ICE	internal combustion engine
LPG	liquefied petroleum gas
MgO	magnesium oxide
MOox	metal oxide
MOred	lower-valence metal oxide
NATO	North Atlantic treaty organization
PAO	poly-alpha-olefin
RF	recovery factor
SAE	Society of Automotive Engineers
SnO_2	tin oxide
USA	United States of America
WS	water splitting

7.1 Introduction

The exhaustion of crude oil resources is apparent globally, which leads the humans to not only ponder over conserving the energy resources but also to find the alternative and replenishable energy rescores to meet the energy requirements without disruption. Since it has been projected that the energy requirement will reach to its hike by 2030, further, it has been estimated that during 2062 and 2094, the accumulated oil reserves would account for 1.4 trillion to 2 trillion barrels while 80 million barrels per day would be the level of depletion [1]. The science of studying the biofuels and synthetic fuel is interdisciplinary that emphasizes on the obtaining cleaner and replenishable

https://doi.org/10.1016/B978-0-12-824001-4.00011-9

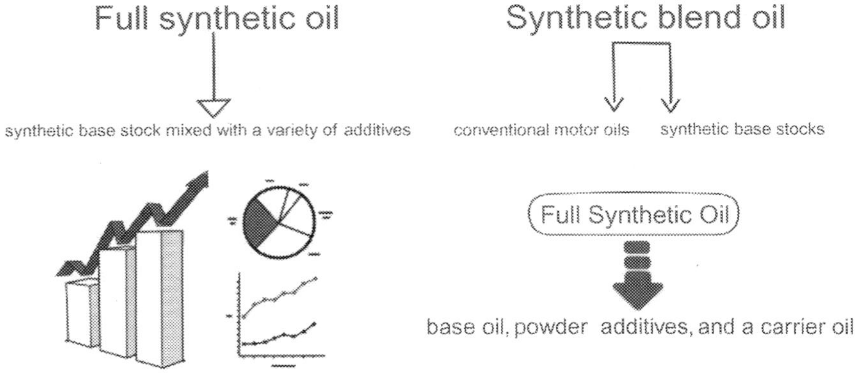

Fig. 7.1 Full synthetic oil vs synthetic blend.

energy resources. Biogas, methane, biodiesel, and ethanol come under the categories of biofuels. On the other hand, synthetic fuels are synfuel and syngas. These fuels are utilized as gasoline and as alternatives to diesel for accommodating the transportation needs and electricity generation. This area of applied science has attracted parallel investment of both the governmental entities and the corporate sector worldwide as shown in Fig. 7.1 full synthetic oil vs synthetic blend [2].

Hint statement

The energy requirements of world are projected to reach 1.4 trillion to 2 trillion during the period of 2062–2096. The consumption per day would be 80 million barrels.

Synthetic fuels are characterized by liquid fuels, i.e., gasoline oil, methanol, diesel oil, low, medium, and high-calorific value gas and solid green fuels [3]. The notion of synthetic fuels is construed in varied and broader aspects, where several kinds of fuels are categorized under the category of synthetic fuels. According to the definition adopted by the International Energy Agency, "synthetic fuels are the liquids produced from natural gas and coal" [4]. Additionally, the Energy Information Administration in its Annual Energy Outlook 2006 stated synthetic fuels as the products that are obtained from natural gas, oil, or through the feedstock of biomass while transforming them into synthetic liquid products or synthetic crude via chemical alteration. Besides, synthesized gaseous fuels, oil sands, oil shales, and liquid fuels are also included as synthetic fuel sources [4].

Hint statement
The science of biofuels and synthetic fuels is an interdisciplinary that deals with producing cleaner and recyclable energy means.

For lubricating numerous internal combustion engines, engine oil is useful. It works to grease the functional parts, cleanses the motor oil, prevents oxidation, and enhances the sealing. Furthermore, it works to lessen the heat of the engine. The production of engine oils and synthetic oil is carried from the petroleum and nonpetroleum and synthetic chemical compounds. The major constituents of engine oil include the organic compounds that carry carbon and hydrogen [5].

The property of viscosity is one of the essential characteristics of the engine oil as it implants the characteristic of greasing [6]. For this purpose, the first-ever lubricant standard J300 was enacted by the Society of Automotive Engineers (SAE) in 1911, which composed of viscosity classification of motor oils [7]. Though this standard has gone through several modifications, it is yet adopted in motor oil applications. Contemporarily, the oil viscosity is recognized through the umber of SAE. If the oil is thinner, its SAE's number would be lower. For example, SAE 10W, where the numeral value is respective to the viscosity on the specified temperature and "W" denotes the oil's appropriateness for extreme lower temperatures [7]. The level of viscosity adds up with an increase in temperature, but lowering the temperature does not affect the viscosity. Subsequently, the optimum level of viscosity is obtained at minimal and increased temperature. Oils with the optimum level of viscosity are termed as multi-grade oils. "20W–40" illustrates viscosity at higher temperatures and a state of shallowness at a minimal temperature [7]. Fig. 7.2 shows synthetic base oils already have multi grads.

Hint statement
Some of the examples of synthetic fuels are low, medium and high-calorific value gas, green solid fuels, gasoline oil, methanol and diesel oil.

Fig. 7.3 shows the viscosity index impresser and synthetic technologies that were adopted earlier were not cost-effective. In the meantime, the projects of developing synthetic products at larger scales are of multibillion-dollar investment. However, it is challenging for the manufactures to produce more quality fuels in comparison with fossil fuels that have lesser hydrogen concentration. Crude oil has a 1.8 hydrogen-to-carbon ratio,

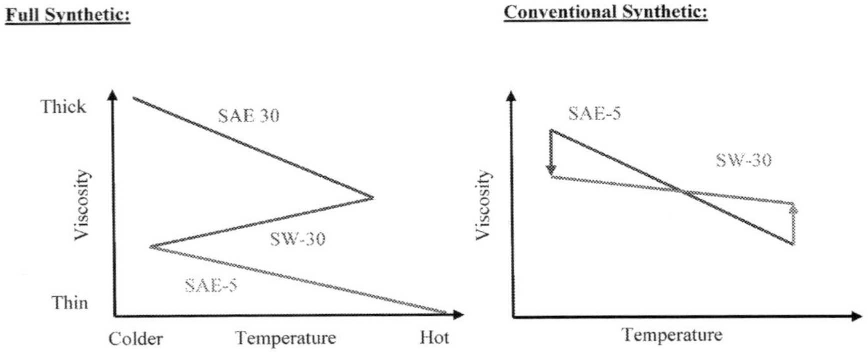

Fig. 7.2 Comparison between full synthetic vs conventional synthetic.

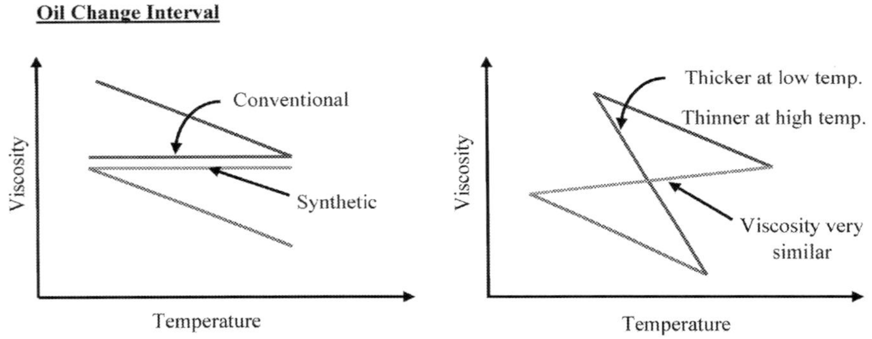

Fig. 7.3 The process of oil change interval.

while shale oil has a 1.5 hydrogen-to-carbon ratio. Tars have 1.2, and coal has less than 1 hydrogen-to-carbon ratio. Fig. 7.4 illustrates the hydrogen-to-carbon ratio of some of the fossil fuels.

Along with the process of advancing the petroleum residuum, synthesizing the quality fuel aims to transform the heavyweight resources in lightweight products and to change the dense compounds in fluid and gaseous forms through concentrating hydrogen or reducing carbon. In contrast with petroleum, these compounds require more resources for tapping style of the natural reservoirs and to bring it to the sites. Although, it is hard to assess the synthetic fuel that is available at the lowest price since the price hinges on the factors of adopting the method of production, level of quality, and on the deposit that it has been tapped. Table 7.1 exhibits the cost of different types of synthetic fuels [8].

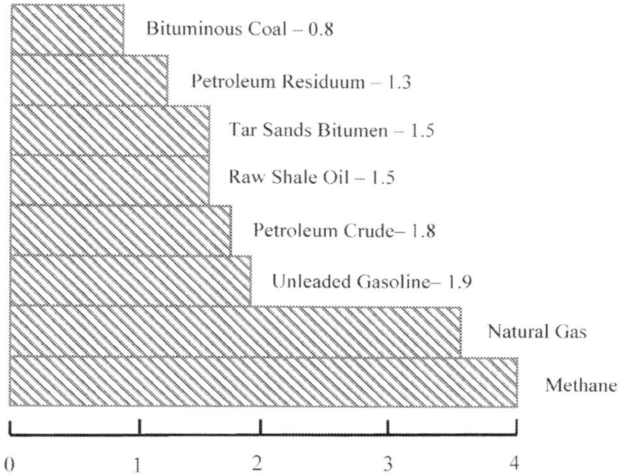

Fig. 7.4 Hydrogen-to-carbon ratio of fossil fuels.

Table 7.1 Cost of different types of synthetic fuels [8].

Raw material	Principal product	Commercial project status	Relative to prices of imports
Oil sands	Refinery feed	Operating	Competitive
Oil shale	Refinery feed	Underway	Competitive to nearly competitive
Coal	Intermediate BTU Gas	Under development	Competitive to nearly competitive
	Methanol	Under development	Competitive to nearly competitive
	Substitute Natural Gas	Under development	Competitive to nearly potentially competitive
	Traditional petroleum products	Operating/Demo/ Pilot	Competitive to nearly potentially competitive
Biomass	Alcohols	Small plants	Competitive only with subsidies
Natural gas	Methanol	Operating for chemicals	Dependent on Gas/Oil price spread
Natural gas	Traditionally petroleum products	Under construction/under development	Higher than Methanol

7.1.1 Advantages of synthetic oil

The three dominant reasons to use synthetic oil include: its potential to ease the technical issues, its ability to not being resource-intensive, and the ability to enhance the functionality of the engine and improves the strength. For instance, Europe has introduced ester base stocks and Poly-alpha-olefin (poly-α-olefin, PAO) to the oil formulators for improving the thermal-oxidative durability. In addition, in the decade of 1970s, the military of the United States of America (USA), world's biggest intergovernmental military alliance the North Atlantic Treaty Organization (NATO) and the public and private companies in the USA had started to use synthetic oil to make functionalized machines amid lower temperatures (e.g., MIL-L46167B) [4]. The usage of synthetic oil offers the advantages of enhanced viscosity at extreme temperatures, enhanced degree of durability, and a lesser degree of loss, which is consequent to evaporation. Also, the synthesized oil is corrosion-resistant, less fragile to the thermal wreckages, and accounts lower tendency of dispensing residue. Another advantage includes higher torque and horsepower on account of a fewer grade at preliminary levels. Moreover, the higher degree of fuel efficacy that ranges between 1.8% and 5% was reported in fleet tests [4].

The formulation of oil, regardless of its nature (mineral or synthetic), is done while mixing the additives in order to fortify the quality of the oil. It reflects that all the synthesized lubricants comprise a base oil that is the blend of various base oils and other additives. When it comes to bringing up the oil (synthetic or mineral), it alludes to the constituents of its base oils. For this purpose, the American Petroleum Institute (API) has devised a chart that serves the information regarding various types of bases oil (Table 7.2) [9].

The molecular make-up is the major factor that differs between the types of oil. The oils that have been placed in groups I and II are generally placed in

Table 7.2 The categorization of base oil by the API.

Group	Saturate wt%	Sulfur wt%	Viscosity index
I	<90 and/or	>0.3	>80 to 120
II	≥ 90 and	≤ 0.03	≥ 80 to ≥ 120
III	≥ 90	≤ 0.03	≥ 120
IV	All poly alpha olefins (PAOs)		
V	All base stocks not included in Groups I-IV		

the category of mineral base oil, and their major component is the crude that is abundant with the molecules of hydrocarbon. In addition to it, the oils that have been included in group II and III base stock, assist in the production of semi-synthetic lubricants which are more economic-friendly. The top three oil groups have their applicability with the inclusion of ester oils, additives (group V), and polyalphaolefins (group IV) to produce semi-synthetic lubricants. Usually, the base oils of group III are termed as synthetic oil if they are hydro-isomerized. Multiple petrochemical entities devised methodologies for feedstocks' acceleratory alteration with hydrogen to produce enhanced mineral lubricating oil. The process was termed as hydro-isomerization or hydro-cracking. In the year of 2005, the production of base stocks (group III) was initiated through the gas-to-liquid method (GTL). Those produced oil encompassed the characteristics of synthetic oil. However, they are considered as semi-synthetic oil lubricants. The formulation of semi-synthetic oil includes mineral oil and a 30% concentration of synthetic oil. Unlike mineral oil, synthetic oil has a steady, molecular size that is responsible for its organized structure. Moreover, synthetic base stocks are produced but not found naturally [9].

Example 7.1

Calculate the recovery factor with parameters—production, reserves, and the water injection of an oil field? If oil production is 23.66 mm^3, water injection = 54.58 mm^3 and reserves are 12.9 mm^3 [10].

Solution:

Recovery factor (RF) = (oil initially in place − oil produced)/oil initially in area).

Oil initially in place = 23.66 + 12.9 = 36.56.

Hence, RF = (36.56 − 23.66)/36.56 = 0.3528 = 35.3% [11].

Pyrolysis, liquefaction, and gasification are the three methods that are adopted for synthesizing coal. In pyrolysis, to obtain liquified coal, the coal is heated without the presence of oxygen (Fig. 7.5). This approach is considered more energy-conservative. Pyrolysis of biomass is the process of thermochemical decomposing wherein complex and large hydrocarbon composites are split into lightweight molecules. Biomass is degraded, which produces bio-oil and char products. The pyrolysis is the initial process of gasification [12] that is undergone without the presence of oxidant agents.

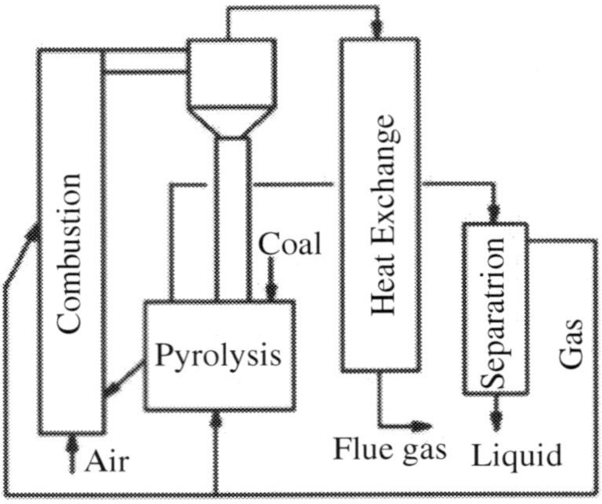

Fig. 7.5 Schematic diagram of the pyrolysis of coal [8].

The range of temperature is 300–1000 °C [13]. The thermochemical process of pyrolysis can be illustrated from the following reaction [14]:

$$C_nH_mO_p(Biomass) \rightarrow \sum_{liquid} C_xH_yO_z + \sum_{gas} C_aH_bO_c + H_2O + C(char)$$

On the other hand, in the process of coal gasification, a chemical reaction takes place among coal, heat, and air through which methane hydrogen and a blend of carbon monoxide are obtained in different quantities (Fig. 7.6). Most often, hydrogen and carbon monoxide are obtained in more quantity in the form of gas, which is used as fuel to industries. However, it concentrates only on 40% of the heat in the form of methane. Additionally, it can be used as a feedstock for synthesizing fluid fuel [8].

Direct liquefaction of coal transforms the coal into liquid form while undergoing heat and hydrogen (Fig. 7.7). Exxon Donor Solvent process is one of the examples of this process which is adopted in modern times. This process entails a hydrogenated solvent that has been recycled and catalyzed to provide hydrogen during the liquefaction process of coal. It gradually takes place at 14 MPa and a temperature of 450 °C with hydrogen [8].

Hint statement

According to the Energy Information Administration, the basic components of synthetic fuels are natural gas, oil, and feed stocks of biomass.

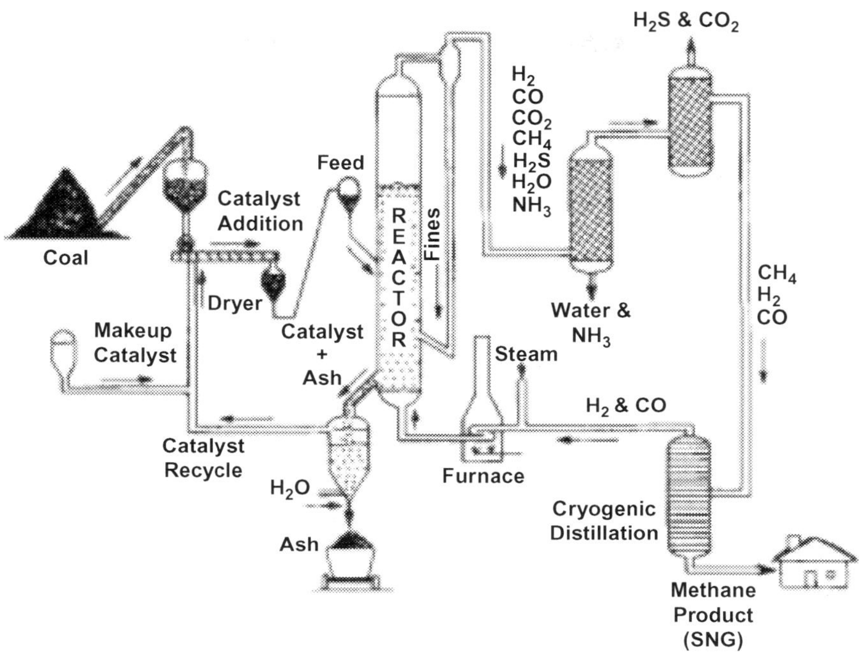

Fig. 7.6 Process of coal gasification [8].

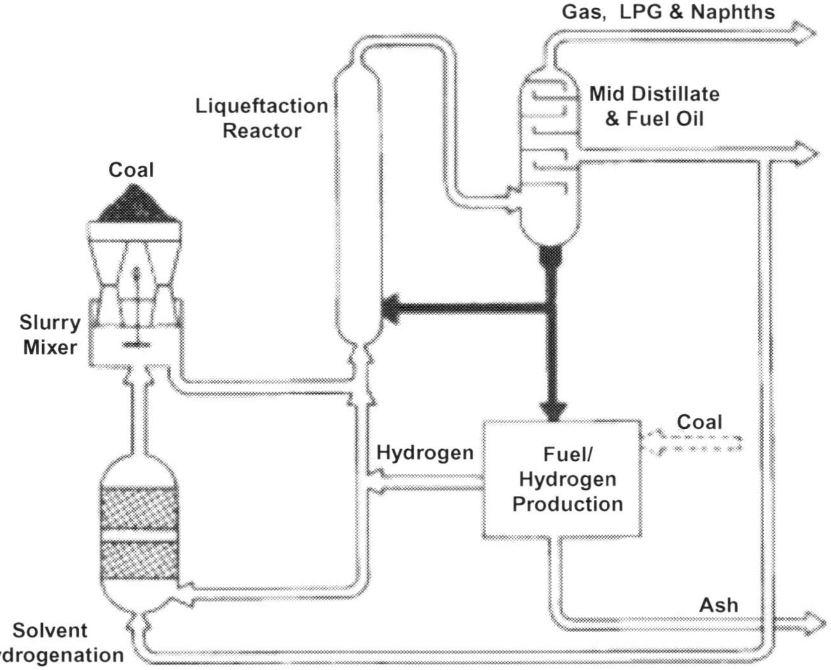

Fig. 7.7 Direct gasification of coal [8].

Example 7.2

Calculate the air to fuel ratio of combustion of Butane [15].

Solution:

Firstly, write the unbalanced theoretical equation of the chemical reaction of butane:

$$C_4H_{10} + a(TA)O_2 + a(TA)(3.76)N_2 \rightarrow bCO_2 + dH_2O + a(TA-1)O_2 + a(TA)(3.76)N_2$$

Afterward, the balancing, following is found:

$C:4=b$

$H:10=2d,\ d=5$

$O:2a=2b+d,\ a=6.5$

General combustion equation of butane at 130% theoretical air is

$$C_4H_{10} + 8.45O_2 + 31.77N_2 \rightarrow 4CO_2 + 5H_2O + 1.95O_2 + 31.77N_2$$

The air to fuel ratio is as follows:

$$AF_{mole} = \frac{8.45 + 31.77}{1} = 40.22\,\text{kg}.$$

7.2 Methods and techniques of synthesizing synthetic oil

Contemporarily, the world is heavily dependent on natural resources to meet the abundant energy requirement. However, there is continuous development in the field of devising the ways and methods to obtain synthetic fuel that are recyclable and can be used as alternatives to fossil fuels. Some of the methods to obtain synthetic fuels are given below:

7.2.1 Thermochemical cycles

Synthetic fuels are obtained through thermochemical processes under higher temperatures. Producing hydrogen through water splitting (WS), carbon dioxide splitting (CDS), the production of syngas via blending CO_2/H_2O splitting are obtained through the thermochemical method. In the meantime, solar-thermochemical cycles, while making use of sunlight, endure the endothermic reactions, which are considered more convenient and useful as compared to photo-electrocatalytic processes [16]. During solar-thermochemical cycles, concerted solar radiation is employed to produce H_2 through breaking the H_2O and producing CO while breaking the CO_2. The catalyze of the thermochemical cycles, metal oxides are used, which is regarded as the most convenient mechanism to generate synthesized fuels. Examining the water splitting and carbon dioxide splitting has brought about numerous reagents that include MgO/Mg, SnO_2/SnO, CeO_2/Ce_2O_3, and Cr_2O_3/CeO_2 [17]. For the production of syngas and hydrogen, thermochemical cycles are considered as appropriate methods. Through thermochemical cycles, water splitting prevents the problem of the dissemination of product gases amid the process of producing hydrogen and oxygen. The required level of heat is comparatively low, which is 1100–1500 [12].

When water is used to obtain fluid, hydrogen is produced as product gas. On the other hand, when a blend of water and carbon dioxide is poured in the container, it results in the production of syngas. Thus, the inputs are heat, water, and carbon dioxide, and outputs are hydrogen, oxygen, and carbon monoxide. Moreover, chemical reactants are redeveloped in a closed circle. Producing hydrogen with no degradation permits the immediate consumption of fuel cells. Generally, the cycles comprise of two or three phases, which include metal oxide redox reactions. In two-steps thermochemical cycles, higher temperature involves, and the efficiency level of the reaction is more than par. The discrepancy of the temperature of oxidation and reduction reactions is caused by the thermodynamic driving force in the two-step thermochemical splitting of water [18]. At the initial phase, metal oxide (MOox) is separated thermally and acquired the metal form, or it sometimes transforms into lower-valence metal oxide (MOred). During this process, oxygen is dispensed (Fig. 7.8). This reactive process is more endothermic with increased temperature [19].

The activation step involves high-temperature reduction [19]:

$$MO_{ox} \rightarrow MO_{red} + \frac{1}{2}O_2$$

The second is the exothermal phase, which involves water splitting and hydrolysis of reduced material oxide wherein production of primary metal oxide and pure hydrogen takes place. Recycling of the pristine metal oxide occurs through the solar reduction reactor by the material cycle [19].

Next step includes H_2 generation, which occurs at lower temperature oxidation in the presence of H_2O [19]:

$$MO_{red} + HO_2 \rightarrow MO_{ox} + H_2$$

Likewise, syngas is produced through the inclusion of H_2O and CO_2 in the reactor flask. The thermochemical cycle of H_2O and CO_2 are regarded as a probable way of making optimal use of carbon dioxide that has been obtained in different ways [19].

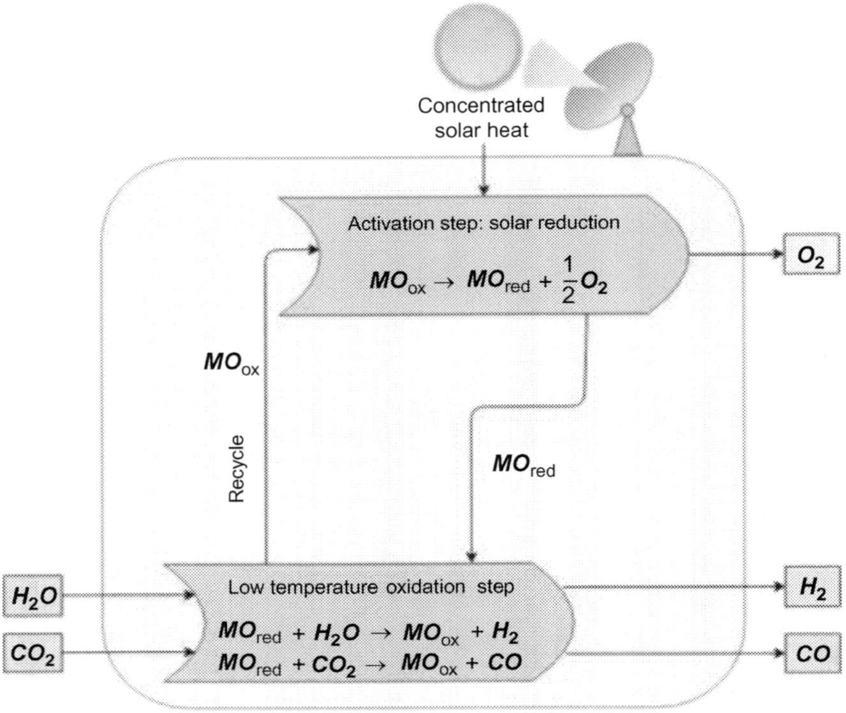

Fig. 7.8 An illustration of two-step thermochemical cycles of H_2O and CO_2 splitting while using metal oxide in a redox system [19].

The step of generating CO occurs with oxidation at a lower temperature [19]:

$$MO_{red} + CO_2 \rightarrow MO_{ox} + CO$$

The above-stated reactions are of oxide types. They are undergoing more investigation in the meantime. However, two reactions take place. One is the hybrid type, and other is hydroxide type [20]:

$$\text{Hybrid type}: MH_2 \rightarrow M + H_2$$
$$(M+)HO_2 \rightarrow MO_2 + \frac{1}{2}O_2$$
$$\text{Hydroxide type}: 2MOH \rightarrow 2M + MO_2 + \frac{1}{2}O_2$$
$$(2M+)HO_2 \rightarrow 2MOH + H_2$$

7.2.2 Gas-to-liquid (GTL)

The method of GTL is an optimal way to exploit natural gas. A chemical reaction undergoes wherein natural gas is converted into fuels such as Liquefied Petroleum Gas (LPG), diesel, and diesel. It comprises of three phases. Firstly, natural gas is converted into hydrogen and carbon monoxide (Syngas). Afterward, the chain reaction between carbon monoxide and hydrogen is undergone to obtain Syncrude [21–24]. Finally, Syncrude is processed to produce the required transporting fuels, which is also known as hydrocracking [1], as shown in Fig. 7.9.

7.2.3 Direct coal liquefaction method

The hydrogen-to carbon ration of coal is lesser than petroleum. When coal is converted into fluid, it stipulates a considerable quantity of hydrogen. However, the exceeding concentration of carbon can also be dispensed. During the process of direct-coal-liquefaction, coal, and hydrogen directly react to produce liquids. The reaction of hydrogen and carbon takes place in the presence of sulfur, nitrogen, and oxygen, which convert them into hydrogen, ammonia, and sulfide (Fig. 7.10). One of the important functions of hydrogen includes hydrogenating the donor solvent. During the hydrogenation, the donor solvent blends hydrogen with coal, which is in dissolved form. It adds up coal's hydrogen-to-carbon ratio (H/C) when it gets a form of liquid. Moreover, the major difference between petroleum and coal is the lower H/C atomic ratio of coal, which is 0.7 and 1.7, respectively (Table 7.3). Subsequently transforming coal into petroleum stipulates the direct addition of hydrogen. The advanced method of coal liquefication

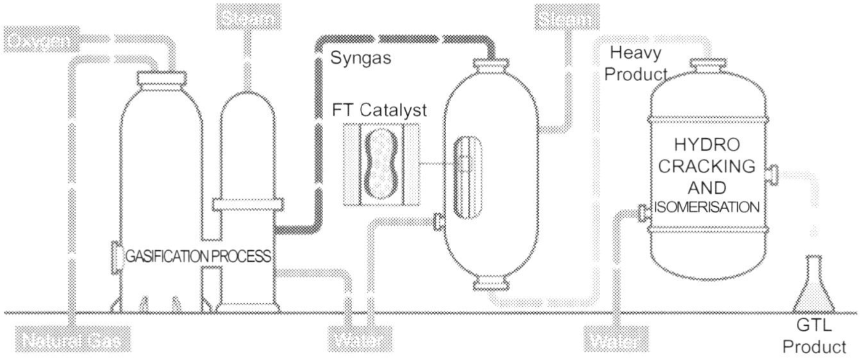

Fig. 7.9 Gas-to-liquids (GTL) technology [25].

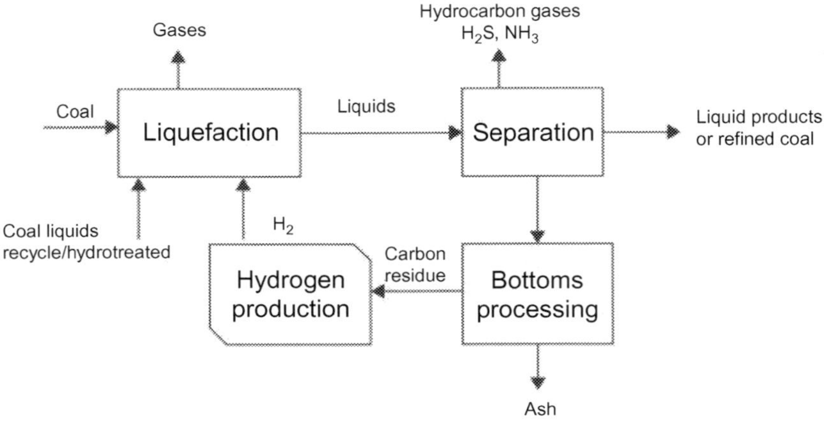

Fig. 7.10 Schematic representation of GTL [26].

to obtain gasoline products creates C5–400°F products only, and these products have no concentration of sulfur hydrocarbon gases, oxygen, and nitrogen [27].

7.2.4 Electrochemical reduction method

Renewable energy resources are more preferred for the generation of electric energy, which includes wind or solar energy. These renewable resources are transferred in an electrochemical reactor to proceed with an electrochemical reduction of CO_2 (ERC) (Fig. 7.11). CO_2 is supplied in the chemical reactor either directly during the gas phase or after the dissolution in a watery solution. When it is supplied directly, the process is known as direct electrochemical reduction gaseous CO_2 (dERC). Additionally, this process is more adaptable. When ERC occurs at the cathode, it leads to the production of value-added products (Eq. 7.1). Additionally, at the anode, O_2 is obtained through the electrochemical oxidation of H_2O (Eq. 7.2). Eq. (7.3) illustrates the complete reaction of forming CH_4. Also, it displays standard potential (E) of the reactive process in relation to the standard hydrogen electrode (SHE) [28].

$$\text{Cathode: } CO_2 + 8H^+ + *8e^- \rightarrow CH_4 + 2H_2O \ (E^\circ = 0.17\ V) \tag{7.1}$$

$$\text{Anode: } 4H_2O \rightarrow 8H^+ + 2O_2 + 8e^- \ (E^\circ = -1.23\ V) \tag{7.2}$$

$$\text{Overall: } CO_2 + H_2O \rightarrow CH_4 + 2O_2 \ (E^\circ = -1.06\ V) \tag{7.3}$$

Table 7.3 Characteristics of coal and petroleum [27].

	High vol. bituminous				Bituminous			
	Lignite	Sub-bituminous	B	A	Med volatile	Low volatile	Anthracite	Petroleum
% C (ash free)	65–72	72–76	78–80	80–87	89	90	93	83–87
% H	4.5	5	5.5	5.5	4.5	3.5	2.5	10–14
%O	30	18	10	10–4	3–4	3	2	0.1–1.5
H/C, atomic	0.77	0.8	0.82	0.8	0.6	0.46	0.32	1.7
%O as OH	15–10	12–10	NA	7–3	1–2	0–1	0	NA
Aromatic C atoms % of total C	50	65	NA	75	80–85	85–90	90–95	NA
Avg. no. Benzene rings/layer	1–2	NA	2–3	2–3	2–3	5	>25	NA
Vol Matter, %	40–50	35–50	NA	31–40	31–20	20–10	>10	NA
Reflectance, %, Vitrinite	0.2–0.3	0.3–0.4	0.6	0.6–1.0	1.4	1.8	4	NA
Calorific value	7000	10,000	13,500	14,500	15,000	15,800	15,200	NA

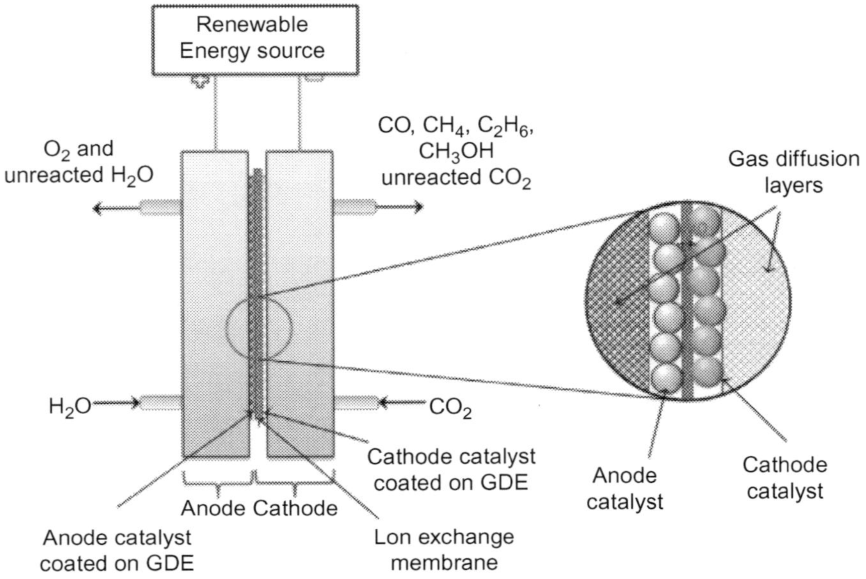

Fig. 7.11 Electrochemical reduction of CO_2 [28].

Through ERC, various products are obtained, such as hydrocarbons, CO, alcohol, and acids. However, such a process includes applied potential and electrocatalysts and electrolyte medium. Moreover, Table 7.4 provides overall reactions of generating the synthetic fuels along with their related thermodynamic requirement of electrical energy (EE) and electrode potential (E°), and such values have been considered by Gibb's free energy [28]. In addition to it, Fig. 7.12 presents the density of synthetic fuels in comparison with conventional energy resources [28].

Table 7.4 Complete reaction to produce synthetic fuels [28].

Complete reaction	E° (V)	EE (MJ/kg)
$2CO_2 + H_2O \rightarrow 2CO + O_2 + H_2O$	−1.33	9.19
$CO_2 + H_2O \rightarrow CH_4 + 2O_2$	−1.06	51.15
$CO_2 + H_2O \rightarrow CH_3OH + 1.5O_2$	−1.20	21.71
$2CO_2 + 3H_2O \rightarrow C_2H_5OH + 3O_2$	−1.14	28.70
$3CO_2 + 4H_2O \rightarrow C_3H_7OH + 4.5O_2$	−1.02	29.53

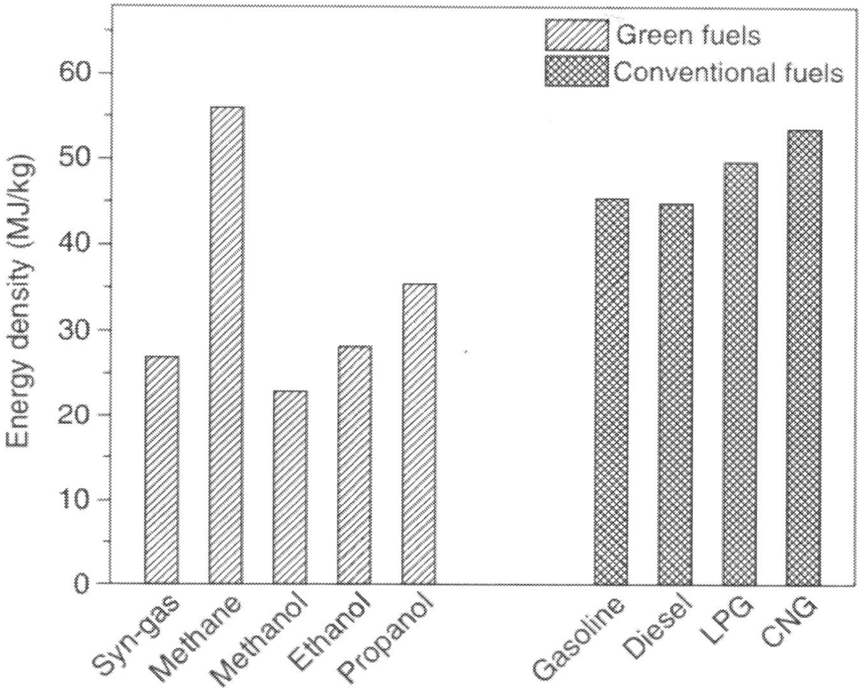

Fig. 7.12 The energy density of synthetic and conventional fuel [28].

7.3 Applications of synthetic oil

The use of crude oil is resource-intensive amid the price hike of mineral oil. Thus, the critical factors of greater importance with regard to the production of synthetic oil include conserving the mineral resources, averting the environmental degradation, particularly air quality deterioration, meeting the energy requirement of the swift expansion of population, social and economic imbalances, the rising conflict of interests among the national and international stakeholders to consolidate the independence and mitigate the financial risks related to fuel [29]. Some of the advantages of synthetic oils are summarized below:

7.3.1 Enhanced engine performance and wellbeing

Several characteristics of methanol, i.e., flame speed, vaporized heat, and rating of heat ($\sim$100) are exploited to improve the performance of the engine.

When it comes to safety, the lesser degree of instability of methanol leads to less inclination of the fuel-driven fire [30].

7.3.2 Clean burning of transportation to protect environment

In contrast with patrol, consumption of methanol causes lesser carbon and other hazardous gaseous concentration into the atmosphere. Subsea engineering and power train manufacturing companies have been inclined to produce and use synthetic fuels to eliminate carbon and footprints from the atmosphere. One of the main applications of synthetic fuel is in the transport sector, where synthetic fuels have more significant potential to replace fossil-fuels completely [30].

7.3.3 Chemical engineering

High-performance octane and cetane are synthesized. For instance, methanol C_1 is a chemical that is used to produce chemicals compound such as propylene that is used in different processes [30].

7.3.4 Generating electricity

Synthetic methanol and methanol are used in traditional electricity-generating power plants. The use of methanol for generating electricity in highly effective electrochemical cells is widely adopted [31–33]. The process is known as Direct Methanol Fuel Cells. At lower temperatures, the cells are used for the electronic devices and also used as an alternative Internal Combustion Engine (ICE) vehicle. Therefore, there is a high likelihood that the suitable blend of the renewable energy mix will ultimately replace the conventional means of generating electricity to meet the energy demands of modernized economies (Fig. 7.13) [30].

Furthermore, synthetic fuels will play a greater role in eliminating the disrupted supply chain of biofuels and contention fuels [34–36]. For instance, dimethyl ether (DME) and methanol are considered as the most appropriate fuels as they account for chemical synthesis in order to obtain synthetic fuels that are used in conventional internal combustion engines with comparatively fewer alterations [37]. Nowadays, China is top-ranked to use methanol in the transportation sector in five different forms of methanol gasoline blends, which are accessible in the market with the names of M100, M85, M15, M10, and M5 [38].

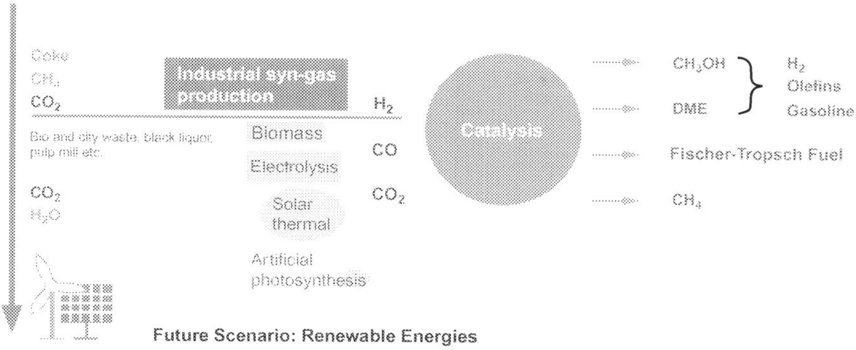

Fig. 7.13 The paradigm shift of fossil fuel to synthetic fuels along with other renewable energy resources [30].

7.4 Conclusion

Extensive and ever-increasing energy requirements of fuels and their resultant consequences for the environment, have led humans to devise the ways to obtain alternative means of energy generation. Synthetic fuels, which mainly include synthetic oil, are the optimal energy resources that have promising advantages and have surpassed the advantages of conventional mineral resources. The core objective of an interdisciplinary science of studying biofuels and synthetic fuels is to obtain the resources that are recycled and economically viable. The usage of synthetic oil is associated with the higher performance of the engine. Also, they are corrosion resistant and are available at a reasonable price comparatively. Moreover, synthetic oil is less vulnerable to thermal wreckage. In the meantime, more resources have been invested in the production of synthetic fuels by governmental and nongovernmental entities. The methods of thermochemical cycles, gas-to-liquid method, direct coal liquefaction method, electrochemical reduction method is commonly adopted to produce synthetic oil. The wider applicability and improved performance of synthetic fuels will greatly help the world to cope with the menace of climate change.

References

[1] U.U. Ntuk, E.N. Bassey, B.R. Etuk, A novel syngas production process design for gas-to-liquid (GTL) technology, Journal of Nigerian Society of Chemical Engineers 27 (1) (2012) 1–8.

[2] Markov, S.A. (2012). Biofuels and Synthetic Fuels. Available at: https://www.researchgate.net/publication/281596949_Biofuels_and_Synthetic_Fuels/link/55ef8f9908aedecb68fdb943/download.

[3] C. Chen, E.B. Kennel, L. Magean, P.G. Stansberry, A.H. Stiller, J.W. Zondlo, Production of Foams, Fibers and Pitches Using a Coal Extraction Process, West Virginia University (US), 2004.
[4] R. Lokapure, Comparative Study of Synthetic Oil and Mineral Oil on the Basis of Physical Characteristics, 2016.
[5] L. Severa, M. Havlíček, V. Kumbár, Acta Univ. Agric. Silvic. Mendelianae Brun 57 (2009) 95–102.
[6] R.M. Stewart, T.W. Selby, The relationship between oil viscosity and engine performance—a literature search, in: The Relationship Between Engine Oil Viscosity and Engine Performance, ASTM International, 1977.
[7] L. Leugner, Natural gas engine lubrication and oil analysis, Practicing Oil Analysis 2 (6) (2003) 30–35.
[8] E.E. David Jr., Strategies in fossil fuel technology: multiple options for unpredicted futures, in: Energy, Resources and Environment, Pergamon, 1982, pp. 22–31.
[9] Anon (n.d.). Synthetic vs Mineral Oil. Available at: http://www.lubricantsonline.co.za/uploads/Synthetic%20vs%20Mineral%20Oil_1.pdf.
[10] Kantharaju, V. (2017). How can I calculate recovery factor with parameters—production, reserves and the water injection of an oil field? Available at: https://www.researchgate.net/post/How_can_I_calculate_recovery_factor_with_parameters-production_reserves_and_the_water_injection_of_an_oil_field.
[11] O.A. Olabode, Effect of water and gas injection schemes on synthetic oil rim models, Journal of Petroleum Exploration and Production Technology (2020) 1–16.
[12] C.N.R. Rao, S. Dey, Solar thermochemical splitting of water to generate hydrogen, Proc. Natl. Acad. Sci. 114 (51) (2017) 13385–13393.
[13] H. Jouhara, D. Ahmad, I. van den Boogaert, E. Katsou, S. Simons, N. Spencer, Pyrolysis of domestic based feedstock at temperatures up to 300 C, Thermal Science and Engineering Progress 5 (2018) 117–143.
[14] P. Basu, Pyrolysis, Biomass Gasification, Pyrolysis and Torrefaction (2013) 147–176.
[15] Combustion Examples, 2006. Available at: https://www.egr.msu.edu/classes/me440/somerton/CombustionExamples.pdf.
[16] J. Yang, D. Wang, H. Han, C.A.N. Li, Roles of cocatalysts in photocatalysis and photoelectrocatalysis, Acc. Chem. Res. 46 (8) (2013) 1900–1909.
[17] S. Mostrou, R. Büchel, S.E. Pratsinis, J.A. van Bokhoven, Improving the ceria-mediated water and carbon dioxide splitting through the addition of chromium, Appl. Catal. A Gen. 537 (2017) 40–49.
[18] E. Rozzi, F.D. Minuto, A. Lanzini, P. Leone, Green synthetic fuels: renewable routes for the conversion of non-fossil feedstocks into gaseous fuels and their end uses, Energies 13 (2) (2020) 420.
[19] H.I. Villafán-Vidales, C.A. Arancibia-Bulnes, D. Riveros-Rosas, H. Romero-Paredes, C.A. Estrada, An overview of the solar thermochemical processes for hydrogen and syngas production: reactors, and facilities, Renew. Sust. Energ. Rev. 75 (2017) 894–908.
[20] T. Kodama, N. Gokon, Thermochemical cycles for high-temperature solar hydrogen production, Chem. Rev. 107 (10) (2007) 4048–4077.
[21] R. Asmatulu, M.A. Shinde, A.R. Alharbi, I.M. Alarifi, Integrating graphene and C60 into TiO2 nanofibers via electrospinning process for enhanced conversion efficiencies of DSSCs, in: Macromolecular Symposia, vol. 365, no. 1, 2016, July, pp. 128–139.
[22] I.M. Alarifi, A. Alharbi, A.G. Potagani, Fabrication and characterization of carbonized polyacrylonitrile nanofibers for composite aircraft and wind turbine manufacturing, Proceedings: 12th Annual Symposium on Graduate Research and Scholarly Projects, Wichita State University, Wichita, KS, 2016, p. 15.
[23] A.R. Alharbi, I.M. Alarifi, W.S. Khan, R. Asmatulu, Highly hydrophilic electrospun polyacrylonitrile/polyvinypyrrolidone nanofibers incorporated with gentamicin as filter

medium for dam water and wastewater treatment, Journal of Membrane and Separation Technology 5 (2) (2016) 38–56.
[24] R. Asmatulu, J. Yeoh, I.M. Alarifi, A. Alharbi, Effects of edge grinding and sealing on mechanical properties of machine damaged laminate composites, in: Nondestructive Characterization and Monitoring of Advanced Materials, Aerospace, and Civil Infrastructure 2016, vol. 9804, International Society for Optics and Photonics, 2016, April, p. 98042G.
[25] Wiersma, R. & Karamagi, M. (2016). Gas-to-Liquids Technology Offers Innovative Solutions for Coatings. Available at: https://www.pcimag.com/articles/102266-gas-to-liquids-technology-offers-innovative-solutions-for-coatings.
[26] C.B. Clifford, C. Song, Direct liquefaction (DCL) processes and technology for coal and biomass conversion, in: Advances in Clean Hydrocarbon Fuel Processing, Woodhead Publishing, 2011, pp. 105–154.
[27] K.K. Robinson, Reaction engineering of direct coal liquefaction, Energies 2 (4) (2009) 976–1006.
[28] K. Malik, S. Singh, S. Basu, A. Verma, Electrochemical reduction of CO2 for synthesis of green fuel, Wiley Interdisciplinary Reviews: Energy and Environment 6 (4) (2017), e244.
[29] E.M. Dickson, E.E. Hughes, Impacts of synthetic liquid fuel development for the automotive market, in: Future Automotive Fuels, Springer, Boston, MA, 1977, pp. 342–366.
[30] M. Ferrari, A. Varone, S. Stückrad, R.J. White, Sustainable synthetic fuels, IASS Fact Sheet, 2014, p. 1, https://doi.org/10.2312/iass.2014.006.
[31] R.M.R. Shagor, I.M. Alarifi, R. Asmatulu, Effects of silanized graphene nanoflakes on mechanical properties of carbon fiber reinforced laminate composites, in: CAMX Conference, Anaheim, CA, 2016, October.
[32] V.S. Swarna, I. Alarifi, V.R. Patlolla, R. Asmatulu, Improving the strengths of metal-metal bonding via inclusion of graphene nanoflakes into adhesive joints, in: CAMX Conference, vol. 11, 2016, October.
[33] I.M. Alarifi, W.S. Khan, A.S. Rahman, Y. Kostogorova-Beller, R. Asmatulu, Synthesis, analysis and simulation of carbonized electrospun nanofibers infused carbon prepreg composites for improved mechanical and thermal properties, Fibers and Polymers 17 (9) (2016) 1449–1455.
[34] A.R. Alharbi, I.M. Alarifi, W.S. Khan, A. Swindle, R. Asmatulu, Synthesis and characterization of electrospun polyacrylonitrile/graphene nanofibers embedded with SrTiO3/NiO nanoparticles for water splitting, J. Nanosci. Nanotechnol. 17 (8) (2017) 5294–5302.
[35] I.M. Alarifi, W.S. Khan, M.M. Rahman, R. Asmatulu, Mitigation of lightning strikes on composite aircraft via micro and nanoscale materials, Advances in Nanotechnology 20 (2017) 39–66.
[36] V. Seewoogolam, I. Alarifi, R. Asmatulu, Highly robust electrospun nanofibers films for the fabrication of MAV wings, in: CAMX Conference, Anaheim, CA, 2016, October.
[37] I. Ridjan, B.V. Mathiesen, D. Connolly, N. Duić, The feasibility of synthetic fuels in renewable energy systems, Energy 57 (2013) 76–84.
[38] Bromberg, L., & Cheng, W.K. (2010). Methanol as an alternative transportation fuel in the US: options for sustainable and/or energy-secure transportation. Available from: http://www.afdc.energy.gov/afdc/pdfs/mit_methanol_white_paper.pdf.

CHAPTER 8

Introduction, properties, and application of synthetic engineering nanomaterials

Abbreviations

AES	Auger electron spectroscopy
EU	European Union
FWHM	full width at the half maximum
ISO	International Organization for Standardization
KBr	potassium bromide
LPCVD	low-pressure chemical vapor deposition
MCP	mechanochemical processing
NCMs	nanocomposites
nm	nanometer
NPs	nanoparticles
NSMs	nanostructured materials
NTs	nanotubes
NWs	nanowires
PEC	photoelectrochemical
SEM	scanning electron microscope
TEM	transmission electron microscopy
XPS	X-ray photoelectron spectroscopy

8.1 Introduction

Nanotechnology has become a significant part of human life due to its usability in the food industry, pharmaceutical, communication electronics, and mobility [1]. Nanotechnology refers to the act of manipulating matter, which is held at the nanoscale that ranges from 1 to 100 nm for obtaining new constituents and tools [2]. Progression in the field of science exhibits the widespread presence of nanomaterials in nature at the cellular level that is a fundamental element of life (Fig. 8.1). It is a significant technology that is applicable for evaluation, creation, and progression of novel and biologically stimulated nanomaterials that have varied approaches of manufacturing, different features, and have a different lifespan [3].

Synthetic Engineering Materials and Nanotechnology
https://doi.org/10.1016/B978-0-12-824001-4.00002-8

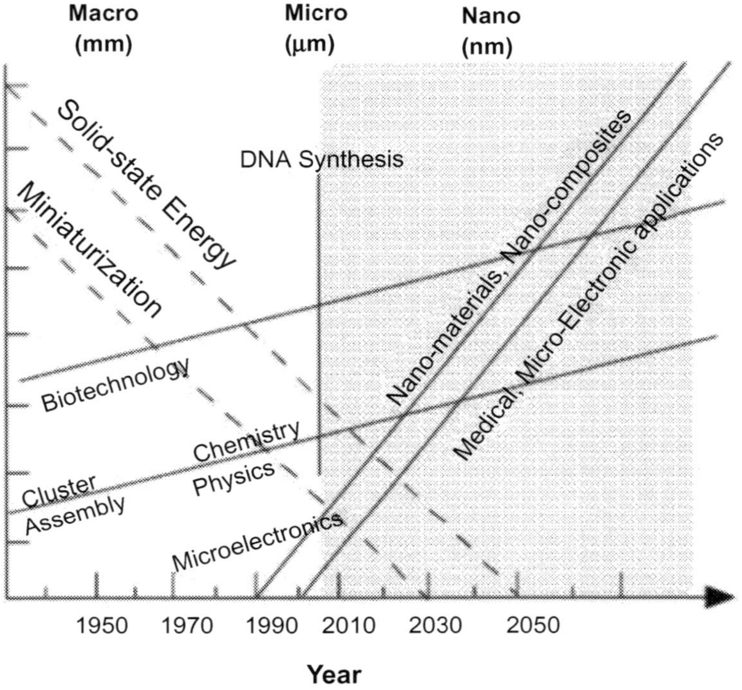

Fig. 8.1 Progression of science and technology [4].

According to the definition of nanomaterials, given by the International Organization for Standardization (ISO), nanomaterials have the inner structure that is measured in nanoscale [5]. Similarly, in the light of the definition by the transnational bloc of European Union (EU), nano-objects have three dimensions at the nanoscale, and the nanoscale provides the range of size that is 1–100 nm [6]. Moreover, they are reckoned as the foundation stone for nanotechnology and neuroscience as well. Additionally, the science of nanostructure is the field of interdisciplinarity studies that has been rapidly advancing globally. Particularly, there is potential to not only alter the trends of manufacturing the products and materials but also to implant the desired functionalities. Till the date, it has dramatically influenced the commercial world, and surely it will increase to a greater extent [4].

Hint statement

Nanotechnology encompasses constructing the submicron constituents through exploiting and controlling the diverse biological, physical, and chemical features of nanoscale materials.

It has further gained the investors' attention across the globe, including profit-oriented entities to governmental bodies that are inclined to obtain new nanodevices, i.e., Nanosensors and Carbon Nanotube. The notion of nanotechnology entails assembling and usability of submicron constituents that is possible through exploiting the distinct biological, physical, and chemical features of nanoscale materials. Moreover, the social and economic contributions of nanotechnology have driven the rise in investment patterns equally among public and private entities across the globe [7].

Hint statement
The term nanoscale reflects the dimensions between 1 and 100 nm.

One nanometer (nm) is equal to the one-billionth part of the mater. It can be further elucidated from the examples that one sheet of paper comprises 100,000 nm, and the diameter of an atom of gold accounts for one-third part of a nanometer. More specifically, the dimensions between 1 and 100 nm are termed as nanoscale [7]. Quantum theory propagates that at the nanoscale, the materials that are placed from 1 to 250 nm, come under the quantum effects of molecules, atoms, and bulk characteristics of materials. Further, the nanoscale is also termed as "No-Man's Land" through which significant dimensions manage various features at the level of nanoscale. Nanoparticles (NPs), nanocomposites (NCMs), nanowires (NWs), and nanotubes (NTs) are different types of nanostructured materials (NSMs) [7].

Hint statement
The method of chemical bonding that is allied with the exterior of the particles of metals, could be stated while viewing the infrared immersion spectra.

Example 8.1

Calculate nanocrystallite size while using the Debye–Scherrer equation by XRD? [8].

Solution:

Scherrer's equation:

Particle Size $= (0.9 \times \lambda)/\ (d \cos\theta)$.

$\lambda = 1.54060\,\mathring{A}$ (in the case of CuKa1) so, $0.9 \times \lambda = 1.38654$

$$\Theta = 2\theta/2\ (\text{in the example} = 20/2)$$

d = the full width at half maximum intensity of the peak (in Rad).
Converting from angle to rad

$$\text{Rad} = (22 \times \text{angle})/(7 \times 180) = \text{angle} \times 0.01746$$

Example: if d = 0.5 angle (θ)

$$= (22 \times 0.5)/(7 \times 180) = 0.00873\ \text{rad}\ [9].$$

8.2 Properties of nanomaterials

The notions of nanomaterials, nanophase, and ultra micro-size materials are interchangeably used that denote the materials that have the grain size in nanometers [9]. Different properties of nanomaterials are encapsulated in the following:

8.2.1 Physical properties

It attributed and recognized the surface contamination, X-ray photoelectron spectroscopy (XPS), and Auger electron spectroscopy (AES) are used for the penetration, which accounts for 0.5–1 nm that encompass the 2.0 m spatial resolution for AES and 0.2 spatial resolution for XPS. Whereas for both the analyzers, the sensitivity account for 0.3%. The diffraction patterns of the powder X-ray and the Transmission Electron Microscopy (TEM) exhibit the organization and composition of the crystal. Moreover, in order to ascertain the mean size of grain R to have knowledge about X-ray wavelength, the following Scherer's equation is used with varied angles [10].

$$\text{FWHM} = 0.94\lambda/R\cos\theta$$

Here, the FWHM reflects the full width at the half maximum of the feature's spectrum for the radian' units in nm. The degree of degradation can be instantly assessed through the electron probe micro-analysis (EPMA) with the help of the Scanning Electron Microscope (SEM). The mode of chemical bonding that is allied with the exterior of the particles of metals could be expressed through viewing the infrared immersion spectra. The nanoparticles that are metallic or in the form of sol gel's grounded powders are entrenched and can be blended with polyethylene glycol powders or the translucent Potassium Bromide (KBr). The blend aims to discern wavelengths that are 400–500 cm^{-1} and 650–200 cm^{-1} sequentially, wherein the volume ratio is 1:10. Subsequently, the nanoparticles are compressed in the disks [11].

8.2.2 Magnetic properties

Rare magnetic and electronic properties of nanomaterials are evidenced at nonzero heat, i.e., in-transition metal oxides, the transition of metal–insulator, and irregular symmetry conditions of high-Tc cuprates that are superconductors [11]. There is a possibility to control the magnetic characteristics of nanomaterials in terms of their shape and size. If the nanoparticles are less than in a specified limit of size, it will be termed as mono-domain particles. Since they exhibit higher magnetic performance at a specified temperature, the temperature is called blocking temperature. When there is an increase in blocking temperature, the particles float in different directions due to the thermal activity [12]. The surface plays an important role in examining the magnetic features of nanomaterials. For further elucidation, the core-shell model is employed that propagates that the structure of the particle is based on magnetically and aligned nucleus and non-aligned shell (Fig. 8.2). The fall of temperature leads to the alignment of the shell [13].

Hint statement

Magnetic characteristics of nanomaterials can be controlled for acquiring desired shape and size.

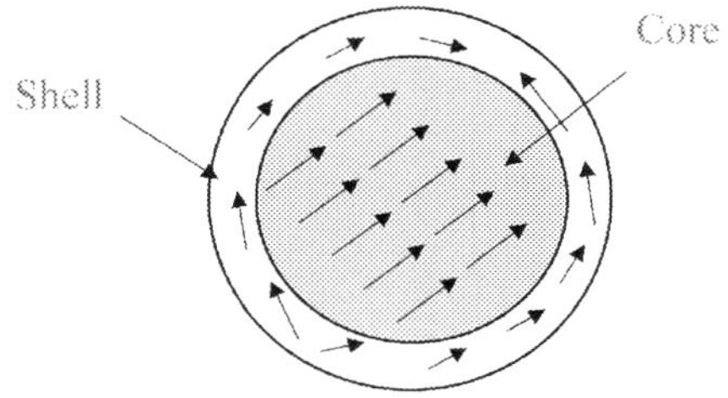

Fig. 8.2 The alignment spherical nanoparticle [13].

8.2.3 Chemical properties

Usage of dispersive mediums is evident for a higher degree of chemical activity of nanomaterials that are exhibited amid the variation in temperature, speediness, the degree of transformation at specified circumstances, and a higher level of certain catalytic and pyrophoric features. Nanoparticles make interaction with passive materials. It can be illustrated from the fact that amid crushing the exterior of nanopowders, there is irretrievable adsorption of passive gases [14].

8.2.3.1 Size effects in chemical process

Amidst the chemical reaction, the effect of size emanates an interface change. Diffusion in hard material is a determining factor of the reaction speed in a large scale varied chemical process where the participation of a solid phase is also involved. Additionally, the interacting particle's radius R is found as bigger than the characteristic diffusion time [14].

$$\delta = \sqrt{Dt}$$

Here D reflects the coefficient of diffusion, and t denotes the time. When $\delta \geq R$, the restraining phase of the method is a chemical reaction [14].

8.2.3.2 Oxidation processes in nanomediums

In line with the rule regarding the oxidation reactivity, at dispersive conditions, the ability of substance for general phase reactivity can be specified. For instance, 180 days is the time duration of nanopowder oxidation in the air for Mo and Fe. In contrast, the size of particles is 70 nm for Fe and 100 nm for Mo (Table 8.1). The process of oxidation in Fe nanopowder

Table 8.1 The temperature of initiation of oxidation of large metal nanopowder.

Metal	Specific surface	Nanoparticle (NP) size (nm)	Temperature of NPS	Large nanopowder particle	Temp. of large powders starting oxidation (°C)
Aluminum	18.5	120	420	<20	870
Iron	4.6	160	310	<45	480
Copper	6.8	105	170	<60	260
Zinc	12.9	65	220	<71	410
Tin	10.2	80	180	<45	270

is more intense than the nanopowder of Mo. Thus, the process of oxidizing Fe will be completed in 180 days, while oxidation of Mo will take place in 140 days (Table 8.1) [14].

Example 8.2

Calculate the percentage of the bulk bonding energy that is lost by atoms at 111 surfaces of the crystal of gold. The sublimation energy of bulk gold accounts for 334 kJ/mol, and $1.5\,J/m^2$ is the surface energy.

Solution:

Au unit cell edge length = 4.08 Å.

Au—Au distance = 4.08/1.414 = 2.88 Å.

A hexagonal array of Au Atoms at the surface area per atom.

$=(2.88\,Å)^2 \times 0.866 = 7.2\,Å^2$

Multiplying by Avogadro's number, the area per mole of Au surface atoms on the (111) crystal face is $4.3 \times 10^4\,m^2$.

Now, multiply the area by the surface energy:

$\left(4.3 \times 10^4 \frac{m^2}{mo}\right) \times \left(1.5\frac{J}{mol}\right) \times \left(\frac{1kJ}{1000J}\right) = 65kJ$ per mole of Au surface atoms.

$\frac{65kJ}{335kJ} \times 100\% = 19\%$bulk binding energy [15].

8.3 Methods of synthesizing engineering nanomaterials

There has been significant growth in the field of nanotechnology. Various methods and technologies are adopted to obtain nanoscale materials and nanodevices. The following are the methods to produce nanomaterials.

8.3.1 Mechanochemical processing (MCP) method

It is an innovative method to synthesize several types of nanopowders. The use of a conventional ball involves a chemical reactor at a lower temperature. The process of ball milling augments the reactive kinetics in a reactant powder mixture. Due to mixing and refining the gain structures, a reaction to the nanometer scale is stimulated amid the milling of chlorides, fluorides, carbonates, oxides, sulfates, and other precursor materials. Subsequently, milling is held for the selected precursor material along with a suitable reactive agent, which results in nanometer-sized grain, which is a single-phase by-product matrix. Afterward, milling, a heat treatment at a lower temperature, is held to make sure the completion of the reaction, which dispenses nonagglomerated nanopowder that has nanoparticles (having a diameter of 1–1000) [16]. For manufacturing crystalline ZnO nanoparticles, the precursors are sodium carbonate zinc chloride ($ZnCl_2$) and (Na_2CO_3). Through milling, both the

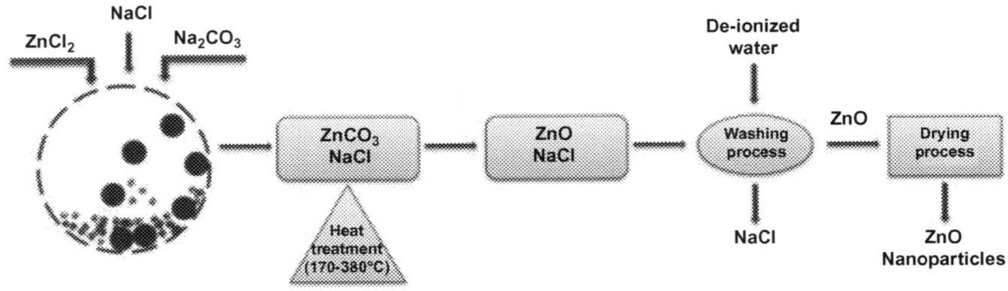

Fig. 8.3 Mechanochemical processing for ZnO nanoparticles [17].

precursor materials, zinc carbonate ($ZnCO_3$), and sodium chloride (NaCl) is produced while a chemical change reaction and ball-powder collisions (Fig. 8.3) [17].

As an inert diluent, NaCl is considered as the precursor, and a nanocomposite and NaCl works as a matrix phase. Following equation displays the mixture of starting materials [17]:

$$ZnCI_c + Na_2C_3 + 8NaCl \rightarrow ZnCO_3 + 1ONaCl$$

The product mix of the nanostructure is given heat at 170–380°C for decomposing $ZnCO_3$ in order to get ZnO [17].

8.3.2 Laser ablation

The process of laser ablation involves eliminating material from a solid or liquid surface through using a laser beam to treat it. The material is provided with the heat at low laser flux through laser energy. Plasma is obtained through a high laser flux material. A laser beam treats the precursor material and vapors during high vacuum settings. From the precursor material, clusters and atoms are dispensed and concentrate to a substrate place (Fig. 8.4) [17].

8.3.3 Chemical reduction method

It is the most commonly adopted method for synthesizing nanoparticles. A significant feature of the chemical reduction method is that it can be processed at room temperature. The fundamental process of the chemical reduction method is exhibited in Fig. 8.5. Numerous reduction agents and their salts, i.e., citric and ascorbic acids, alkali hydrides such as $LiB(C_2H_5)_3H$, $NaB(C_2H_5)_3H$, and $NaBH_4$ and other capping agents have been found effective in eliminating metal aggregation of nanoparticles [17].

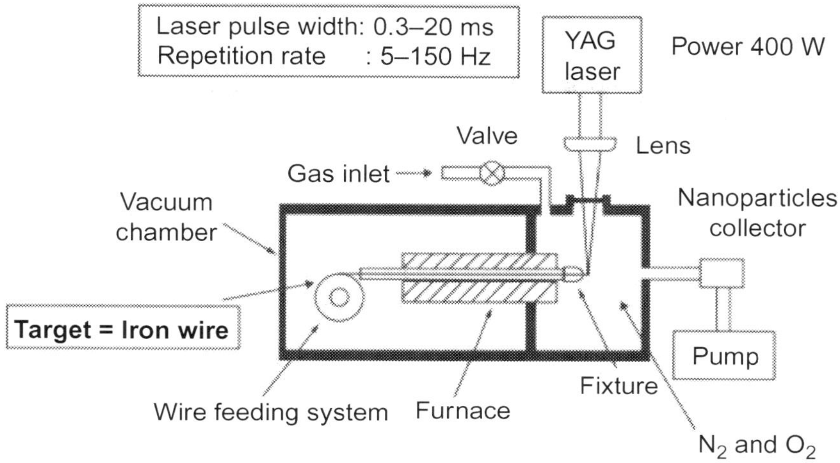

Fig. 8.4 Laser ablation method for synthesizing Fe_2O_3 nanoparticles [17].

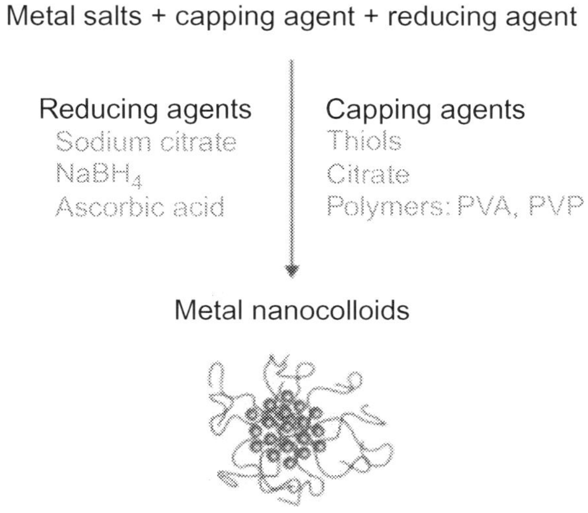

Fig. 8.5 Electrochemical reduction of nanomaterials [17].

8.4 Techniques of synthesizing engineering nanomaterials

Several techniques have been adopted for the synthesis of nanomaterials that vary in shape, orientation, and size. These techniques are broadly categorized as top-down approaches and bottom-up approach [18].

8.4.1 Top-down approach

Top-down techniques are employed to process nanoparticles through energy-intensive ball milling from bulk materials. These processes take place in a chemically inactive atmosphere, which results in highly reactive nanoparticles that are enabled to create agglomerates. In the presence of reactive gas, there is a likelihood of the occurrence of additional reaction, which may be used for coating nanoparticles. For instance, thin-film deposition, etching, and nanolithography techniques are used to remove the undesired material. Following are some of the top-down techniques:

8.4.1.1 Etching

It involves eliminating material that is mostly used in microfabrication and also for synthesizing nanoparticles. Through plasma, electric arc-discharge method, and chemicals, etching is processed. During chemical etching, a chemical reaction occurs along with a substrate compound in order to obtain an etch profile. The selection of processing is made for chemical etching of substrate compounds. Another name for plasma etching is reactive ion etching, wherein the use of plasma aims for ionizing gas [16]. The etching attributed as unmasked and photochemical is employed to manufacture systematic arrays of shape that are in the range of nanometer. For instance, crystalline silicon wafers are etched electrochemically for creating porous silicon layers. The process employs a blend of ethanol and hydrofluoric acid (Fig. 8.6) [16].

8.4.1.2 Electrospinning

It is an evolving technology for manufacturing fiber-based nano polymer. The foundation of this technique is on spinning a dilute polymer fluid amid an electric field that carries high-voltage. By this process, a 1000 V charge is gained by a suspended drop of the polymer. The droplets create a Taylor

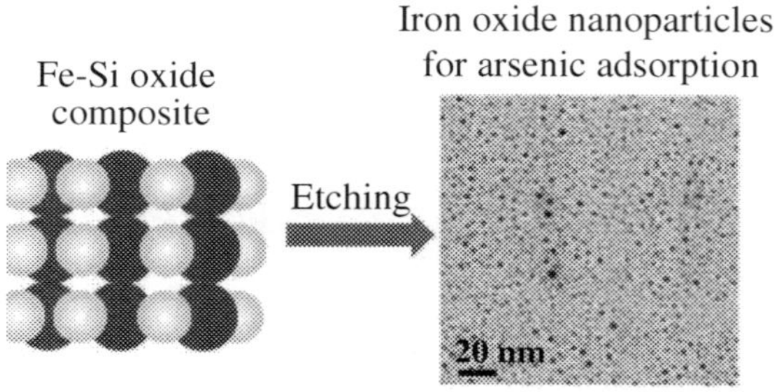

Fig. 8.6 Etching for synthesizing iron oxide nanoparticles [19].

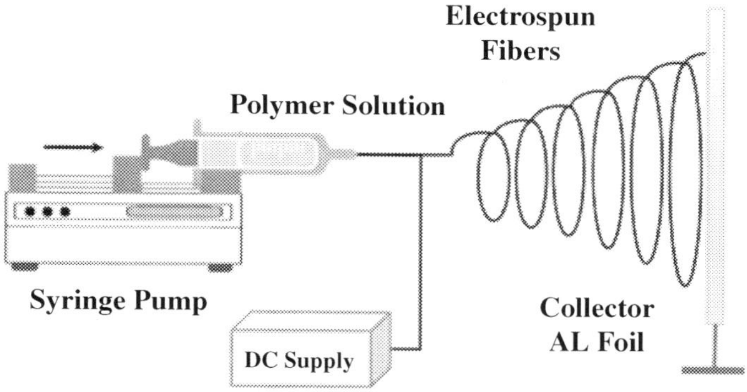

Fig. 8.7 Schematics of the electrospinning process for nanofiber synthesis [20].

cone where a jet of polymer dispenses via exterior as a result of tensile forces that are produced through an interface of an electric field. Also, the polymer jet comprises electrical charge. Consequently, a bulk of polymer firs are produced. Moreover, the jet can be turned into the direction of the grounded surface, and it is amassed as a constant web of fiber that varies in size to <100 nm to few micrometers. For obtaining solid threads, neither the coagulation chemistry nor extreme hot temperature is required from the solution [16] (Fig. 8.7).

8.4.2 Bottom-up techniques

While adopting the physical or chemical deposition process, nanomaterials are produced from the precursor materials in gas, liquid, or solid phase in bottom-up techniques. Different bottom-up techniques are summarized below:

8.4.2.1 Chemical vapor deposition

The approach of gas-phase synthesizing is gaining popularity as it facilitates a unique method to manage the control process for obtaining the required nanostructures that are controlled via chemical composition. The systems that encompass multicomponent are convenient in terms of preparation. Generally, many of the synthesis paths are attributed towards producing minor clusters that can be amassed to create nanoparticles through condensation. The process of condensation takes place upon supersaturating the vapor. During this process, standardized nucleation involves gas-phase, which is exploited for manufacturing nanoparticles (Fig. 8.8) [16].

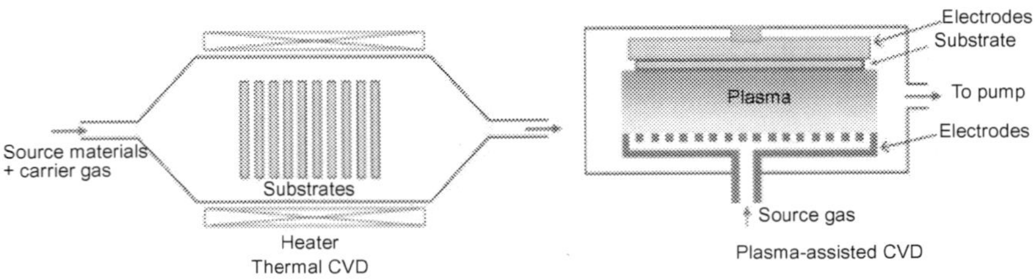

Fig. 8.8 Chemical vapor deposition [16].

Some of the examples of nanomaterials that have greater commercial importance and are obtained through chemical vapor deposition are given in the following:

- Polycrystalline silicon is composed of silane (SiH_4) and trichlorosilane ($SiHCl_3$), and follows the given chemical reaction:

$$SiH_3Cl \rightarrow Si + H_2 + HCl$$
$$SiH_4 \rightarrow Si + 2H_2$$

Low-Pressure Chemical Vapor Deposition (LPCVD) systems are involved in this reaction. It may include primary saline feedstock, a saline solution that comprises 70%–80% nitrogen concentration. The required temperature range is stated between 600 and 650°C, while the range of favorable pressure is 25 and 150 Pa that has production rate per minute from 10 to 20 nm [16].

- Silicon between dioxide is known as an oxide in the field of semiconductors. It can be obtained via various methods. Generally, the source gases are oxygen, silane, nitrous oxide (N_2O), and dichlorosilane ($SiCl_2H_2$). Following is the reaction [16]:

$$SiH_4 + O_2 \rightarrow SiO_2 + 2H_2$$
$$SiCl_2H_2 + 2N_2O \rightarrow SiO_2 + 2N_2 + 2HCl$$
$$Si(OC_2H_5)_4 \rightarrow SiO_2 + \text{byproducts}$$

8.5 Applications of synthetic engineering nanomaterials

There are a plethora of nanomaterials applications from electronics to nanobiological systems and in pharmaceuticals [21]. Following are the major applications of engineered nanomaterials in different areas:

8.5.1 Ecological remediation

It is the novel usage of nanomaterials as they work for deteriorating uncleanness from soil and water. They are also used to enhance air quality while adopting cost-effective methods. Their miniature size is very useful for improving ecological conditions where both the methods; in situ and ex situ are employed to diminish the environmental pollution and hazards. Carbon-based nanomaterials, nanocomposites (NCPs), and metal and metal oxide nanoparticles are the most effective absorbents for environmental pollution. Moreover, Titanium oxide nanoparticles are used for removing micropollutants. Other engineered nanomaterials for ecological well-being are nano-sized cerium oxide, zero-valent iron, aluminum oxides, ferric oxide, etc. For instance, heavy metals and chlorinated compounds are deteriorated through electron donation by zero-valent iron. Besides, the shell structure of the iron nanoparticles boosts the remedial process of contamination (Fig. 8.9) [21].

8.5.2 Pharmaceutical industry

The advent of the year 1965 brought the usage of nanomaterials in medicine due to their various useful properties. The major areas of nanomaterials include biosensors and diagnosis, tissue engineering, and molecular engineering [23–25]. A significant property of nanomaterials is their greater capacity of infiltrating. Also, they do not have a negative impact on healthy tissues [26]. Moreover, the biological and physiochemical features of nanoparticles can be modified for their different applications. For cancer therapy, as regenerative medicine and for biomolecular detection, nanomaterials are used [27] (Fig. 8.10).

8.5.3 Manufacturing electronic devices

Since nanomaterials contain chemical, optical, and electric properties, they are used to manufacture the cores of recording media and transformers, memory devices, biosensors antenna cores, density storage, sensors, and

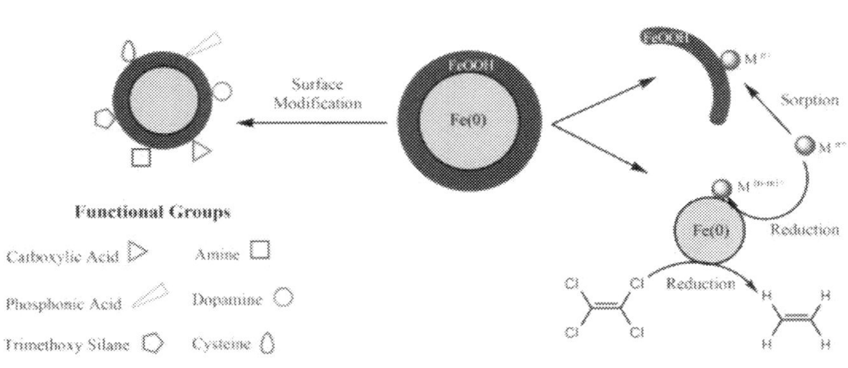

Fig. 8.9 Removing contamination by using iron nanoparticles [22].

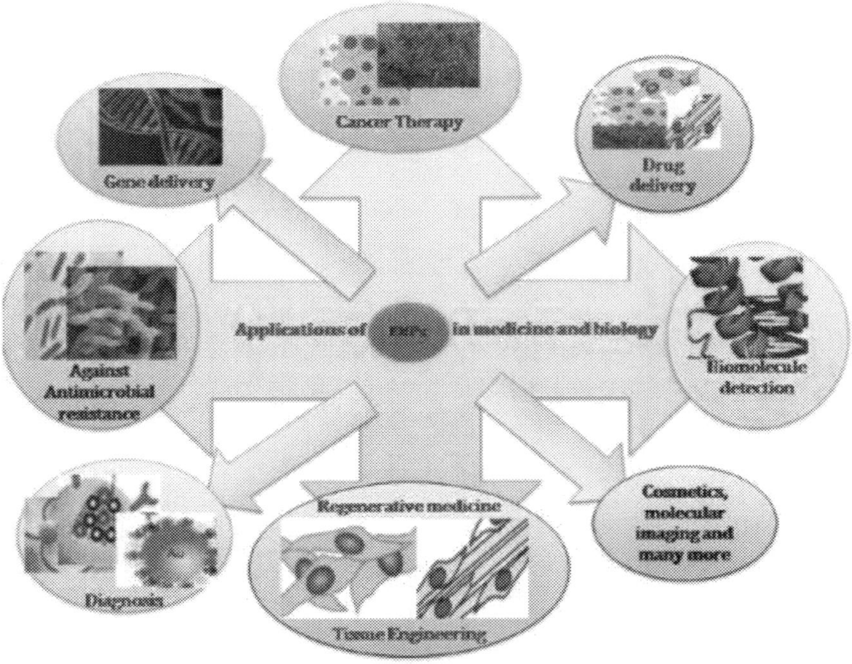

Fig. 8.10 Use of nanomaterials in pharmaceuticals [27].

computer units [28]. The physiochemical features are implanted into nanomaterials to obtain imaging, optical and mechanical characteristics for their applicability in ecology, commercial, and medical fields [29]. Health fitness products are also produced for consumers. Various technologies are imminent for use in food packaging. Recent developments in Resonant energy transfer systems have been introduced wherein nanomaterials are used for organic dye molecules that are delivering promising benefits [30].

8.5.4 Energy harvesting

Energy reservoirs globally are fast depleting that emanates the need to acquire other sustainable means for energy production [31–33]. In this instance, nanomaterials are being used for energy production due to their bigger surface area [34–36], catalytic nature, and optical behavior. Through photoelectrochemical (PEC) and electrochemical water splitting, nanomaterials are used for generating energy [34].

8.6 Conclusion

The advent of nanotechnology is reckoned as a revolution in science and technology, which has made possible the manufacturing of novel and unique materials and compounds that have caused the maximization of the profitability and enhanced functionality in all spheres of life. A one-billionth part of matter which is named as a nanometer has gained drastic attention of governmental and profitable bodies for manufacturing enhanced nanosensors and other useful materials and desired equipment through exploiting the unique chemical, physical, and magnetic properties of nanomaterials. Mechanochemical processing methods, laser ablation, and chemical reduction methods are employed for acquiring synthesized nanomaterials. On the other hand, the techniques through which the process of synthesizing nanomaterials is held include etching, electrospinning, and chemical vapor deposition. The various applications of synthesized nanomaterials are evident in the environmental remediation process, in making medicine and drugs, in producing electrical goods and tools and in building the tools and equipment that are used in harvesting energy. Other applications of nanomaterials are imminent in the food and packaging industry.

References

[1] M. Rossi, F. Cubadda, L. Dini, M.L. Terranova, F. Aureli, A. Sorbo, D. Passeri, Scientific basis of nanotechnology, implications for the food sector and future trends, Trends Food Sci. Technol. 40 (2) (2014) 127–148.
[2] M.C. Roco, C.A. Mirkin, M.C. Hersam, Nanotechnology Research Directions for Societal Needs in 2020: Summary of International Study, 2011.
[3] Anon (2017). Engineered Nanomaterials: Impact & Safety Aspects. Available at: http://www.nfp64.ch/SiteCollectionDocuments/White-Paper-NFP64-E.pdf.
[4] U. Gangopadhyay, S. Das, S. Jana, P. Ghosh, State of art of nanotechnology, Int. J. Eng. Res. Dev. 3 (6) (2012) 95–112.
[5] ISO, International Organization for Standardization. Nanotechnologies Vocabulary Part 1: Core Terms. ISO/TS 80004-1:2010, 2010.
[6] ISO, International Organization for Standardization. Technical Specification: Nanotechnologies Terminology and Definitions for Nano-objects Nanoparticle, Nanofibre and Nanoplate. ISO/TS 80004-2:2008, 2008.
[7] Wang, G. (2018). Nanotechnology: the new features. arXiv preprint arXiv:1812.04939.
[8] Sudhan, P.N. (2015). How do I calculate nanocrystallite size by Debye-Scherrer equation using XRD? Available at: https://www.researchgate.net/post/How_do_I_calculate_nanocrystallite_size_by_Debye-Scherrer_equation_using_XRD.
[9] M.S. El-Shall, A.S. Edelstein, Formation of clusters and nanoparticles from a supersaturated vapor and selected properties, in: Nanomaterials: Synthesis, Properties and Applications, 1996, pp. 13–20.
[10] B.D. Cullity, Elements of X-Ray Diffraction, Addison Wesley, MA, 1978.

[11] J.T. Lue, Physical properties of nanomaterials, Encycl. Nanosci. Nanotechnol. 10 (1) (2007) 1–46.
[12] P.M. Kulal, D.P. Dubal, C.D. Lokhande, V.J. Fulari, Chemical synthesis of Fe2O3 thin films for supercapacitor application, J. Alloys Compd. 509 (5) (2011) 2567–2571.
[13] N.M. Čitaković, Physical properties of nanomaterials, Vojnotehnički glasnik 67 (1) (2019) 159–171.
[14] Z. Abdullaeva, Nano-and Biomaterials: Compounds, Properties, Characterization, and Applications, John Wiley & Sons, 2017.
[15] Anon (2020). Surface Energy. Available at: https://chem.libretexts.org/Bookshelves/Inorganic_Chemistry/Book%3A_Introduction_to_Inorganic_Chemistry/11%3A_Basic_Science_of_Nanomaterials/11.05%3A_Surface_Energy.
[16] N. Kumar, S. Kumbhat, Introduction, in: N. Kumar, S. Kumbhat (Eds.), Essentials in Nanoscience and Nanotechnology, 2016, https://doi.org/10.1002/9781119096122.ch1.
[17] P.J.P. Espitia, N.D.F.F. Soares, J.S. dos Reis Coimbra, N.J. de Andrade, R.S. Cruz, E.-A.A. Medeiros, Zinc oxide nanoparticles: synthesis, antimicrobial activity and food packaging applications, Food Bioproc. Tech. 5 (5) (2012) 1447–1464.
[18] P. Held, Prof. Beer Pal Singh (doctoral dissertation), Jaypee University of Information Technology, Waknaghat, 2017.
[19] W. Cheng, W. Zhang, L. Hu, W. Ding, F. Wu, J. Li, Etching synthesis of iron oxide nanoparticles for adsorption of arsenic from water, RSC Adv. 6 (19) (2016) 15900–15910.
[20] I.M. Alarifi, Fabrication and characterization of electrospun polyacrylonitrile carbonized fibers as strain gauges in composites for structural health monitoring applications, Doctoral dissertation, Wichita State University, 2017.
[21] A.M. Markeb, Environmental Applications of Engineered Nanomaterials; Synthesis and characterization, Doctoral dissertation, Universitat Autonoma de Barcelona (UAB), 2017. Available at: https://www.tdx.cat/bitstream/handle/10803/454768/amaam1de1.pdf?sequence=1&isAllowed=y.
[22] F.D. Guerra, M.F. Attia, D.C. Whitehead, F. Alexis, Nanotechnology for environmental remediation: materials and applications, Molecules 23 (7) (2018) 1760.
[23] I.M. Alarifi, A. Alharbi, W.S. Khan, R. Asmatulu, Electrospun carbon nanofibers for improved electrical conductivity of fiber reinforced composites, in: Electroactive Polymer Actuators and Devices (EAPAD) 2015, vol. 9430, International Society for Optics and Photonics, 2015, April, p. 943032.
[24] M.A. Shinde, I. Alarifi, A. Alharbi, R. Asmatulu, Electrospun TiO2 nanofibers incorporated with graphene nanoflakes for energy conversion, in: Smart Materials and Nondestructive Evaluation for Energy Systems 2015, vol. 9439, International Society for Optics and Photonics, 2015, March, p. 94390Z.
[25] A. Alharbi, I.M. Alarifi, W.S. Khan, R. Asmatulu, Electrospun strontium titanata nanofibers incorporated with nickel oxide nanoparticles for improved photocatalytic activities, in: Smart Materials and Nondestructive Evaluation for Energy Systems 2015, vol. 9439, International Society for Optics and Photonics, 2015, March, p. 94390F.
[26] L.A. Kolahalam, I.K. Viswanath, B.S. Diwakar, B. Govindh, V. Reddy, Y.L.N. Murthy, Review on nanomaterials: synthesis and applications, Mater. Today Proc. 18 (2019) 2182–2190.
[27] G.R. Rudramurthy, M.K. Swamy, Potential applications of engineered nanoparticles in medicine and biology: an update, J. Biol. Inorg. Chem. 23 (8) (2018) 1185–1204.
[28] R.S. Yadav, J. Havlica, J. Masilko, L. Kalina, J. Wasserbauer, M. Hajdúchová, Z. Kožáková, Impact of Nd3+ in CoFe2O4 spinel ferrite nanoparticles on cation distribution, structural and magnetic properties, J. Magn. Magn. Mater. 399 (2016) 109–117.

[29] H. Dong, B. Wen, R. Melnik, Relative importance of grain boundaries and size effects in thermal conductivity of nanocrystalline materials, Sci. Rep. 4 (1) (2014) 1–5.
[30] Y.M. Lei, W.X. Huang, M. Zhao, Y.Q. Chai, R. Yuan, Y. Zhuo, Electrochemiluminescence resonance energy transfer system: mechanism and application in ratiometric aptasensor for lead ion, Anal. Chem. 87 (15) (2015) 7787–7794.
[31] A. Alharbi, I.M. Alarifi, W.S. Khan, R. Asmatulu, Synthesis and analysis of electrospun SrTiO3 nanofibers with NiO nanoparticles shells as photocatalysts for water splitting, in: Macromolecular Symposia, vol. 365, no. 1, 2016, July, pp. 246–257.
[32] K.B. Mahat, I. Alarifi, A. Alharbi, R. Asmatulu, Effects of UV light on mechanical properties of carbon fiber reinforced PPS thermoplastic composites, in: Macromolecular Symposia, vol. 365, no. 1, 2016, July, pp. 157–168.
[33] I.M. Alarifi, A. Alharbi, W.S. Khan, A.S. Rahman, R. Asmatulu, Mechanical and thermal properties of carbonized PAN nanofibers cohesively attached to surface of carbon fiber reinforced composites, in: Macromolecular Symposia, vol. 365, no. 1, 2016, July, pp. 140–150.
[34] I.M. Alarifi, A. Alharbi, W.S. Khan, R. Asmatulu, Thermal and electrical properties of carbonized pan nanofibers for improved surface conductivity of carbon fiber composites, in: CAMX/SAMPE Conference, 2015, pp. 27–29.
[35] I.M. Alarifi, A. Alharbi, W.S. Khan, A. Swindle, R. Asmatulu, Thermal, electrical and surface hydrophobic properties of electrospun polyacrylonitrile nanofibers for structural health monitoring, Materials 8 (10) (2015) 7017–7031.
[36] I.M. Alarifi, A. Alharbi, O. Alsaiari, R. Asmatulu, Training the engineering students on nanofiber-based SHM systems, in: American Society for Engineering Education (ASEE), Zone III Conference, 2015.

CHAPTER 9

Ceramic nanomaterials

Abbreviations

Al_2O_3	alumina
$C_{12}H_{22}O_6$	sucrose
$C_{12}H_{22}O_6$	sucrose
$C_2H_5NO_2$	glycine
$C_2H_5NO_2$	glycine
C_2H_6O	ethanol
$C_2H_6O_2$	furfuryl alcohol
$C_3H_8O_2$	2-methoxyethanol
$C_5H_8O_2$	acetylacetone
$C_6H_{12}N_4$	hexamethylenetetramine
$C_6H_{12}O_{11}$	glucose
C_6H_6	benzene
$C_6H_8O_7$	citric acid
$CaCl_2$	calcium chloride
$CaCO_3$	calcium carbonate
CH_2O	formaldehyde
CH_4N_2O	urea
CH_6N_4O	carbohydrazide
CMNCs	ceramic matrix nanocomposites
CuO NCs	copper oxide nano composites
HA	hydroxyapatite
MgO	magnesium oxide
MMNCs	metal matrix nanocomposites
Na_2SiO_3	sodium silicate
PMNCs	polymer matrix nanocomposites
SHS	self-propagating high-temperature synthesis
SiO_2	silica
TEOS	tetraethyl orthosilicate
TiO_2	titanium oxide
ZrO_2	zirconia

9.1 Introduction

Nanomaterials have various contributions in the 21st century and marked a growth rate of 25% yearly due to their multipurpose functionalities. Because of distinct restructuring features, they have grabbed research attention across the globe [1]. Ceramic nanomaterials are made through

Synthetic Engineering Materials and Nanotechnology
https://doi.org/10.1016/B978-0-12-824001-4.00015-6

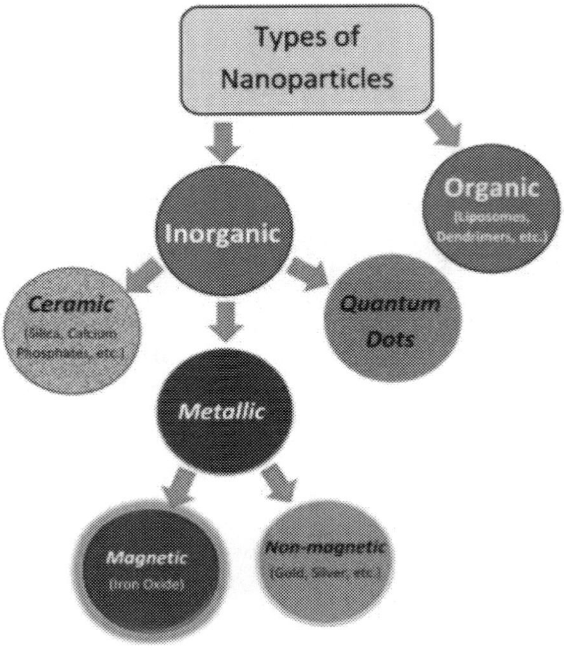

Fig. 9.1 Classification of inorganic nanoparticles [3].

inorganic mechanisms with penetrable features. As it is easier to obtain the nanoparticles according to the desired morphology and size, and permeability, a spike has been observed to exploit the ceramic nanoparticles forh drug delivery. Also, recent researches are more inclined to produce biocompatible ceramic nanoparticles, i.e., alumina and silica [2]. A general categorization of nanoparticles is shown in Fig. 9.1. Recently, biocompatible ceramic nanomaterials have revolutionized the drug delivery system and made it possible to manufacture modern materials. Some of the examples of nanoceramics materials are silica (SiO_2), alumina (Al_2O_3), zirconia (ZrO_2), calcium carbonate ($CaCO_3$), hydroxyapatite (HA), and titanium oxide (TiO_2) [3].

Hint statement

Ceramic nanoparticles comprise materials that contain both the characteristics of metals and nonmetals.

Inorganic nanoparticles' composition is based on the composites that are carbon-free; however, calcium carbonate is one to name a few composites that are exempted from this criterion. Inorganic materials are classified as ceramic and metallic. Ceramic nanoparticles are made of the materials that contain both the characteristics of metals and nonmetals. The ratio of thermal and electrical conduction is not higher in ceramic nanomaterials. Conversely, the elasticity modulus is higher, along with increased tensile strength and corrosion resistance. The kind of bonding among the atoms of their constituent materials bring about these features in ceramics nanomaterials.

On the other hand, metallic nanoparticles are composed of metals wherein the bond among atoms is not strong, which causes the unfettered mobility of electrons, which makes them exceptional conductors [3]. The 1980s period was marked with unearthing the nanomaterials, and the method to produce nanomaterials was the sol–gel method. In this method, nanoparticles were blended in a fluid and gel. The method further went through modifications where the use of heat and pressure were included [4].

Hint statement

The type of bonding among the atoms of the major component of materials, determines the characteristics of ceramics nanomaterials.

Example 9.1

Calculate the surface area if the size of the particle is reduced [5].

Solution:

One particle of volume $(V_1) = 4\pi/3$ (volume units).

Radius of the particle (r1) = 1 (length unit).

Area of the particle (A1) = 4π (area units).

The subscripts are showing the number of particles.

The area per unit volume for one particle:

$$\frac{A_1}{V_1} = \frac{4\pi r^2}{\frac{4}{3}\pi r^3} = \frac{3}{r}$$

When this particle is broken down into two spherical particles with equal volume, therefore, by the consistency of volume, it can be achieved:

$$\frac{4}{3}\pi\left(2r_2^3\right) = \frac{4\pi}{3} \Rightarrow r_2 = \left(\frac{1}{2}\right)^{\frac{1}{3}} = 0.79A_2 = 24\pi\left(\frac{1}{2}\right)^{\frac{2}{3}} = 2(7.92) = 15.84\,m^2$$

For denoting any general "n" (number of particles), the applicable numbers are given as:

$$r_n = \left(\frac{1}{n}\right)^{\frac{1}{3}}, V_n = \left(\frac{4}{3}\pi\right)\left(\frac{1}{n}\right), A_n = n(4\pi)\left(\frac{1}{n^{1/3}}\right)^2 = 4\pi(n)^{1/3} \text{ and } \frac{A_n}{V_n} = 3(n)^{1/3}$$

Ceramic nanomaterials demonstrate exceptional features due to their lesser density, stiffness, and resistance toward oxidation when they are relatively engineered with increased heat [6]. For instance, alumina-based nanocomposites' electrical and thermal characteristics hinge on the concentration by conductive reinforcing and dispersal degree. The interfacial extents resistibility on nanocomposites is extraordinary owing to the nanoparticles' high surface area. For the intent to ascertain the thermal conductivity of ceramic composites, a variety of methods, i.e., Johnson's model and Hamilton and Crosser's model, are employed [7]. Nonetheless, the methods mentioned above cannot provide the exact figures related to ceramic nanoparticles' thermal conductivity. The fact is that it is not easier to assess the interfacial resistibility of nanoceramics [8]. More specifically, thermal conductivity is denoted as

$$\Lambda = \alpha\varphi\, C_p$$

where λ represents thermal diffusivity, C_p stands for specific heat capacity, while φ refers to nanocomposites' bulk density [9]. Generally, thermal conductivity is calculated via a flash method that carries unvarying radiation that contains a specimen sand trivial pulse of energy. Amid a short time span, inviolable and viability of exact reproduction of results are few to name the prominent features of this method [9].

Hint statement

During 1980s, the nanomaterials were unearthed. This decade is also referred to the initiation of sol–gel method.

Adding up the nanoparticles in alumina ceramics increases the possibility of contacting due to high surface areas, leading to higher resistance concerning microparticles. Electric resistance is illustrated as [9]:

$$R = \frac{V}{I}$$

Here, V demonstrates the voltage of the surface, whereas I denotes the current that is applied. Once the resistibility is assessed, it is used to gauge the electric resistance [9].

$$\varphi = \frac{RA}{t}$$

wherein A refers to a cross-sectional area while t shows the viscosity of the materials [10].

Hint statement

The electric features of ceramics facilitate the transference of energy at 100% efficiency rate.

Example 9.2

Calculate the number of particles through the volume consistency if a spherical particle is of (r1 =) 1 mm radius is diminished in size to (r_n =) 10 nm (radius) particles [5].

Solution:

$$\frac{4}{3}\pi\left(10^{-3}\right)^3 = n\frac{4}{3}\pi\left(10^{-8}\right)^3 \Rightarrow n = 10^{15}\ nm\ \frac{A_n}{A_1} = 10^{15}\frac{4\pi\left(10^{-8}\right)^3}{4\pi\left(10^{-3}\right)^2} = 10^5$$

⇒ Thus, the rise in the areas of surface ratio through the factor of 10^5.

9.2 Methods of synthesizing nanoceramics

There is a rapid advancement in devising the methods of synthesizing ceramic materials to enhance their chemical and physical properties and applicability [11]. Different methods of obtaining nanoceramics are as follows:

9.2.1 Sol–gel method

It is the most commonly used method to obtain pristine particles of silica. This method effectively controls the particles of nanoparticles' division, shape, and size by systematically adjusting the reaction settings [12].

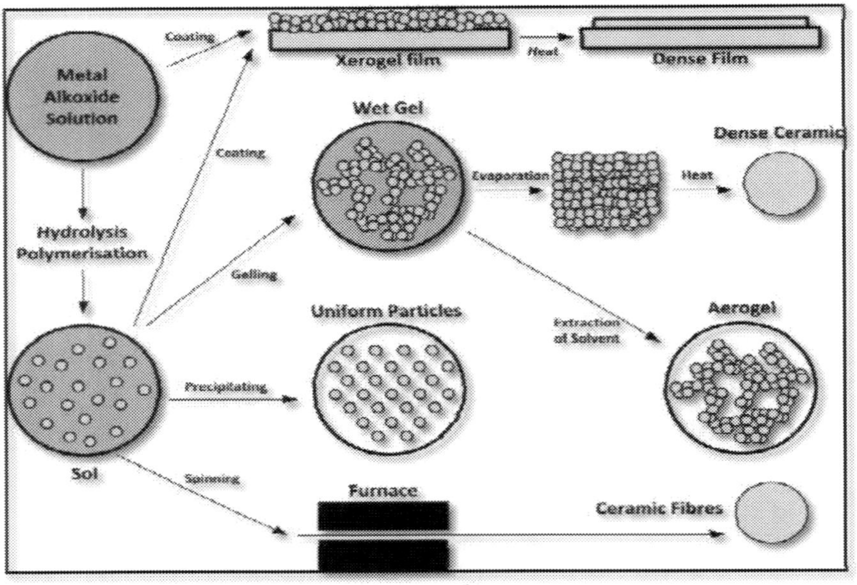

Fig. 9.2 The synthesizing ceramics process of nanomaterials [12].

Sol–gel undergoes a wet chemical process in order to obtain hybrid organic and inorganic materials, mainly oxides and hybrid materials, based on oxides. In this process, hydrolyzing and condensing of metal alkoxides such as tetraethyl orthosilicate (TEOS), $Si(OC_2H_5)$ or inorganic originators, i.e., sodium silicate (Na_2SiO_3) occurred along with mineral acid such as hydrochloric acid (HCL) or it may include base, in the form of catalyst (NH_3) (Fig. 9.2) [12].

The following are the reactions of TEOS that delineates the creation of silica particles [12].

$$Si(OC_2H_5) + H_2O \xrightarrow{\text{Hydrolysis}} Si(OC_2H_5)OH + C_2H_5OH$$

$$= Si - O - H + H - O - Si\equiv$$

$$\xrightarrow{\text{Water condensation}} \equiv Si - O - Si\equiv + H_2O$$

$$\xrightarrow{\text{Alcohol condensations}} \equiv Si - O - Si\equiv + C_2H_5OH$$

9.2.2 Self-propagating high-temperature synthesis (SHS) method

This method is employed to produce various materials, including metallic carbides, borides, oxides, and long-range-ordered alloy compounds.

More importantly, it facilitates the process of pipes' inner coating and deteriorates radioactivity. The reaction in SHS is higher exothermic. The allowed temperature is dispensed when the blend coating is burnt; thus, no warmth is required from external sources. The reactions involve rapid heating and chilling, and two processes include burning. Initially, a circulation wave initiates the reaction that leads to the reaction mixture. Subsequently, the need to produce single-phase composites emanates. At large, heat is provided to the reactants in furnace oil when the combustible heat is acquired. The process is termed as a thermal flash. An indispensable feature of the SHS method is the completion of obtaining higher combustible temperatures in a short period, which becomes viable by the liberation of heat amidst exothermic reactions [13]. Fig. 9.3 shows (a) the reaction chamber and the process of SHS, while (b) demonstrates the effectiveness of Magnesium oxide (MgO) to create layers of graphene [14].

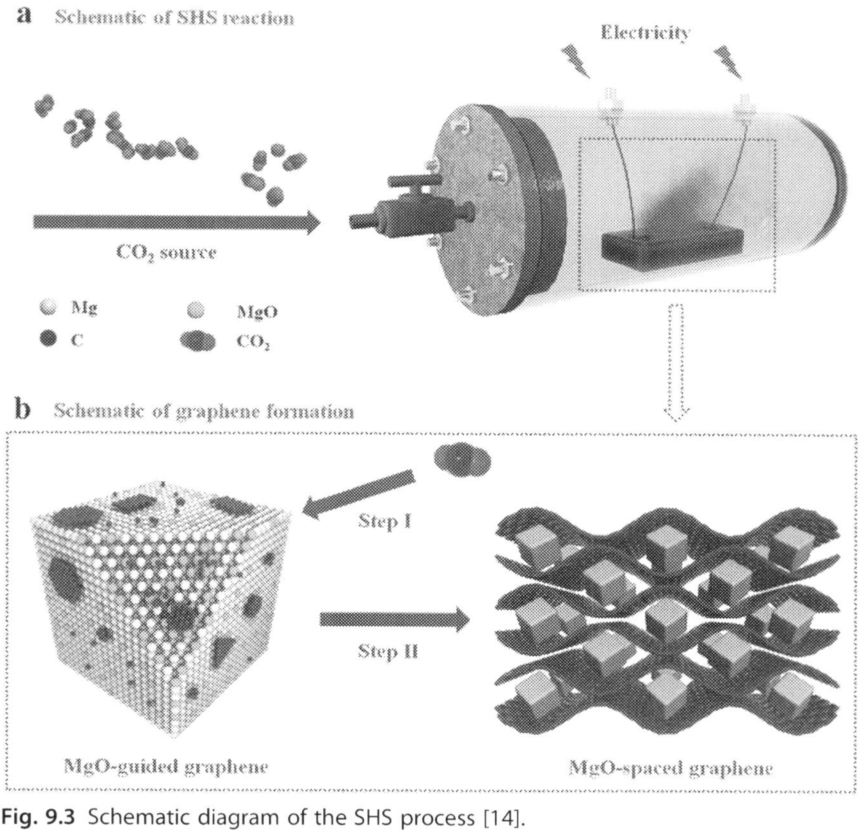

Fig. 9.3 Schematic diagram of the SHS process [14].

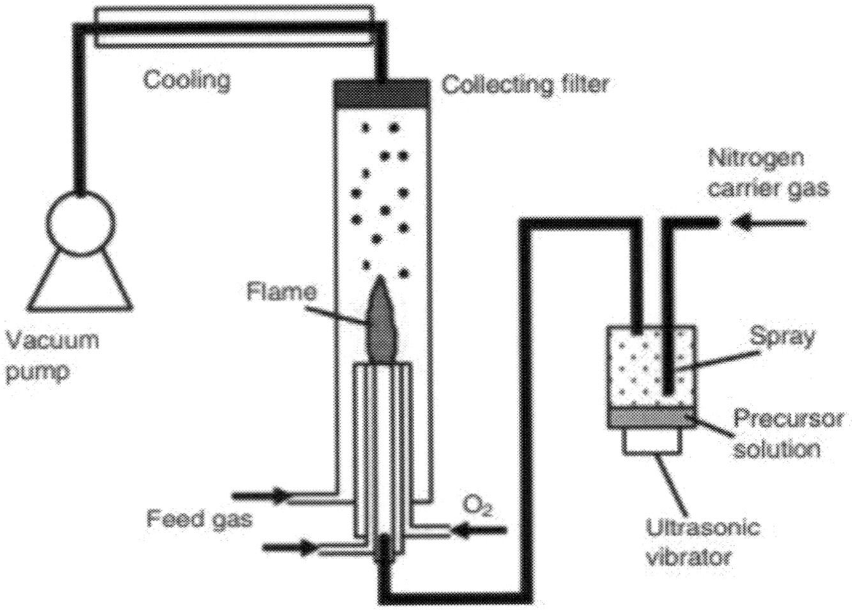

Fig. 9.4 The mechanism of spray pyrolysis [13].

9.2.3 Spray pyrolysis

This method is very useful for the synthesis of greater quality and stabilized ceramic powder. It is also known as aerosol decomposition, solution aerosol thermolysis, and phase transition. The primary material used in this process is chemical precursors that comprise suspension or salts. Aerosol droplets are obtained through atomization or nebulization of the primary solution or suspension. As a droplets evaporating result, its concentrated droplets are made, and the sleet of particles is caused by thermolysis. It creates microporous particles and dense particles (Fig. 9.4) [13].

9.2.4 Chemical vapor condensation (CVC) method

There is a wide range of nanosized powder in order to manufacture ceramic nanomaterials. This method is assumed as one of the techniques of gas compression wherein liquid chemical precursor is essential. During the process, the area unit of chemical precursor will undergo gasification. It happens before alteration in the process of combustion while using fuel-oxidant that blends likewise methane-air. The method signals the rapid cessation of precursor carrier stem. The temperature range of the flame is 1200–3000 K,

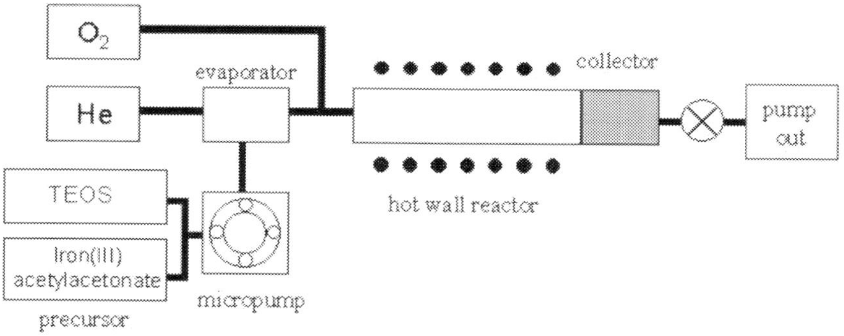

Fig. 9.5 Schematic CVC process [16].

which adds up a gas-phase chemical reaction. Usually, the extreme temperature rises at the originating place of flame when cumulations merge and separate from the flame. Consequently, the flame's temperature reduces, and particle merges form bulk agglomerates of initial particles. This process brings about the spread of ceramics nanoparticles, i.e., SiO_2 and TiO_2 [15]. Fig. 9.5 presents a schematic diagram for the CVC process [16].

9.3 Techniques of synthesizing nanoceramics

9.3.1 Solution combustion technique

This technique occupies greater significance and is based on modern materials' combustion synthesis, particularly for preparing nanocrystalline ceramics and compounds. The use of citric acid is involved in it, and metal nitrate is used with citric acid as a fuel agent. Their usage delivers various advantages to obtain nanocrystalline ceramic powders and amorphous. Moreover, the type of fuels may vary based on oxides. For instance, for connected compounds and aluminum oxides, carbamide is used [13]. Table 9.1 presents the classification of fuel and oxidizer and resultant solvents [17]. Whereas, Fig. 9.6 exhibits the solution combustion synthesis of nanopowder [18].

9.3.2 Alginate template technique

Techniques of biomineralization have been popular for producing organic and inorganic nanomaterials. In saccharide, the alginate template technique contains di-and trivalent cations that create a rigid cross-linking gel. Fig. 9.7

Table 9.1 Classification of fuel and oxidizer for solution preparation [17].

Oxidizer	Fuel	Solvent
Metal nitrates: Me^{V} $(NO_3)_v$ nH_2O	Urea (CH_4N_2O) Glycine ($C_2H_5NO_2$) Sucrose ($C_{12}H_{22}O_6$) Glucose ($C_6H_{12}O_{11}$) The citric acid ($C_6H_8O_7$)	Water (H_2O) Hydrocarbons: Kerosene Benzene (C_6H_6) Alcohols: Ethanol (C_2H_6O)
Ammonium nitrate (NH_4NO_3)	Hydrazine-based fuels: Carbohydrazide (CH_6N_4O) Oxalyldihydrazide ($C_2H_6N_4O_2$)	Methanol (CH_4O) Furfuryl alcohol ($C_2H_6O_2$)
Nitric acid	Hexamethylenetetramine ($C_6H_{12}N_4$) Acetylacetone ($C_5H_8O_2$)	2-Methoxy ethanol ($C_3H_8O_2$) Formaldehyde (CH_2O)

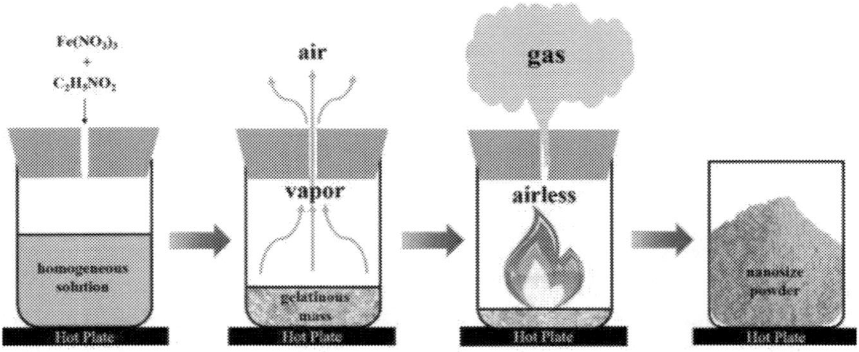

Fig. 9.6 Solution combustion synthesis [18].

displays the alginate technique template, wherein an alginate solution is dispensed. The drops are poured through a large syringe in the suspension of calcium chloride. In the solution of $CaCl_2$, beads are morphologically transformed [13]. In recent times, iron oxide in micrometer scale size is blended with calcium, which results in magnetic alginate in a cross-linked gel. Besides, alginic acid combines with polysaccharides that work as templates to produce different magnetic substrate [13].

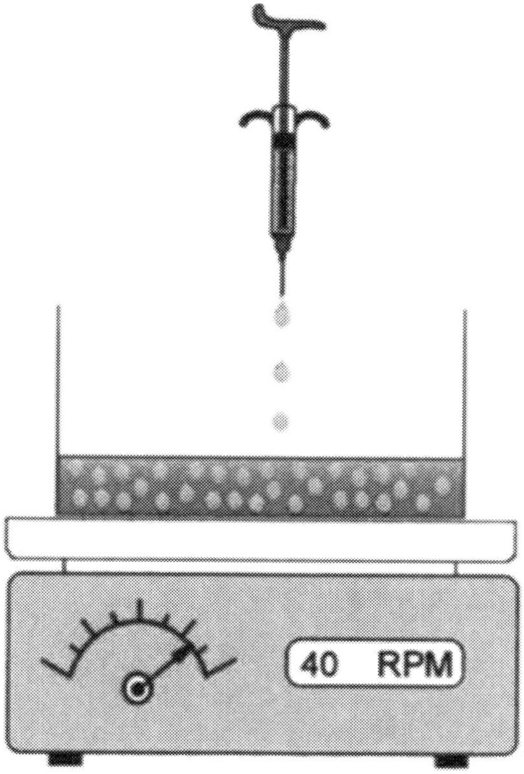

Fig. 9.7 Alginate template technique [13].

9.3.3 Microwave-assisted technique

Since the 1940s, microwave technique has been dispensing several benefits in different areas, particularly in the food industry. But the advent of the year 1986 had brought its usage in chemistry for manufacturing manifold inorganic materials. Through adopting this technique, many materials are produced with enhanced kinetics. The higher reactive rates enable to production nanostructures by nano-powders amid metastable increased temperature. There are the effects of heat due to the molecules' interaction having a higher frequency (2.45 GHz) and the radiation of nanoparticles that lead to steady and swift thermal reactions. Furthermore, using microwave tools in order to synthesize the bulk of products is more convenient (Fig. 9.8) [13].

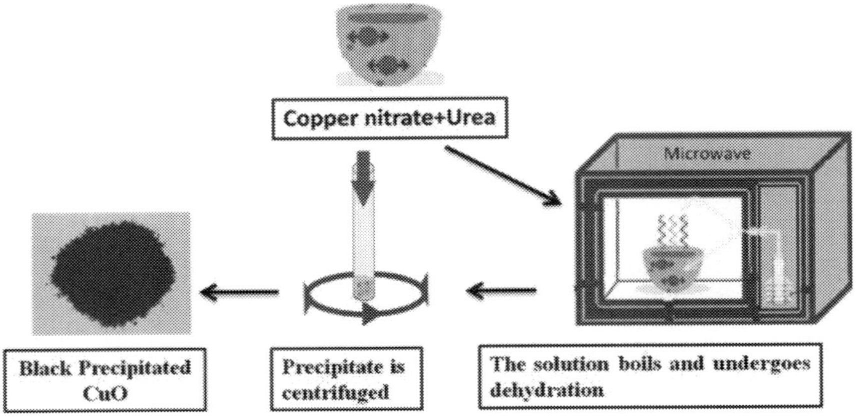

Fig. 9.8 Synthesis of copper oxide nanocomposites (CuO NCs) by microwave-assisted technique [19].

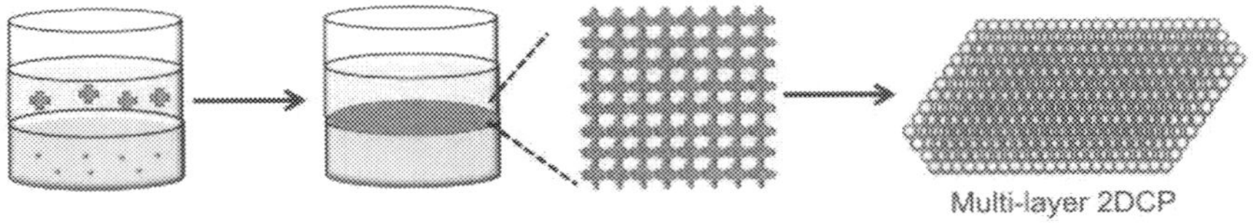

Fig. 9.9 Schematic diagram of liquid–liquid interface for two-dimensional crystalline polymers [20].

9.3.4 Liquid–liquid Interface technique

The liquid–liquid interface technique comprises a single synthesis that facilitates the synthesizing of nanocrystal arrays amidst ambient settings. This technique involves the reaction between organometallic compounds that are liquefied in the organic layer while diminishing an oxidized reactive agent. The composites are created in the ultrathin nanocrystalline film having interface level and comprise of nanocrystals. The films of ultrathin nanocrystalline are of nanocrystals that have a coating with organic compounds. Fig. 9.9 shows the process of synthesizing two-dimensional crystalline polymers by the liquid–liquid interface technique [13].

9.4 Application of nanoceramics

Nanoceramics are used in medicine, i.e., for the regeneration and repair of bones. Other potential nanoceramics applications are storing and supplying energy, transportation and communication infrastructure, and the construction industry. The electric characteristics of ceramics make them very beneficial for transferring energy with 100% [4]. The most significant applicability for ceramic nanomaterials is their functionality in extreme settings and higher temperatures. For instance, they are used in nozzle tips as combustion materials internally in the automotive sector. Additionally, in the aerospace sector and energy sector for heavy-duty gas turbines, nanoceramics are employed in the form of a hot gas path and combustion components [6].

Based on their kinds of matrix, various nanomaterials are classified into three groups: metal matrix nanocomposites (MMNCs), ceramic matrix nanocomposites (CMNCs), and polymer matrix nanocomposites (PMNCs). Fig. 9.10 exhibits the aerospace applications of nanocomposites, particularly ceramics matrix nanocomposites [1].

Moreover, in drugs and medicine, there is the use of calcium phosphate as a drug carrier, antigens, proteins, and enzymes, and also in delivering non-viral gene, they are used as promising material. Other applications of calcium phosphate are given below:

- They act as suitable deliverables for drugs, having a lesser extent to invasiveness.
- They can be fabricated at a lower cost.
- The timespan of calcium phosphate is long; they take for biodegradation.

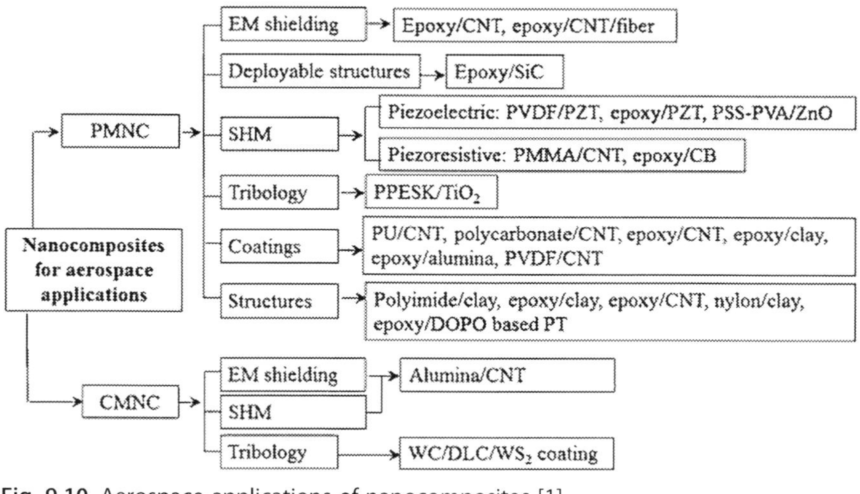

Fig. 9.10 Aerospace applications of nanocomposites [1].

- They are not inclined to alter their porosity and maintain their stable nature at temperature and pH modifications.
- Similar to the components of the tissues, they encompass size, chemistry, and crystalline organization.
- Through adopting the fabrication methods, their biocompatibility and bioavailability can augment [3].

Besides, some of the nanoceramics and their applications are presented in Table 9.2 [21].

Table 9.2 Nanoceramics and their applications [21].

Ceramics raw material		Particle size (nm)	Usability
Oxide-based	Al_2 $Al(OH)_3$	200–10	They are used in ultra-polishing and as stabilizers to foams. They work for mechanical reinforcing for the polymeric composites and have their applications as ceramic binders in refractory
	$MgO/Mg(OH)_2$	200–80	They are used in the hydrogenation of ethanol and act as an antiflame in carbon nanotubes and polymers, and they also provide them catalytic support
	SiO_2 microsilica	300–50	For paper fibers, they perform the binder agent's function and for water clarification and refractory castable. In manufacturing filler and polymer films, they are used as a binder agent
	TiO_2 (rutile), ZnO	100–5	They are used to prepare white

Table 9.2 Nanoceramics and their applications [21]—cont'd

Ceramics raw material		Particle size (nm)	Usability
			pigment and skin sun blocker
	TiO_2 (anatase)	200–10	They work as semiconductors and self-cleaning surfaces. Also, they have included in bactericide hydrophobic coatings
	ZrO_2, CeO_2	100–20	They are used to manufacture the cells of hydrogen and fuel
	Clays (montmorillonite/ hydrotalcite)	300–100 × 1–10	They perform the function as catalysts and useful n drug delivery. Moreover, in nanocomposites and inks, they are major constituents and work as resisting means for polymeric films
Non-oxide based	Carbon black	200–50	It is used to prepare black pigment and is used in reinforcing for ink and rubber composites
	Expanded graphite	300–100 × 1–50	It is a significant component for obtaining graphite pigment and works as an excessive temperature lubricant. Also, it is a source of acquiring graphene
	Graphene	300–100 × 0.50	It works as a semiconductor and is

Continued

Table 9.2 Nanoceramics and their applications [21]—cont'd

Ceramics raw material		Particle size (nm)	Usability
			used in mechanical reinforcement
	Nanotubes and fullerenes	1000 × 5, 1	They are used for synthesizing lubricants and manufacturing sensors, conductors, and semiconductors. Other applications are mechanical reinforcement and drug delivery
	Diamond and SiC	500–50	They work as ultrapolisher and for surface-coating

Apart from it, ceramics as semiconductors and in the form of nanofishes are used in the textile industry in order to implant certain features such as oil and water repellency, thermal resistance and flame retarding, etc. Moreover, UV blockers are made from the nanoparticles of titanium oxide and zinc oxide [22]. Other applications of nanoceramics are evident in dentistry and orthopedics. The advent of bio-compatible ceramics constituents entails enhanced biomedical functionality that is very advantageous in the healthcare sector across the globe. These materials play a pivotal role in the repairability and healing of various human body parts. For instance, the biomaterials in the form of nanoparticles are composed of calcium phosphate, which is used for scaffolds, cement, and coatings, particularly in dentistry and orthopedics.

Furthermore, replacing the bones of the knee, hips' tendons, and muscles, the use of ceramics nanoparticles has increased. While in dentistry, it is used in regenerating in periodontal disorders, and it works in maxillofacial regeneration, healing the jawbone, and is used as bone filler after the surgical treatment of tumors. Other biocompatible nanoceramics are hydroxyapatite (HAP) that have evident applicability in repairing enamel and mechanical furcation. It also works for restricted dispensation of bone morphogenetic protein and gene delivery [23].

9.5 Conclusion

Ceramic nanomaterials are modern materials with wide-ranging applications worldwide. They are composed of inorganic materials that encompass permeable properties. A spike has been recorded in the area of producing biocompatible ceramic nanomaterials. It's mainly biocompatible nanoceramics that have dramatically improved drug delivery, and other sub- healthcare sectors have in organ regeneration, scaffolds preparation, and bone repair. Nanoceramics have the features of both organic and inorganic simultaneously. They do not have a higher degree of thermal and electrical conduction. At higher temperatures, nanoceramics demonstrate resistance toward corrosion. The sol–gel method is one of the foremost methods of producing nanoceramics since its dawn in the 1980s. However, other methods that are adopted to prepare nanoceramics are self-propagating high-temperature synthesis and spray pyrolysis. Whereas some of the techniques that are employed in manufacturing nanoceramics include the chemical vapor condensation method and alginate template technique. Apart from their drug and medicine productivity, they are also used in the textile industry, electronics industry, and ultraviolet-resistant treatment preparation.

References

[1] V.T. Rathod, J.S. Kumar, A. Jain, Polymer and ceramic nanocomposites for aerospace applications, Appl. Nanosci. 7 (8) (2017) 519–548.

[2] S. Singh, V.K. Pandey, R.P. Tewari, V. Agarwal, Nanoparticle based drug delivery system: advantages and applications, Indian J. Sci. Technol. 4 (3) (2011) 177–180.

[3] C Thomas, S., Kumar Mishra, P., & Talegaonkar, S. (2015). Ceramic nanoparticles: fabrication methods and applications in drug delivery. Curr. Pharm. Des., 21(42), 6165–6188.

[4] A.P. Tiwari, S.S. Rohiwal, Synthesis and bioconjugation of hybrid nanostructures for biomedical applications, in: Hybrid Nanostructures for Cancer Theranostics, Elsevier, 2019, pp. 17–41.

[5] Subramaniam, A. & Balani, K. (n.d.). Nanostructures and Nanomaterials: Nanostructures and Nanomaterials: Characterization and Properties Characterization and Properties. Available at http://home.iitk.ac.in/~anandh/MSE694/Introduction_to_Nanomaterials-3.pdf.

[6] G. Mikhail, Ceramic nanomaterials for high temperature applications, Res. Dev. Material Sci. 2 (2) (2017), https://doi.org/10.31031/RDMS.2017.02.000535. RDMS.000535.

[7] D.P.H. Hasselman, L.F. Johnson, Effective thermal conductivity of composites with interfacial thermal barrier resistance, J. Thermoplast. Compos. Mater. 21 (6) (1987) 508–515.

[8] H. Gleiter, Nanostructured materials: basic concepts and microstructure, Acta Mater. 48 (1) (2000) 1–29.

[9] F.M. Smits, Measurement of sheet resistivities with the four-point probe, Bell Syst. Tech. J. 37 (3) (1958) 711–718.
[10] M. Zulkarnain, A.M. Husaini, M. Mariatti, I.A. Azid, Particle dispersion model for predicting the percolation threshold of nano-silver composite, Arab. J. Sci. Eng. 41 (6) (2016) 2363–2376.
[11] I.A. Rahman, V. Padavettan, Synthesis of silica nanoparticles by sol-gel: size-dependent properties, surface modification, and applications in silica-polymer nanocomposites—a review, J. Nanomater. 2012 (2012).
[12] K.J. Klabunde, J. Stark, O. Koper, C. Mohs, D.G. Park, S. Decker, D. Zhang, Nanocrystals as stoichiometric reagents with unique surface chemistry, J. Phys. Chem. 100 (30) (1996) 12142–12153.
[13] S.V. Ganachari, N.R. Banapurmath, B. Salimath, J.S. Yaradoddi, A.S. Shettar, A.M. Hunashyal, G.B. Hiremath, Synthesis techniques for preparation of nanomaterials, in: Handbook of Ecomaterials, Springer, Cham, 2017, https://doi.org/10.1007/978-3-319-48281-1_149-1.
[14] C. Li, X. Zhang, K. Wang, X. Sun, G. Liu, J. Li, Y. Ma, Scalable self-propagating high-temperature synthesis of graphene for supercapacitors with superior power density and cyclic stability, Adv. Mater. 29 (7) (2017) 1604690.
[15] A. Arafat, J.C. Jansen, A.R. Ebaid, H. Van Bekkum, Microwave preparation of zeolite Y and ZSM-5, Zeolites 13 (3) (1993) 162–165.
[16] J.H. Yu, C.W. Lee, S.S. Im, J.S. Lee, Structure and magnetic properties of SiO_2 coated Fe_2O_3 nanoparticles synthesized by chemical vapor condensation process, Rev. Adv. Mater. Sci. 4 (2003) 55–59.
[17] X. Yu, J. Smith, N. Zhou, L. Zeng, P. Guo, Y. Xia, R.P. Chang, Spray-combustion synthesis: efficient solution route to high-performance oxide transistors, Proc. Natl. Acad. Sci. 112 (11) (2015) 3217–3222.
[18] X. Wang, M. Qin, F. Fang, B. Jia, H. Wu, X. Qu, A.A. Volinsky, Effect of glycine on one-step solution combustion synthesis of magnetite nanoparticles, J. Alloys Compd. 719 (2017) 288–295.
[19] N.M. Shaalan, M. Rashad, M.A. Abdel-Rahim, CuO nanoparticles synthesized by microwave-assisted method for methane sensing, Opt. Quant. Electron. 48 (12) (2016) 531.
[20] L. Wang, H. Sahabudeen, T. Zhang, R. Dong, Liquid-interface-assisted synthesis of covalent-organic and metal-organic two-dimensional crystalline polymers, NPJ 2D Mater. Appl. 2 (1) (2018) 1–7.
[21] M.H. Wakamatsu, R. Salomao, Ceramic nanoparticles: what else do we have to know? InterCeram 59 (1) (2010) 28–33.
[22] Y.W.H. Wong, C.W.M. Yuen, M.Y.S. Leung, S.K.A. Ku, H.L.I. Lam, Selected applications of nanotechnology in textiles, AUTEX Res. J. 6 (1) (2006) 1–8.
[23] S. Balasubramanian, B. Gurumurthy, A. Balasubramanian, Biomedical applications of ceramic nanomaterials: a review, Int. J. Pharm. Sci. Res. 8 (12) (2017) 4950–4959.

CHAPTER 10
Carbon-based nanomaterials

Abbreviations

AFM	atomic force microscopy
C_2H_2	acetylene
CH_4	methane
CNTs	carbon-based nanotubes
CVD	chemical vapor deposition
FFI	fast frame imaging
GO	graphene oxide
GQDs	graphene quantum dots
MWCNT	multiwalled carbon nanotubes
OES	optical emission spectroscopy
ORR	oxygen reduction reaction
PEG	polyethylene glycol
PLGA	poly-lactic-co-glycolic acid
PLIF	laser induced fluorescence
RGO	reduced graphene oxide
SCAs	synthetic carbon allotropes
SWCNT	single-walled carbon nanotubes
TEM	transmission electron microscopy

10.1 Introduction

Carbon is one of the most abundantly found elements on the earth. Arguably, carbon has wider applicability among all the elements that are included in the periodic table. Since, it has the capability of binding with its own as well as with all the elements. Moreover, Deoxyribonucleic acid's base material is carbon, reflecting that carbon is the element to give the basis to life. The electron configuration of carbon is 1s2, 2s2, 2p2 that creates different kinds of crystalline and disorganized materials due to its existence in three varied hybridizations, which are sp3, sp2, and sp1 [1]. However, fullerenes serendipitous discovery is led to the commencement of synthetic carbon allotropes (SCAs) or carbon nanomaterials [1]. They comprise entirely coalesced π-electron systems and can be taken as restrained composites in the dimensions of zero, one, or two (Fig. 10.1).

Synthetic Engineering Materials and Nanotechnology
https://doi.org/10.1016/B978-0-12-824001-4.00016-8

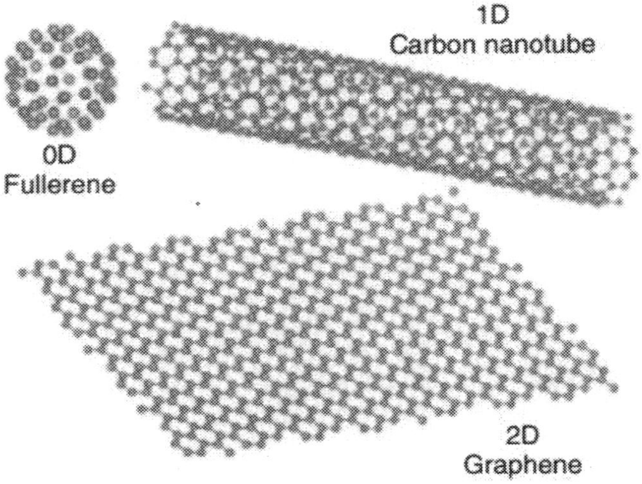

Fig. 10.1 Allotropes of carbon nanomaterials in different dimensions [2].

Hint statement

Carbon nanotubes (CNTs) are nano-structured allotropic cylindrical carbon.

The broader surface area of carbon offers dynamic places for interaction with materials that have varied forms of the chemical by wastewater. They have different allotropic kinds, i.e., diamond, graphite, fullerenes, carbon nanotubes, graphitic carbon nitrate, and graphene (Fig. 10.2) [2].

Hint statement

For enhancing regeneration of injured hard muscles, three-dimensional graphene-based HA nano-composites are used.

Carbon nanotubes in their pristine forms are composed of various hexagonal graphene sheets that are made of the atoms of carbon. They are molded in the form of tubes and are reckoned as nonreactive. For instance, single-walled carbon nanotubes (SWCNT) require a temperature of 500°C to be oxidized and for its burning. Since synthesizing carbon-based nanotubes (CNTs) typically stipulates the existence of catalytic metals during its manufacturing. Thus it has various contamination, which is higher than multiwalled carbon nanotubes (MWCNT) in (CNTs) that are used in the industry. The atomic structure of CNT was defined in tube chirality, which

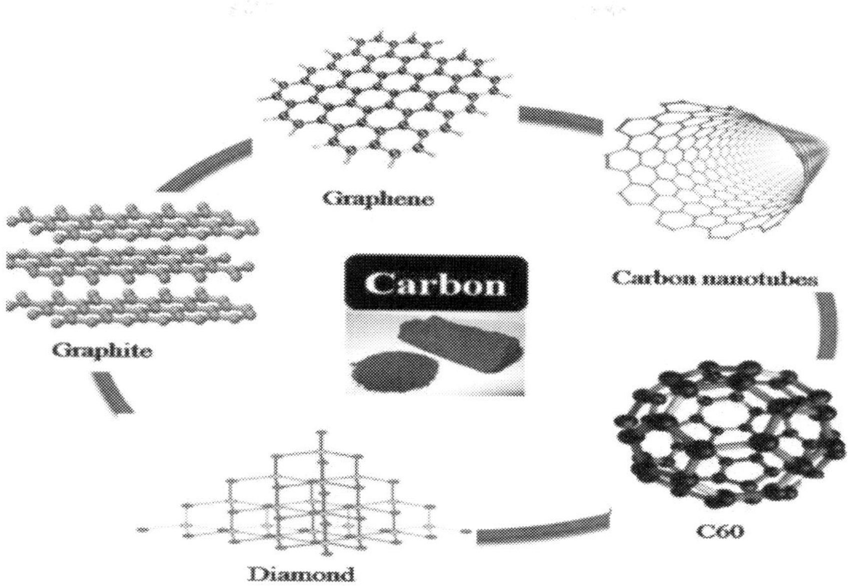

Fig. 10.2 Allotropic kinds of carbon-based nanomaterials [3].

is described through the positioning of the sheet of graphene at the time of synthesizing carbon nanotubes. The most widely recognized conformations are armchair and zig–zag conformations, which typically occur with a blend of various conformations. Synthesis of CNTs occurs in different ways concerning length, diameter, atomic structure defects, and surface chemistry. The thickness of CNTs is subjected to the chirality of the tubes and layers of graphene. Typically, the exterior diameter of SWCNT is found as 1–3 nm. However, CNTs have a length in micromeres, but it can be found with the length of a few hundred nanometres and tens in micrometers. The tubes that are 50 μm lengthy are commonly found and carry hundreds of micrometers. Moreover, amid the synthesis of CNTs, some kinds of defects may be encountered; for example, the tube may collapse in the form of bamboo closures inside the tube, which can be traced through electron microscopy [4].

Hint statement

The prerequisites of synthesizing carbon-based nanomaterials in larger quantity, are simpler equipment, mild pressure and temperature.

10.2 Methods of synthesizing carbon-based nanomaterials

Carbon-based nanomaterials can be formed via various methods. Carbon arch discharge, chemical vapor deposition, and laser ablation are the most commonly adopted methods to synthesize carbon-based nanomaterial. These methods are described in the following:

10.2.1 Chemical vapor deposition (CVD)

For the synthesis of carbon-based nanomaterials, more straightforward tools, mild pressure, and temperature are required, which make them more appropriate for producing carbon-based nanomaterials in large quantities. Chemical vapor deposition is processed through decomposing hydrocarbon to obtain carbon. Chemical vapor deposition (CVD)'s reactive agents are composed of a reactive chamber where the tubes are occupied with hydrocarbons and inactive gas (Fig. 10.3). For the production of SWCNT, methane is used whereas, in synthesizing MWCNTs, acetylene or ethylene are used. The substrate is given heat up to 850–1000 °C in synthesizing SWCNT whereas, the temperature is added up to 550–700 °C for manufacturing MWCNT. Carbon is manufactured through hydrocarbon decomposition and gets dissolved in the catalyst of the metal nanoparticle. Once acquiring a specified amount of carbon, semifullerene cap is formed as an initial structure for manufacturing a nanotube in a cylindrical shell, created by a steady drift of carbon from a hydrocarbon source. Ultimately, the process of eliminating catalysts through the edges of nanotubes and purifying occur for high-quality manufacturing nanotubes [5].

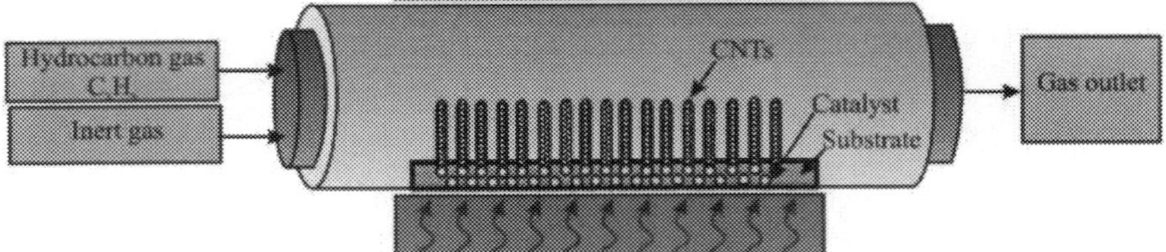

Fig. 10.3 Schematic representation of chemical vapor deposition (CVD) process [5].

10.2.2 Carbon arc discharge

Generally, this process is conducted with a graphite node that has a diameter of 0.65 and a graphite cathode, which has a diameter of 0.95 cm. A regulated power supply gives power to the arc. A ballast resistor is included with the electric circuit of the arch, in order to reduce relaxation oscillation related to a variance resistance of the arch, a reduced resistibility averting gradually with the dispensation of arc to examine its electrodes. The intersection at the center of the axis at the right angles between the electrodes is the place where planar laser-induced fluorescence (PLIF), optical emission spectroscopy (OES), fast frame imaging (FFI) are installed. The foundations of the OES setup are laid on an imaging spectrograph (Chromex is 250) attached with a fiber-coupled low-resolution spectrometer, a camera, and a linear CCD (Ocean Optics 2000+) (Fig. 10.4) [5].

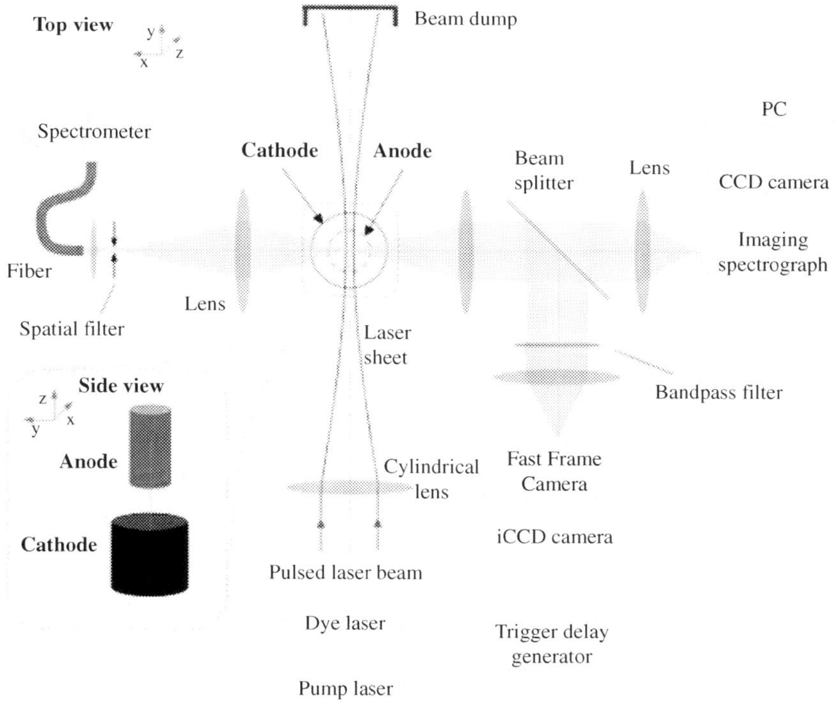

Fig. 10.4 Carbon arc discharge [5].

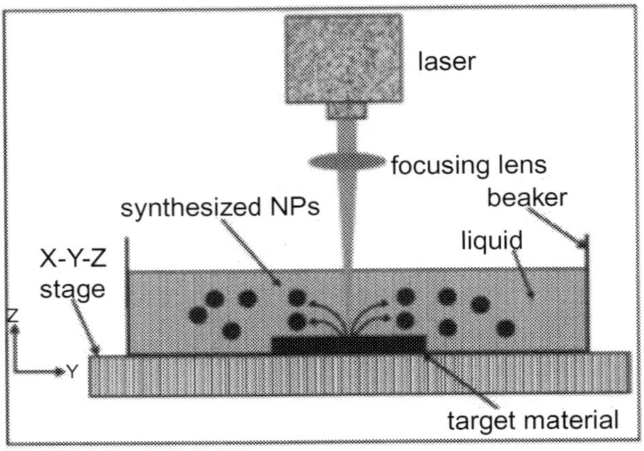

Fig. 10.5 Laser ablation method [6].

Hint statement

Graphene-based HA nanocomposites are reshaped in form of bulks, coatings, powders, and scaffolds due to their usability in repairing of bones fissures.

10.2.3 Laser ablation method

The laser ablation method is used to produce graphene oxide (GO) through graphite oxidation. Fig. 10.5 shows the process of laser ablation wherein (GO) was liquefied in water. The platinum plate was also engrossed in the solution of GO. The setup of laser ablation comprises a lens, a liquid cell, and a high-power laser. Through Nd: YAG Q-Switch laser beam, the platinum plate was vaporized. The laser beam's reputation rate, wavelength, and energy were 40 Hz, 532 nm, and 1200 mJ sequentially. Further, the nanoparticles were diffused in the solution of GO while stirring, during which the platinum plate dissolved. The time required in the process of ablation was 5–30 min. Furthermore, to reshape the particle size of platinum nanoparticles, transmission electron microscopy (TEM), and a UV visible spectrometer were used [7].

Example 10.1

Evaluate the work function difference between copper, cupric oxide, and cuprous oxide nanoparticles? [8]

Solution:

The required temperature to obtain Cupric oxide via heating cuprous oxide (Cu_2O) or copper in the air = 1000–1100°C.

Thus, the Cupric oxide is formed as $Cu_2O + 1/2O_2 = 2CuO$ [9].

10.2.4 Single- or double-emulsion method

Through single emulsion (oil–water) or double emulsion (water–oil–water) methods, the nanoparticles of the poly-lactic-co-glycolic acid (PLGA) are synthesized. PLGA are significant components of hydrophilic or hydrophobic drugs at the nanoscale. PLGA is liquefied in the solution of organic phase oil. Further, it is processed through emulsification along with a stabilizer or water. The hydrophobic drugs are then blended with the oil phase wherein hydrophilic drugs are emulsified in the polymer solution before creating particles. There is the formation of the trivia droplets of the polymer through the facilitation of high-intensity sonication bursts. Furthermore, the emulsification is mixed in a large fluid phase while stirring well. The stirring facilitates the evaporation of the solvent. The hard-edged nanoparticles are amassed and cleansed through the centrifugation (Fig. 10.6) [11].

10.2.5 Emulsion-solvent evaporation method

This is the most commonly adopted method for synthesizing nanoparticles. It comprises of two phases; firstly, it stipulates emulsifying polymer solution in a liquid phase. In the second phase, the polymer's solvent is processed through evaporation, along with the precipitation of the polymer as nanospheres. Afterward, nanoparticles are amassed through the ultracentrifugation and cleansed by the distilled water, in order to eliminate the stabilizer waste or any other drug residue. Due to the alteration in this method, it is termed as high-pressure emulsification and solvent evaporation method. The modification is characterized by preparing emulsion concerning the homogenizing through high pressure, backed by dispensing the organic solvent via stirring. The volume is controllable via regulating the stirring rate, the kind of dispersion agent, and the thickness of organic and watery phases and heat (Fig. 10.7) [11].

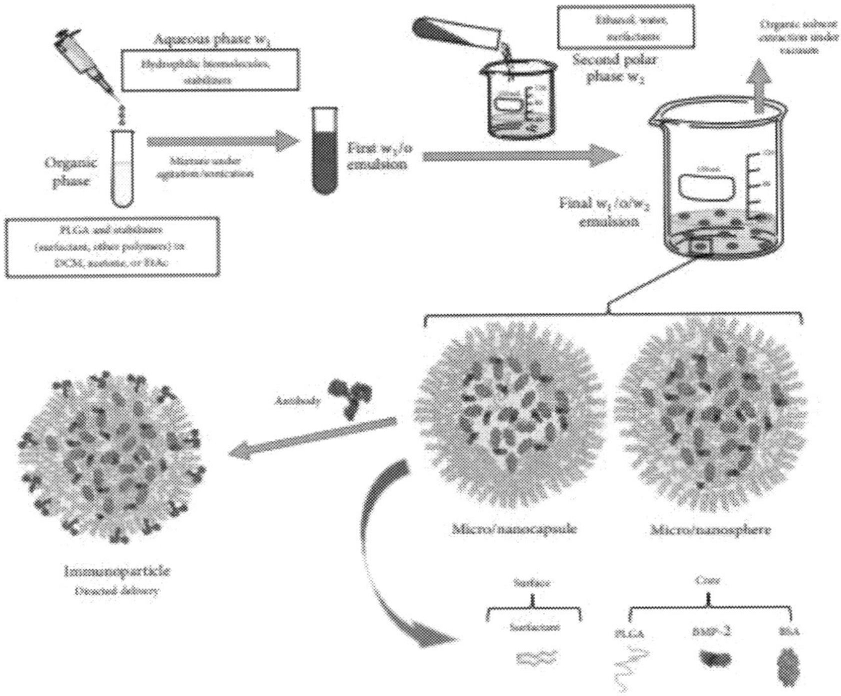

Fig. 10.6 Double emulsion process to acquire PLGA micro/nanoparticles [10].

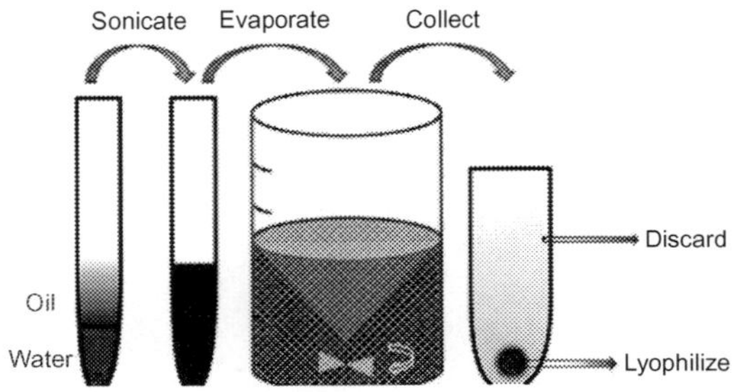

Fig. 10.7 Schematic of nanoparticle fabrication [11].

10.3 Techniques of synthesizing carbon-based nanomaterial

10.3.1 Transmission electron microscopy

In this microscopy technique, an electronic beam is diffused by an ultra-thin specimen that mingles while passing through sampling. An electronic beam is central to study magnetic fields, and electrons are created in a spiral path. Further, an image is observed due to the transmission of electrons by the specimen, which is enlarged and central to an objective lens and visible on the image screen. The nanotubes defined characteristics, i.e., morphology and diameter, and dissemination, TEM analysis micrographs, are arranged. Moreover, to distinguish a single wall by multi-wall nanotubes and examine their organization, reaction among the nanotubes, and other components, TEM is the most appropriate technique. Fig. 10.8A displays carbon nanotube having dissimilar diameters. While, in Fig. 10.8B, a metal nanotube is present with an overall diameter of 81.3 nm, which is [12].

10.3.2 Atomic force microscopy (AFM)

Owing to the convenience and durability, atomic force microscopy (AFM) is considered as a suitable technique to examine the viscosity and the number of layers. Generally, for measurement, a cantilever with a shrill tip of ~5–10 nm has used that probes around the surface of graphene materials. Subtle modification in frequency and vibration amplitude of the tip are amassed owing to elusive changes in the surface heterogeneity that can be employed to evaluate the sample's topography. It facilitates measuring the viscosity of the films of graphene. While considering the fact that the thickness of the one atom of graphene is 0.35 nm, the number of layers can be calculated. For instance, Yang et al. [13] chalked out a model catalyst that was graphene-based Fe/N/C (FeN-MLG) for investigating the dynamic sites of pyrolyzed Fe/N/C, which the most appropriate cheaper catalyst for oxygen reduction reaction (ORR). The structures that have more heterogeneous organizations are created amid the pyrolysis in traditional catalysts that fathom dynamic sites. Yang et al. [13] employed AFM as indispensable equipment to describe synthesizing the anticipated catalyst (Fig. 10.9). However, in Fig. 10.9A and B exhibits the AFM image of creased FeN-MLG, while Fig. 10.7C shows the conforming height profile and the viscosity of the layer of FeN-MLG is ~0.5 nm [13].

10.3.3 Scanning electron microscopy

The technique involves the electronic microscopy and deliberates manifold benefits in morphological evaluation. Meanwhile, it offers inadequate knowledge regarding the size division and actual mean of the population.

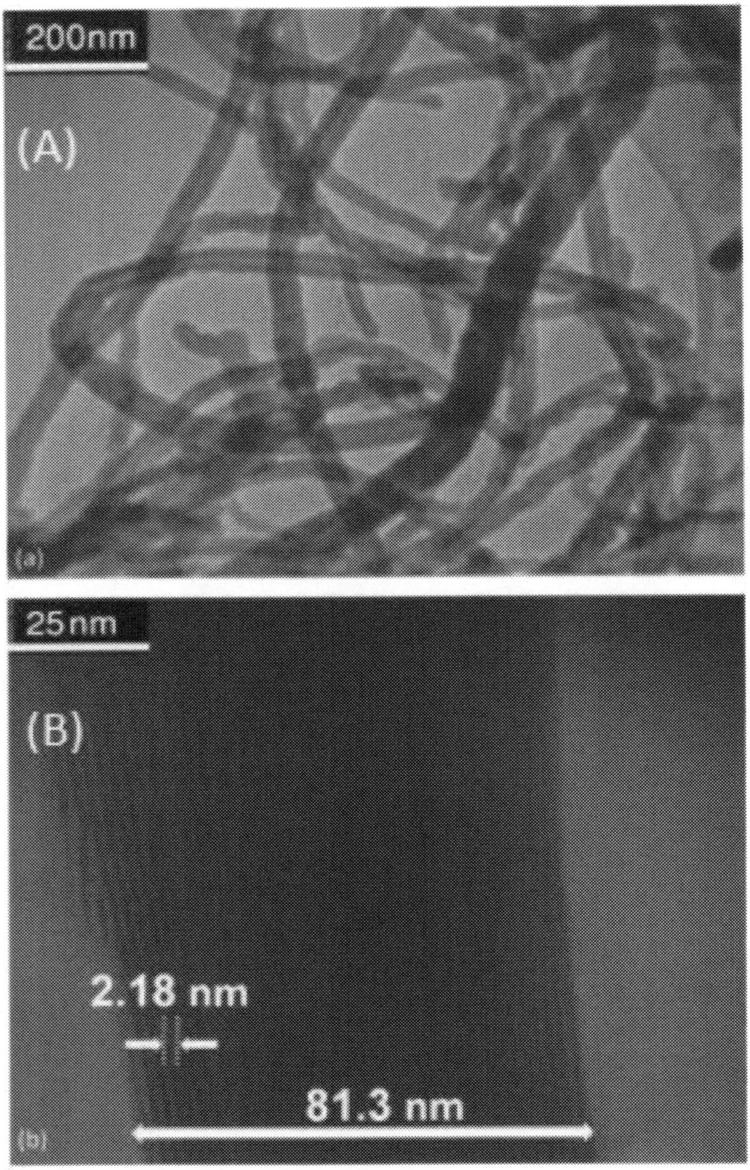

Fig. 10.8 TEM micrographs: (A) carbon nanotubes and (B) carbon nanotubes carry wall of a metal oxide [12].

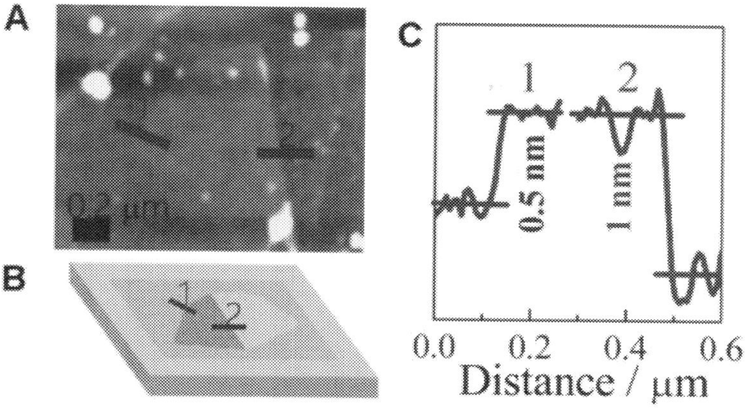

Fig. 10.9 AFM dimension of creased FeN-MLG. (A) AFM appearance (B) morphological representation of bright spot (C) profiles of height, shown by the *solid lines* 1 and 2, sequentially, displays the viscosity of FeN-MLG, which is 0.5 nm [13].

Aiming to the characterization of the SEM, it is imperative to convert the solution of the nanoparticles into a dried powder. It is then to pour into a holder that carries a coating of the appropriate metal, i.e., gold in the form of a sputter coater. Furthermore, the sample is skimmed through a reasonably central beam of electrons. The model's surface features are acquired through the ancillary electrons that emanate from the surface's exterior. The nanoparticles must-have the capability to tolerate the vacuum (Fig. 10.10) [15].

10.3.4 Flame synthesis

The flame synthesis technique involves fuels, typically ethylene (C_2H_4), acetylene (C_2H_2), and methane (CH_4) that have a chemical reaction with the oxidizing agents (O_2 from the air). The reaction produces the gaseous mixture that carries saturated and unsaturated hydrocarbons (C_2H_4, C_2H_6, C_2H_2), and radicals carry carbon dioxide (CO_2), carbon monoxide (CO), water vapor (H_2O), and hydrogen (H_2). The formation of hydrocarbons and carbon monoxide composes the gaseous precursor mixture that provides solid carbon that is concentrated on the catalyst particles to create the nanostructured carbons. The metal catalysts that are implanted in the

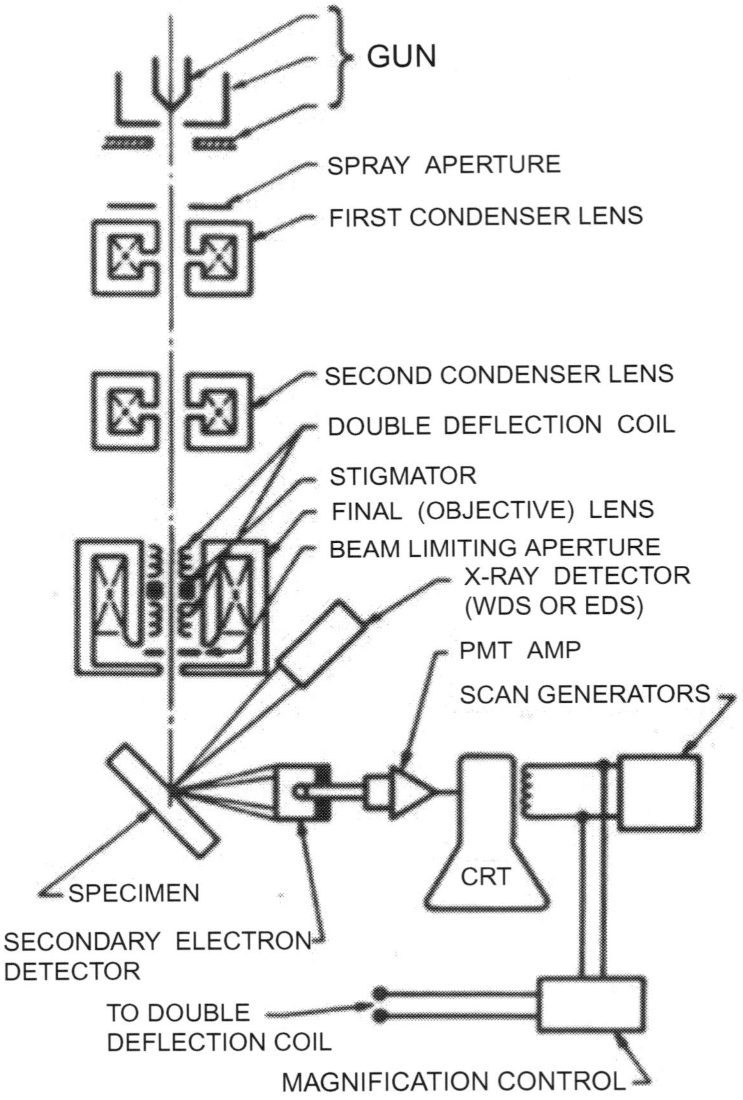

Fig. 10.10 Schematic diagram of SEM [14].

flame, which may be found either in the form of a substrate coating or as aerosol particles, provide essential reactive places for harmonizing the solid carbon's disposition. The carbon nanotubes' organization hinges on the carbon deposition rate and the particle size of the catalyst (Fig. 10.11) [17].

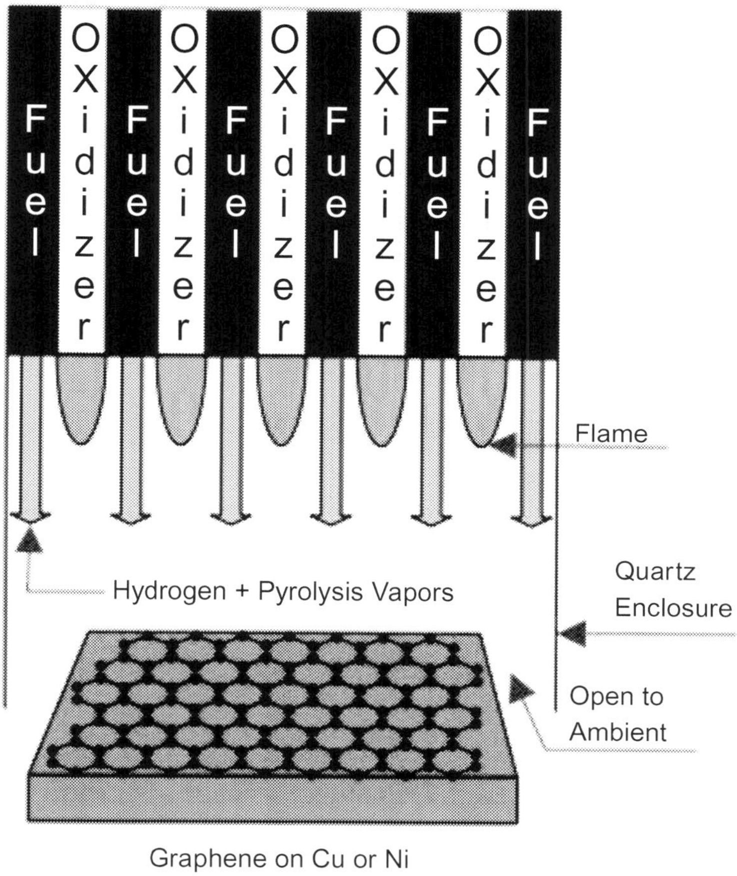

Fig. 10.11 Flame synthesis [16].

Example 10.2

Calculate the heat of fusion of a 2 nm Ag nanocrystal [18].

Solution:

Suppose, to melt a silver nanocrystal, and it has 2 nm in diameter; it requires about 1/2 of the atoms on the surface. The lower surface area of a spherical liquid droplet is lower than the faceted crystal. Suppose the surface area of a cube is 1.24 greater than the surface area of a sphere of the same volume. When the surface area's decrease is approximately 20% upon liquefying, the surface energy would be 1/4 of the bulk bonding of the atoms of energy. Thus, the diameter of the nanocrystal would be:

$$\Delta H^0_{fusion} \Delta H^0_{fusion,\,bulk} - \left(\frac{1}{2}\right)(0.25)\left(\frac{1}{2}\right)\Delta H^0_{vap} = \Delta H^0_{fusion,\,bulk} - 0.025\Delta$$

Here, ΔH°vap denotes the heat of vaporization that is the accumulated bonding energy of a bulk crystal's atoms. Since, silver has ΔH fusion, bulk = 11.3 kJ/mol, and ΔH°vap, bulk = 250 kJ/mol. Therefore, calculate the heat of fusion of a 2 nm Ag nanocrystal would be:

$$\Delta H^{0}_{fusion} \approx 11.3 - (0.025)(25) = 5.1\frac{kj}{mol}$$

10.4 Application of carbon-based nanomaterials

Researchers and several studies have discovered carbon-based nanomaterials, i.e., graphene and carbon nanotubes. Graphene is a nanomaterial which occupies numerous commercial and scientific applications [19]. Carbon nanotubes (CNTs) are allotropic cylindrical carbon that is nanostructured. The properties possessed by cylindrical carbon molecules have applicability in optics, electronics, medicine, and pharmaceutical [19]. Since graphene is a hybridized structure, having a single atomic layer, it has other materials that possess few layers. These are GO, nanolayers, reduced GO, etc., known as graphene-family. The graphene (Fig. 10.12A) family consists

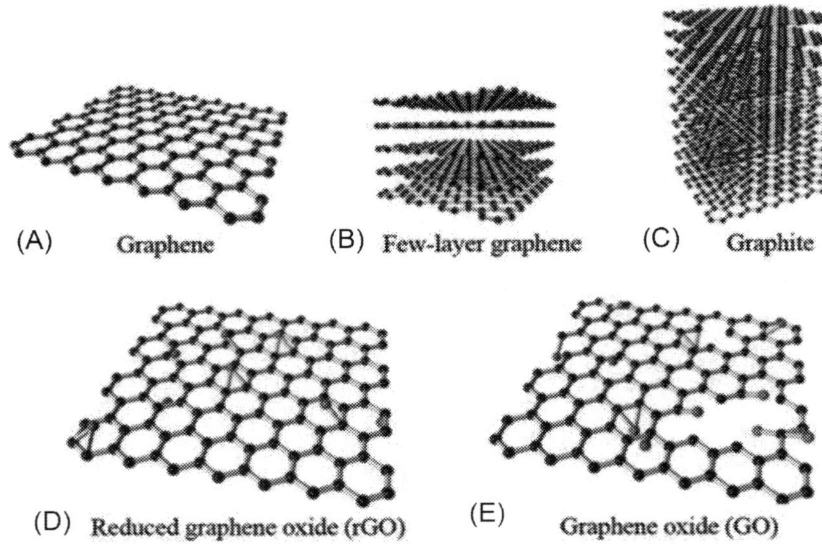

Fig. 10.12 Nanomaterials of the graphene family [19].

of family nanomaterials that are few-layer graphene (Fig. 10.12B), graphite (Fig. 10.12C), reduced GO (Fig. 10.12D), and GO (Fig. 10.12E). They vary based on their structures, number of layers, and characteristics [20].

Graphene-based HA nanocomposites are transformed into bulks, coatings, powders, and scaffolds used to repair bone fissures. Also, they are the major constituent of the coating that is used to coat an orthopedics metallic implant, which adds up to the binding ability of bone, for the regeneration of larger defects or bone injuries. These graphene-based HA nanocomposites are three-dimensional increase injured hard muscles [21]. The novel features of CNTs, diamond-like films, and diamonds have promising contributions toward manufacturing nanostructured biosensors and electrochemical sensors capable of performing analytically compared to conventional electrochemical sensing mechanisms [22].

All the carbon family materials have unmatchable properties that are useful in drug delivery, tissue engineering, imaging, biosensing, cancer therapy, and diagnosis [23]. CNTs are known as hollow cylinders and are composed of graphite sheets. They are subdivided into SWCNT and MWCNT. CNTs act as new era's nanoprobes due to their excellent mechanical, structural, and electronic features. Due to the higher aspect ratio, high chemical stability, rapid electronic transference, and high conductivity, they are most appropriate for biosensing. Also, the carbon allotropes are promising components to be used in drug delivery molecules. Active cells due to their pristine morphology expedites nonintrusive infiltration to the biological membranes [24]. Carbon nanotubes are used in anticancer medicines and work as proteins for chemotherapy [25]. Table 10.1 represents various carbon-based nanomaterials used in cancer therapy.

Apart from the applications in medicine, carbon nanomaterials are also used as a remedial agent against contamination. Fig. 10.13 provides the absorption capability of carbon-based nanomaterials [34]. Since, there are adverse impacts due to the presence of the heavy metal ions in water. Thus, it is imperative to eradicate these elements from water. Nanomaterials have promising applications in removing the harmful materials from water as they have a higher surface area and ability to modify the elements chemically and regenerate them. Some examples of the polluting agents removed by the nanomaterials are arsenic, metal ions, biological contamination, and organic materials [43].

Table 10.1 Applications of carbon-based nanomaterials in cancer therapy.

Carbon-based nanomaterials	In vitro	Therapy
NY-ESO, CpG-ODNs with MWCNT	Dendritic cells	Immune response [26]
Magnetic ferrite nanoparticles filled	CNTSKOV3 cells	Imaging and therapy [27]
Polyethylene glycol (PEG) functionalized MWCNTS	U87, U373MG, NHA	Brain tumor therapy [28]
CNT	–	Microbeam radiation therapy [29]
MWSCNT	PANC-1	Pancreatic cancer [30]
MWSCNT	HeLa	Photothermal therapy [30]
SWCNT	4T1	Chemo-photothermal therapy [21]
(PEG-g-PDMA-HA)@rGO	MDAMB-231, A549	Photothermal [31]
GO decorated Ru(II)-PEG complex	A549	Photodynamic-photothermal
Iron oxide-GO	HeLa	Chemophotothermal [32]
GO (^{188}Re)-modified Fe_3O_4/ silica	–	Chemophotothermal [33]
(HA)-modified Q-Graphene	A549, MRC-5	Chemotherapy [34]
CuS-GO	HeLa	Chemophotothermal [35]
RGO-PEG	U87	Chemophotothermal photodynamic [36]
GQD-Ce6-HA	A549	Photodynamic [37]
UCNP-GQD	4T1	Photodynamic [32]
7Gd-encapsulated Graphene Carbon	SSC-7	Photodynamic [32]
GQDs	BT-474, MCF-7	[38]
GQDs	PANC-1, A-549, HePg2	[39]
GQDs	Rg2	Chemotherapy [40]
GQDs	La29,HaCaT, Mia-Pa-Ca-2	Photothermal photodynamic [41]
GQDs	SW620, HCT116	Radiotherapy [42]

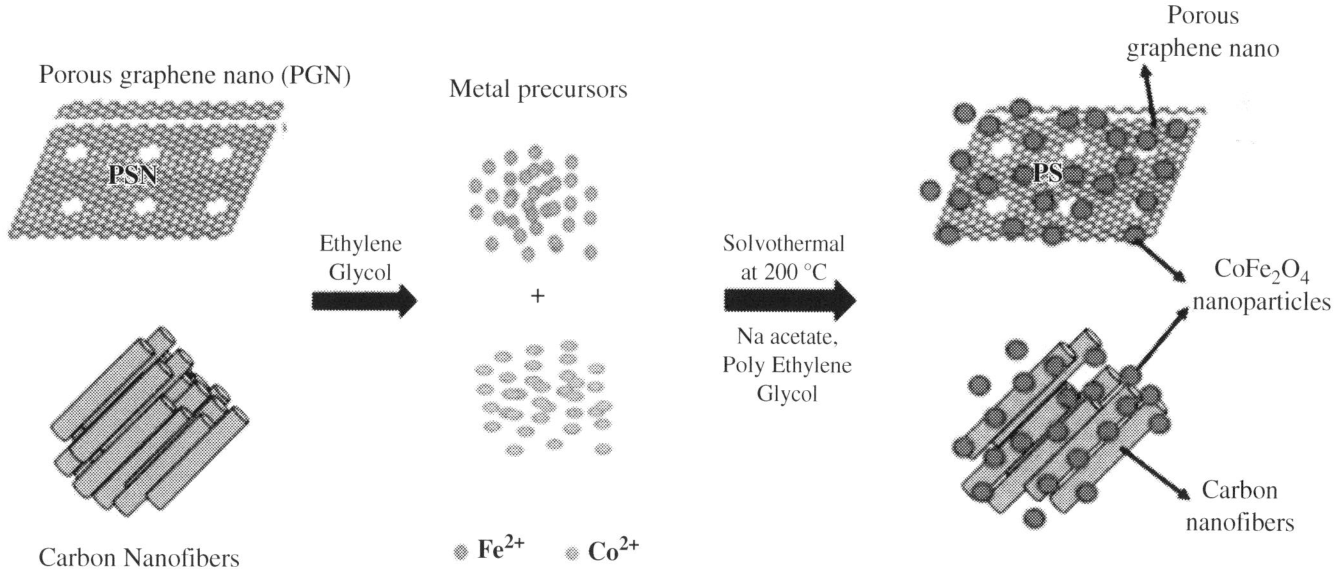

Fig. 10.13 Absorption of heavy metal ions from water while using carbon nanomaterials [43].

10.5 Conclusion

Carbon is a fundamental component of DNA that is reckoned as a life building component. It is one of the abundant elements found on the earth. The electronic configuration of carbon is the determining factor that forms various other hybridized and disorganized components. The methods employed to manufacture carbon-based nanomaterials are the laser ablation method, arch discharge, and chemical vapor disposition methods. On the other hand, the techniques to form carbon-based nanomaterials are TEM and AFM. Carbon-based nanomaterials, particularly carbon nanotubes, have their promising contribution in numerous areas of life. Most importantly, they are used in manufacturing anticancer drugs and work as proteins in chemotherapy. Other applications are included ecological remediation, obtaining biosensing tools, and tissue engineering.

References

[1] N. Kumar, S. Kumbhat, Essentials in Nanoscience and Nanotechnology, vol. 486, Wiley, Hoboken, NJ, 2016.

[2] X. Wan, Y. Huang, Y. Chen, Focusing on energy and optoelectronic applications: a journey for graphene and graphene oxide at large scale, Acc. Chem. Res. 45 (4) (2012) 598–607.

[3] T. Jayaraman, A.P. Murthy, V. Elakkiya, S. Chandrasekaran, P. Nithyadharseni, Z. Khan, M. Jagannathan, Recent development on carbon based heterostructures for their applications in energy and environment: a review, J. Ind. Eng. Chem. 64 (2018) 16–59.

[4] Anon (n.d.). Carbon Nanotubes. Available at: https://monographs.iarc.fr/wp-content/uploads/2018/06/mono111-01.pdf.

[5] O. Zaytseva, G. Neumann, Carbon nanomaterials: production, impact on plant development, agricultural and environmental applications, Chem. Biol. Technol. Agricult. 3 (1) (2016) 17.

[6] Y.H. Chen, C.S. Yeh, Laser ablation method: use of surfactants to form the dispersed Ag nanoparticles, Colloids Surf. A Physicochem. Eng. Asp. 197 (1–3) (2002) 133–139.

[7] A.R. Sadrolhosseini, M. Habibiasr, S. Shafie, H. Solaimani, H.N. Lim, Optical and thermal properties of laser-ablated platinum nanoparticles graphene oxide composite, Int. J. Mol. Sci. 20 (24) (2019) 6153.

[8] Khatun, S. (2020). Nanoparticles. Available at: https://www.researchgate.net/post/What_are_the_work_function_difference_between_copper_cupric_oxide_and_cuprous_oxide_nanoparticles.

[9] Alkathy, M. (2020). Re: What are the work function difference between copper, cupric oxide and cuprous oxide nanoparticles? Available at: https://www.researchgate.net/post/What_are_the_work_function_difference_between_copper_cupric_oxide_and_cuprous_oxide_nanoparticles/5f49afb744dd6f372175fdf0/citation/download.

[10] I. Ortega-Oller, M. Padial-Molina, P. Galindo-Moreno, F. O'Valle, A.B. Jódar-Reyes, J.M. Peula-García, Bone regeneration from PLGA micro-nanoparticles, Biomed. Res. Int. 2015 (2015).

[11] R.L. McCall, R.W. Sirianni, PLGA nanoparticles formed by single-or double-emulsion with vitamin E-TPGS, J. Vis. Exp. 82 (2013), e51015.

[12] M.A. Asadabad, M.J. Eskandari, Transmission electron microscopy as best technique for characterization in nanotechnology, Synth. React. Inorg. M. 45 (3) (2015) 323–326.
[13] X.D. Yang, Y. Zheng, J. Yang, W. Shi, J.H. Zhong, C. Zhang, S.G. Sun, Modeling Fe/N/C catalysts in monolayer graphene, ACS Catal. 7 (1) (2017) 139–145.
[14] Swapp, S. (n.d.). Scanning Electron Microscopy (SEM). Available at: https://serc.carleton.edu/research_education/geochemsheets/techniques/SEM.html.
[15] S.L. Pal, U. Jana, P.K. Manna, G.P. Mohanta, R. Manavalan, Nanoparticle: An overview of preparation and characterization, J. Appl. Pharm. Sci. 1 (6) (2011) 228–234.
[16] N.K. Memon, D.T. Stephen, J.F. Al-Sharab, H. Yamaguchi, A.M.B. Goncalves, B.H. Kear, M. Chhowalla, Flame synthesis of graphene films in open environments, Carbon 49 (15) (2011) 5064–5070.
[17] J.P. Gore, A. Sane, Flame synthesis of carbon anotubes, in: Carbon Nanotubes-Synthesis, Characterization, Applications, vol. 1, 2011, p. 16801.
[18] Anon (2020). Surface Energy. Available at: https://chem.libretexts.org/Bookshelves/Inorganic_Chemistry/Book%3A_Introduction_to_Inorganic_Chemistry/11%3A_Basic_Science_of_Nanomaterials/11.05%3A_Surface_Energy.
[19] L. Crisan, B.V. Crisan, S. Bran, F. Onisor, G. Armencea, S. Vacaras, C. Dinu, Carbon-based nanomaterials as scaffolds in bone regeneration, Part. Sci. Technol. (2019) 1–10.
[20] N. Chatterjee, H.J. Eom, J. Choi, A systems toxicology approach to the surface functionality control of graphene–cell interactions, Biomaterials 35 (4) (2014) 1109–1127.
[21] M. Li, P. Xiong, F. Yan, S. Li, C. Ren, Z. Yin, Y. Cheng, An overview of graphene-based hydroxyapatite composites for orthopedic applications, Bioact. Mater. 3 (1) (2018) 1–18.
[22] A.C. Power, B. Gorey, S. Chandra, J. Chapman, Carbon nanomaterials and their application to electrochemical sensors: a review, Nanotechnol. Rev. 7 (1) (2018) 19–41.
[23] G. Hong, S. Diao, A.L. Antaris, H. Dai, Carbon nanomaterials for biological imaging and nanomedicinal therapy, Chem. Rev. 115 (19) (2015) 10816–10906.
[24] T. Panczyk, P. Wolski, L. Lajtar, Coadsorption of doxorubicin and selected dyes on carbon nanotubes. theoretical investigation of potential application as a pH-controlled drug delivery system, Langmuir 32 (19) (2016) 4719–4728.
[25] Y. Hwang, S.H. Park, J.W. Lee, Applications of functionalized carbon nanotubes for the therapy and diagnosis of cancer, Polymers 9 (1) (2017) 13.
[26] P.C.B.D. Faria, L.I.D. Santos, J.P. Coelho, H.B. Ribeiro, M.A. Pimenta, L.O. Ladeira, R.T. Gazzinelli, Oxidized multiwalled carbon nanotubes as antigen delivery system to promote superior CD8+ T cell response and protection against cancer, Nano Lett. 14 (9) (2014) 5458–5470.
[27] X. Liu, I. Marangon, G. Melinte, C. Wilhelm, C. Ménard-Moyon, B.P. Pichon, S. Begin-Colin, Design of covalently functionalized carbon nanotubes filled with metal oxide nanoparticles for imaging, therapy, and magnetic manipulation, ACS Nano 8 (11) (2014) 11290–11304.
[28] B.N. Eldridge, B.W. Bernish, C.D. Fahrenholtz, R. Singh, Photothermal therapy of glioblastoma multiforme using multiwalled carbon nanotubes optimized for diffusion in extracellular space, ACS Biomater Sci. Eng. 2 (6) (2016) 963–976.
[29] L. Zhang, H. Yuan, C. Inscoe, P. Chtcheprov, M. Hadsell, Y. Lee, O. Zhou, Nanotube X-ray for cancer therapy: a compact microbeam radiation therapy system for brain tumor treatment, Expert Rev. Anticancer Ther. 14 (12) (2014) 1411–1418.
[30] Z. Sobhani, M.A. Behnam, F. Emami, A. Dehghanian, I. Jamhiri, Photothermal therapy of melanoma tumor using multiwalled carbon nanotubes, Int. J. Nanomedicine 12 (2017) 4509.
[31] C. Fiorica, N. Mauro, G. Pitarresi, C. Scialabba, F.S. Palumbo, G. Giammona, Double-network-structured graphene oxide-containing nanogels as photothermal agents for the treatment of colorectal cancer, Biomacromolecules 18 (3) (2017) 1010–1018.

[32] D.Y. Zhang, Y. Zheng, C.P. Tan, J.H. Sun, W. Zhang, L.N. Ji, Z.W. Mao, Graphene oxide decorated with Ru (II)–polyethylene glycol complex for lysosome-targeted imaging and photodynamic/photothermal therapy, ACS Appl. Mater. Interfaces 9 (8) (2017) 6761–6771.

[33] L. Deng, Q. Li, S.A. Al-Rehili, H. Omar, A. Almalik, A. Alshamsan, N.M. Khashab, Hybrid iron oxide–graphene oxide–polysaccharides microcapsule: a micro-matryoshka for on-demand drug release and antitumor therapy in vivo, ACS Appl. Mater. Interfaces 8 (11) (2016) 6859–6868.

[34] Y. Yang, Y. Liu, C. Cheng, H. Shi, H. Yang, H. Yuan, C. Ni, Rational design of GO-modified Fe3O4/SiO2 nanoparticles with combined rhenium-188 and gambogic acid for magnetic target therapy, ACS Appl. Mater. Interfaces 9 (34) (2017) 28195–28208.

[35] Y. Luo, X. Cai, H. Li, Y. Lin, D. Du, Hyaluronic acid-modified multifunctional Q-graphene for targeted killing of drug-resistant lung cancer cells, ACS Appl. Mater. Interfaces 8 (6) (2016) 4048–4055.

[36] L. Han, Y.N. Hao, X. Wei, X.W. Chen, Y. Shu, J.H. Wang, Hollow copper sulfide nanosphere–doxorubicin/graphene oxide core–shell nanocomposite for photothermo-chemotherapy, ACS Biomater Sci. Eng. 3 (12) (2017) 3230–3235.

[37] J. Liu, K. Liu, L. Feng, Z. Liu, L. Xu, Comparison of nanomedicine-based chemotherapy, photodynamic therapy and photothermal therapy using reduced graphene oxide for the model system, Biomater. Sci. 5 (2) (2017) 331–340.

[38] N.R. Ko, M. Nafiujjaman, J.S. Lee, H.N. Lim, Y.K. Lee, I.K. Kwon, Graphene quantum dot-based theranostic agents for active targeting of breast cancer, RSC Adv. 7 (19) (2017) 11420–11427.

[39] Z. Fan, S. Zhou, C. Garcia, L. Fan, J. Zhou, pH-Responsive fluorescent graphene quantum dots for fluorescence-guided cancer surgery and diagnosis, Nanoscale 9 (15) (2017) 4928–4933.

[40] Y. Su, Y. Hu, Y.U. Wang, X. Xu, Y. Yuan, Y. Li, M. Li, A precision-guided MWNT mediated reawakening the sunk synergy in RAS for anti-angiogenesis lung cancer therapy, Biomaterials 139 (2017) 75–90.

[41] M. Thakur, M.K. Kumawat, R. Srivastava, Multifunctional graphene quantum dots for combined photothermal and photodynamic therapy coupled with cancer cell tracking applications, RSC Adv. 7 (9) (2017) 5251–5261.

[42] J. Ruan, Y. Wang, F. Li, R. Jia, G. Zhou, C. Shao, S. Ge, Graphene quantum dots for radiotherapy, ACS Appl. Mater. Interfaces 10 (17) (2018) 14342–14355.

[43] C. Santhosh, R. Nivetha, P. Kollu, V. Srivastava, M. Sillanpää, A.N. Grace, A. Bhatnagar, Removal of cationic and anionic heavy metals from water by 1D and 2D-carbon structures decorated with magnetic nanoparticles, Sci. Rep. 7 (1) (2017) 1–11.

CHAPTER 11

Metal oxide nanomaterials

Abbreviations

EMT effective mass theory
KW kilowatt
LIB lithium ion batteries
MRI magnetic resonance imaging
MSD magnetic storage devices
NO_x nitrogen oxide
PEEK polyether ether ketone
SO_x sulfur oxides

11.1 Introduction

The metal oxides are considered to be significant due to their highly captivating characteristics and single class material. Among the transition metal oxides are considered to be most valuable. Unlike the other nanomaterials, the oxide nanomaterials have enhanced lattice parameters. Appropriate synthesis procedures can be implemented on oxide compounds of metal elements; these metal elements have diverse oxide compounds because of their varying valences. The following equation explains the synthesis of metal oxide nanomaterials through the process of hydrolysis:

$$M(NO_3)_x + xH_2O = M(OH)_x + xHNO_3 \quad (11.1)$$

$$M(OH)_x = MO_{\frac{x}{2}} + x/2\,H_2O \quad (11.2)$$

The metal nanoparticles are considered to be the essential genre for the research because of their varied application in multiple productions and technological, industrial sectors. Their presence in multiple shapes and flexibility of different sizes increases their application. Fig. 11.1 provides evidence for the significance of metal oxides and their adequacy in research and development. Fig. 11.1 represents the number of studies on metal oxides during the last 15 years.

Synthetic Engineering Materials and Nanotechnology
https://doi.org/10.1016/B978-0-12-824001-4.00007-7

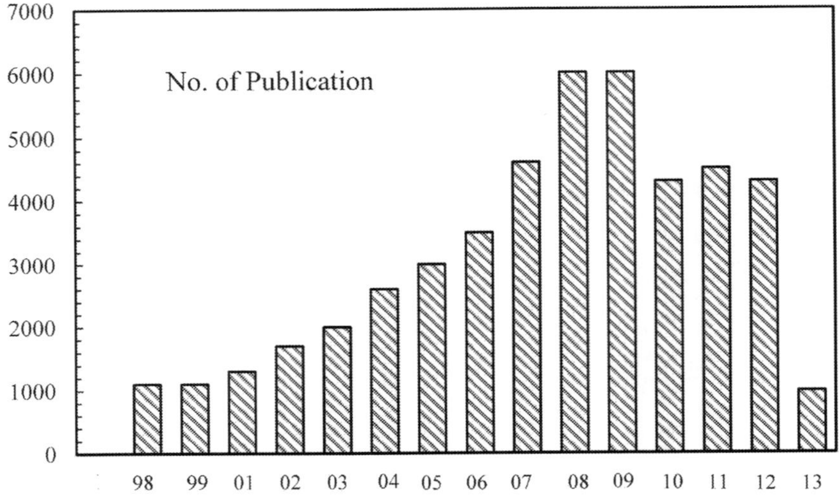

Fig. 11.1 Number of publications in 15 years [1].

The contrasting electronic structure of metal oxides enables them to reflect metallic, semiconducting, or insulating characters. In general conditions, the metal oxides formed from the metals present at the leftmost side of the Periodic Table are insulating in nature, including MgO, CaO, Al_2O_3, and SiO_2. Those forms from the metals placed in the middle of the periodic Table are semiconductor or metallic, for instance, ZnO, NiO, TiO_2, CuO, Fe_2O_3, and Cr_2O_3. Table 11.1 illustrates the different nature of metal oxide nanoparticles depending upon their parent metal.

The nanomaterials of metal oxides have diverse applications in the preparation of electronic items and circuits. Moreover, these are widely consumed in different chemical reactions as a catalyst. The chemical and petrochemical industry produces products worth millions by consuming metal oxide nanomaterials as raw material, which secures its significance. Along with this, the nanomaterials are utilized to remove environmental pollutants from land, air, and water. The oxides of CO, nitrogen oxide (NO_x), and sulfur oxide (SO_x), generally produced as a by-product of fossil fuel combustion, are eliminated from the environment using metal oxide nanomaterials. The computer and semiconductor industry significantly deepened upon metal oxides nanomaterials, as the majority of chips inside computation devices are designed by utilizing these materials.

Table 11.1 Nature of metal oxide nanomaterials [1].

Classification	Name	Position of metal in periodic table
Insulator	MgO	Group III
Insulator	CaO	Group IV
Semiconductor	CeO_2	Group VI
Semiconductor	TiO_2	Group V
Insulator	ZrO_2	Group IV
Semiconductor	VO_2	Group IV
Semiconductor	Cr_2O_3	Group VI
Semiconductor	WO_3	Group IV
Semiconductor	MnO	Group IV
Semiconductor	Mn_3O_4	Group IV
Semiconductor	MnO_2	Group IV
Semiconductor	FeO	Group IV
Semiconductor	Fe_3O_4	Group IV
Semiconductor	Fe_2O_3	Group IV
Semiconductor	RuO_2	Group V
Semiconductor	CO	Group IV
Semiconductor	Co_3O_4	Group IV
Semiconductor	NiO	Group IV
Semiconductor	CuO	Group IV

11.2 Properties of nanoparticles

In recent years, the research and application of metal oxide nanomaterials have extensively increased due to their adequacy to be practically implemented in different fields [2]. There are different electronic characters of metal oxide, that is, conductors, semiconductors, and insulators. The general properties of nanomaterials are unique and moldable according to a diverse range of fields. The materials' chemical and physical properties include higher density and confined size of corners and edges on the surface.

Hint statement
Metal oxides have diverse electrical properites, it can be conductor, semiconductor, or insulator.

11.2.1 General properties

The metal oxide nanomaterials are considered to be more potent for practical implementation in the field of technology. Therefore, it is considered an essential element of research in chemistry, biology, biomedical, agriculture, medicine, information technology, electronics, information technology, and the energy sector. The change in cell parameters causes structural changes in nanomaterials; as the nanoparticle's size decreases, the number of edges and corners increases in the cell. These seize variations generally include metal oxides of CuO, ZnO, SnO_2, Al_2O_3, MgO, ZrO_2 AgO, and CeO_2. The increasing size also enables atoms to create stress/strain on the adjoining structural perturbations [3]. The specified nanomaterials' specified size enables it to alter magnetically, chemical, and electronic conducting properties [4]. The chemical, physical, and electronic properties of nanomaterials depend on size and shape, and the properties vary according to the size of metal oxide [5]. The nanomaterials of iron oxide hold its significance as it is consumed in the production of Magnetic Resonance Imaging (MRI), Magnetic Storage Devices (MSD), and other contrasting agents [6]. The size dependency was observed in γ-Fe_2O_3 nanomaterials oxide. The ferromagnetic behavior was observed in 55 nm nanoparticle, whereas the 12 nm Nanoparticle reflects super magnetic behaviors [7]. The super magnetic features are affected by the size; the decrease in size also reduces the nanoparticle's overall magnetic anisotropy [8]. For the attainment of required properties, the altered size metal oxides are essential.

11.2.2 Optical properties

The optical conductivity is considered to be an essential property of metal oxides. This property can be obtained by experimenting with reflectivity and absorption measurements. Whereas the reflectivity procedure is affected by the nanomaterial size as individual oxides showed varied characteristics. The changes are dependent upon the range of photon wavelengths of the particles. The particles having lower photon wavelength absorption features are solid [9]. The quantum-seize confinement enables the absorption of light to be discrete and size-dependent. The contrast and transition between the quantized electronic and electron and whole discrete causes linear and non-linear properties in the nano-crystalline semiconductors. Based on the relationship among the nanoparticle radius (R) and the exaction Bohr radius, as explained in the Eq. (11.3):

$$RB = \in h^2/\mu e^2 \tag{11.3}$$

μ represents the reduction of mass and ε constant dielectric of semiconductors. The effect of quantum confinement is sorted into three major categories, which correlate to $R \gg RB$, $R \approx RB$, and $R \ll RB$, respectively

[10]. The size effect on the metal oxide nanomaterial's optical properties can be significantly elaborated using Effective Mass Theory (EMT). However, multiple EMA theory deficiencies can be catered by applying the free section collision model [11] or other theories based on length-strength correlation [12]. Eq. (11.2) reflects the direct bandgap, Fe_2O_3 [13], and the indirect bandgap, CdO_{67}. However, Cu^2O, CeO_2, ZnO, and TiO_2 [14, 15]. The limited number of limitations from the R-2 behavior reflects that the theory has overestimated the blue shift; it can be justified by electronic calculation using simple quantum mechanical methods. Whereas the marked deviations are not accounted for the physical and chemical phenomena. The q factor relaxation model is consumed to compare experimental data with the phonon confinement predicted by theory. According to the model, Raman intensity can be explained as mentioned in the below (Eq. 11.4);

$$I(\omega) \cong \int BZ \exp\left(-\frac{q^2L^2}{8}\right) \frac{d^3 q}{\left[\omega - \omega(q^{\rightarrow}] + \left[\frac{\Gamma_0}{2}\right]^2\right.} \tag{11.4}$$

The particle size distribution is represented by $p(L)$, q represents units of π, whereas the $\omega(q)$ is the phonon dispersion and Γ_0 intrinsic linewidth.

11.2.3 Mechanical properties

The low mechanical properties include yield and hardness, whereas high mechanical properties include superplasticity and observations related to temperature. Generally, the information regarding metal oxides nanomaterials is pivotal in ductility and superplasticity. The sintering with 600 k lower temperatures reflects significant improvement. For the conventional materials, the yield stress (σ), and harness (H) is catered by the Hall–Petch (H–P) equation:

$$\frac{\sigma}{H} = \frac{\sigma_0}{H_0} + kd^{-1/2} \tag{11.5}$$

Initial constants represent the friction stress and hardness, the grain size or primary article by d, and corresponding slope by K. The effect of the Hall–Petch equation on the nanoparticles are considered to be particle or grain boundaries functioning as an adequate difficulty for slip transfers and dislocations.

11.2.4 Magnetic properties

The magnetic properties of metal oxide nanomaterials are considered to be significant. This is because of their widespread applications in different fields. This may include the application in the field of fluids, water decontamination, and environmental remediation. The size dependency of nanoparticles is also implicated in magnetic properties; it is considered that nanoparticles ranging in between 10 and 20 nm are the ideal ones to reflect the metal oxides nanomaterial's magnetic property [16].

Hint statement
Metal oxide nanomaterials have enhanced metallic properites. Which increases their application.

Due to the lower scale, magnetic properties dominate other properties and can function effectively in the scenario. The magnetic property is dominant generally due to the uneven electron distribution in the particle. The synthesis methods and techniques are also significantly impactful on the magnetic properties. For instance, nanoparticles synthesize by micro-emulsion, thermal decomposition, and flame spray are more magnetic than other nanomaterials [17].

11.3 Methods of synthesis of metal oxide

11.3.1 Bottom-up

Atoms are the basic block and fundamental particles that gain the preparation method Bottom-up attain the interest of many chemistries and material sciences applications, as shown in Fig. 11.2. The method Bottom-up has two wings chemical and physical, which are the founding stone of science as it plays a significant part in the result of the quantitative method. In three states of matter, solid, liquid, and gas, chemical synthesis carriers are related to any three states of matter.

Hint statement
Bottom-up technique is segmented into physical and chemical methods and these methods are the core in material science.

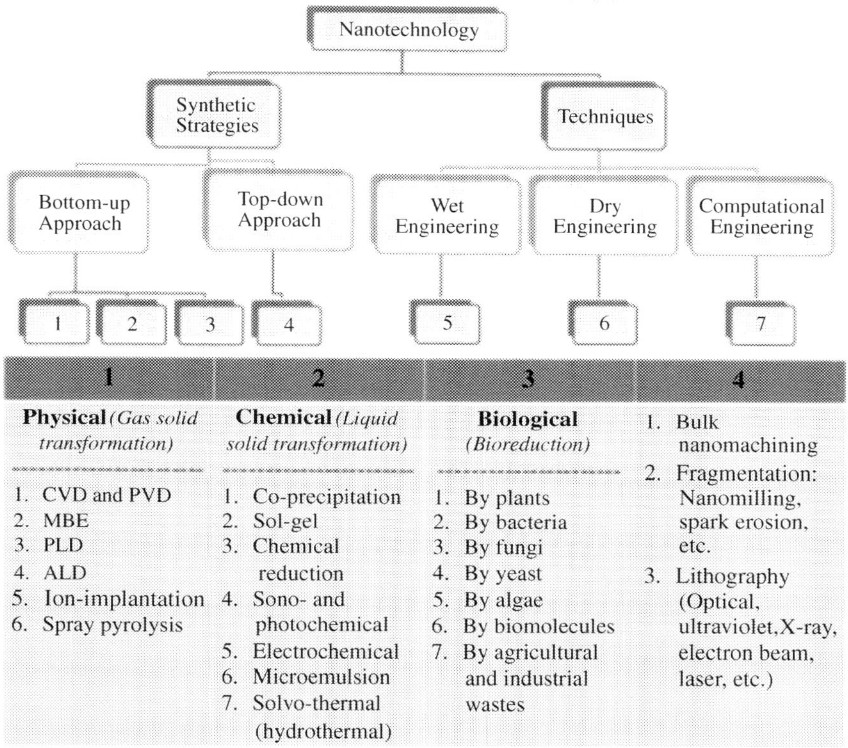

Fig. 11.2 Synthesis methods for bottom-up approach.

The reaction involved in a solid-state is the method of iterative using metallurgical technique and then sent to the diffusion of atoms through the high temperature of heat to form the reaction's product. The result of increased heat leads to abnormal growth, and the initial product is with large grain size until the response with growth limiter. Periodically it is difficult to obtain nanoscale systems through solid-state. The process of Synthesis. In the liquid and gaseous state, the diffusion rate is larger compared to solid-state systems; the process of synthesis can be implied at a low temperature and restricted the abnormal growth factor. The focus is only on the liquid state and gaseous state. The schematic branch of metal oxide nanomaterials synthesis is provided. The Mott-Schottky equation can determine the n-type characteristics of metal oxide nanomaterials that are synthesized by utilizing the bottom-up approach (Fig. 11.2);

$$\frac{1}{C_2} = 2/(\varepsilon\varepsilon_0 N_D)\left(E - E_{tb-K} T_{b/q}\right) \tag{11.6}$$

11.3.2 Hydrothermal/solvothermal approach

The obtained Nanoparticle is of high temperature due to the solvent's higher temperature present in the reaction. The higher temperature solvent can be harmful to the environment and consumes a greater amount of funds; therefore, the hydrothermal process is introduced. The usage of general solvents is reduced due to the lesser temperature requirement. The pressure inside the closed chamber increases as the solvent is heated. This enables the solvent to boil at a higher boiling point than normal. The stage in between liquid and gas is achieved, which is termed as the supercritical stage. With the increasing temperature, the solvent's viscosity and dielectric constant also increase. With the increase of pressure, the temperature became dominant.

The preparation of metal oxide nanomaterials from the hydrothermal process requires a sequential mechanism shown in Fig. 11.3. The reaction between the precipitant ion and solution method is performed in an autoclave. A dilutant is consumed for amalgamation of all the components involved in the reaction; for this reaction's conduction, an acidic or basic medium is required. For the morphology manipulation, and the additives are introduced in the reaction, and the mixture is heated at a high temperature ranging from 100 to 1000 °C.

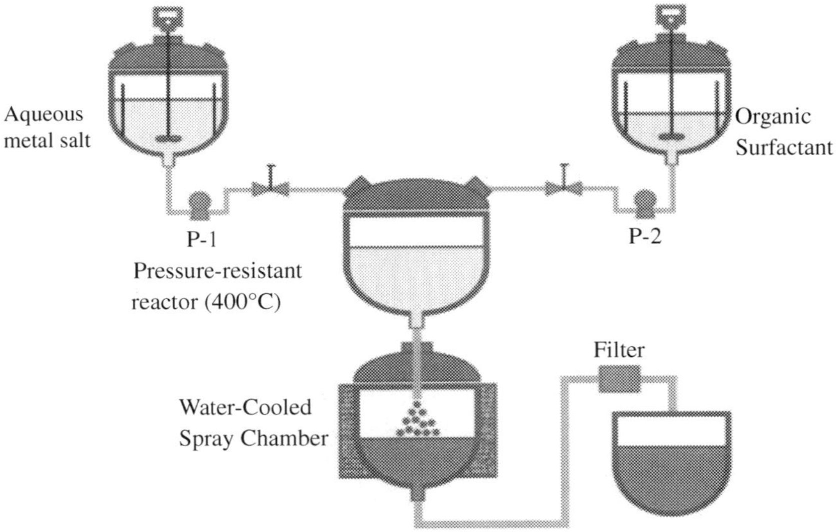

Fig. 11.3 Hydrothermal approach equipment.

11.3.3 Sol–gel approach

In common, this is the most conventional and standard method for the production of metal oxide nanoparticles. The derivatives of alkali metals that react with the alcoholic solution are hydrolyzed to form the corresponding hydroxide. The reaction in this process is called nucleophilic reaction, and the general form of the equation is given as

$$M(OR)_n + xH_2O \leftrightharpoons M(OR)_{n-x}(OH)_x + xROH \qquad (11.7)$$

The formation of the anhydrous cluster of metal hydroxide results from water loss and the solvent's removal. It's obtained gelation, and dense porous gel, the derivatives of hydroxide are linked in point to point topology of the network. The three-dimensional structure of bones is in the different kinds of gel, which are point to point connection of pores. The amorphous powder of hydroxide is obtained by drying the solvents and properly drying the required gel. Moreover, contacting the hydroxide with the heat results in the ultrafine powder of hydroxide. This process starts with a nanosized unit's involvement, which in chemical reaction undergoes at the nanometer scale. The product obtained initially is also the Nano realm.

11.4 Techniques for the synthesis of metal oxides

11.4.1 Induction thermal plasma

The technique utilizes in the synthesis of metal oxide nanoparticles can vary from material to material, as shown in Fig. 11.4. It has a wide range of techniques that can be implemented for the synthesis. The technique of thermal plasma induction is elaborated below. The technique's pictorial representation clearly illustrates the experimental setup to synthesize metal oxide nanoparticle fabrication using plasma. The apparatus utilized in the experiment comprises a plasma torch, the chamber area for the synthesis of nanoparticles, and the filter unit. The temperature of more than 10,000,000 is generated in the plasma torch utilizing induction heating at 4 MHz along with the controlled input power of 20 Kilowatt (KW). Noble gases are consumed to transport raw powder at 3 L/min and inner gas at 5 L/min. At 60 L/min, a mixture of noble gas with oxygen is consumed to form a plasma. A mix of Lithium and Cobalt, generally Li_2CO_3 with a diameter of 3.5 μm and 3–10 μm in mean diameter particles of metal oxides, is inserted in the powder feeder's carrier gases. A diverse range of metals is pre-tested for evaluating the adequate preparation of nanoparticles using a plasma induction

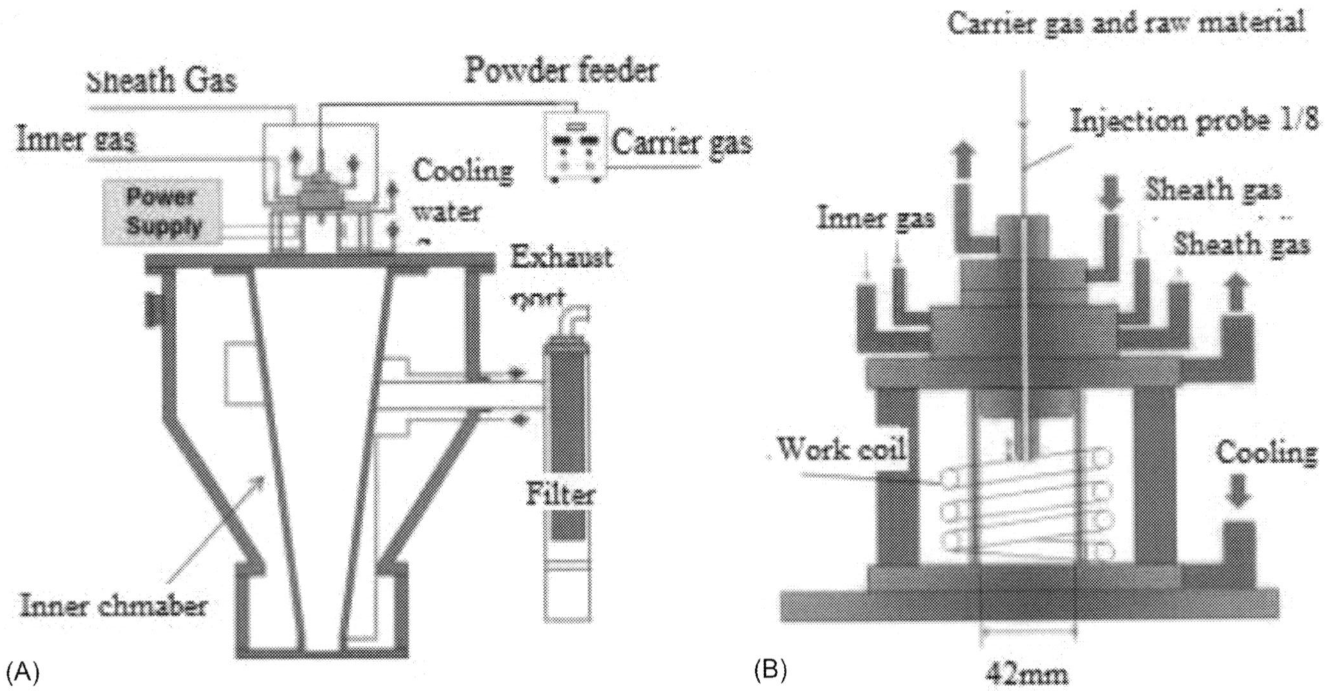

Fig. 11.4 (A) Induction thermal plasma equipment and (B) plasma torch [18].

Plasma Conditions				
Input power			20 kW	
Frequency			4 MHz	
Pressure			101.3 KPa	
Sheath gas			Ar: 57.5 L/min, O_2: 1.5 L/min	
Inner gas			Ar: 5 L/min	
Carrier gas			Ar: 3 L/min	
Discharge time			5 min	
Feed rate			400 mg/min	
Raw Materials				
System Raw	Li-Mn	Li-Cr	Li-CO	Li-Ni
powders	Li_2CO_3, MnO_2	Li_2CO_3, Cr	Li_2CO_3, Co	Li_2CO_3, Ni

Fig. 11.5 Conditions of the experiment.

technique. These metals include; Mn, Cr, Co, and Ni; the stoichiometric ratios of Li_2CO_3 with metal were 0.5, whereas the amorphous elements were feed at the rate of 400 mg/min. The experimental analysis is presented in the tabular form below. Fig. 11.5 shows the induction thermal plasma process; the yields and the mole fraction are corresponding.

Hint statement

Induction thermal technique is considered to be the most relevant technique in extraction of metal oxides.

11.4.2 Electrospinning

Electrospinning is considered the most convenient and widely consumed technique for producing metal oxide nanoparticles. The technique secures its significant position generally because of its ability to fabricate fibers in larger quantities [19, 20]. The consumption of the electrospinning technique can adequately produce nanofibrous mats and scaffolds. The electrospinning equipment functions so that; a narrow tube pump divides the solution by using spinneret into a high-voltage electric field produced in between the spinneret and collector grounded.

The materials consumed in the technique were; Poly Vinyl Pyrrolidone of 13,00,000 MW. The purified DCM was consumed in the technique as a

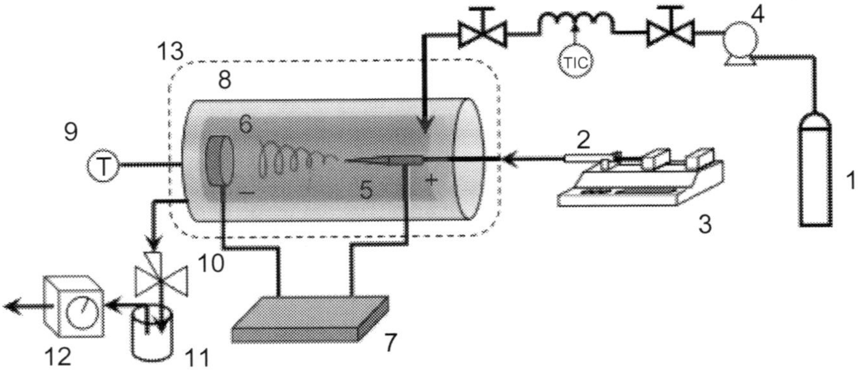

Fig. 11.6 Electrospinning apparatus: 1, Carbon dioxide cylinder; 2, syringe; 3 syringe pump; 4, Carbon dioxide pump; 5, nozzle; 6, collector; 7, high-voltage source; 8, spinning autoclave; 9, temperature sensor; 10, BPR; 11, solvent trap; 12, flow meter; 13, heater [21].

solvent along with TiO_2 nanoparticles. The polymer feed solution was DCM with a 4 wt% concertation with pure carbon dioxide.

The equipment consumed in the technique of electrospinning is illustrated in Fig. 11.6. The apparatus is mainly composed of dense Co_2, Non-conductive Polyether Ether Ketone (PEEK), high-voltage power, autoclave, high-pressure pump, and syringe. Along with the backpressure regulator and syringe prepared from stainless steel. The difference between the collector to the nozzle was 8 cm. In the technique, a PEEK vessel is heated up to a temperature of 30 °C. The PEEK capillary tube was consumed to introduce dense Co_2 with the pressure of 5 MPa into the PEEK vessel. The standard of the temperature and pressure are estimated based on previous researches conducted in this field. A manometer can be utilized for the maintenance of constant pressure in the equipment. By utilizing a high-pressure stainless steel syringe, the solution of feed is injected into the PEEK vessel. The solution is placed in the high-pressure syringe pump using PEEK capillary tube. The ideal flow rate of the feed solution is $0.05\,mL\,min^{-1}$. The electrostatic force is generated by the application of a high-voltage power supply. The estimated time for the production of nanomaterials using this technique is 15 min.

11.4.3 Solution combustion technique

In recent years, the metal combustion technique has received extensive attention because of its adequate control over the structure of reactive materials on length scales nanoparticles; it is considered the most potent long-term technique for synthesizing nanomaterials [22]. The technique is considered the convenient, simple, and diverse process in which different oxidizers and fuels are consumed and the self-sustained inhomogeneous solution. Multiple combustion modes are utilized for the reaction based on the type of precursors [23]. For the synthesis of metal oxide nanomaterials, the adequate precursor is considered metal nitrate, and for the fuel, glycine or citric acid is ideal. As the ratio of the solution was based on the principles and calculation instructed in the propellant chemistry. According to which, the metal nitrate to fuel ratio was at unity. The molar ratio in the combustion reaction is considered as an equivalent stoichiometric ratio. Therefore, it is considered that oxygen present in metal nitrate can be oxidized fully or consume the fuel entirely. Thus, the gases Co_2, H_2O, and N_2 are produced as the chemical reaction by-product, which exempts the need for oxygen as an external supply. The solution heated up until the temperature of 300°C when a fumes large amount produced in the reaction that indicated the reaction completion. An amorphous form is produced, which is later crushed and grounded. Fig. 11.8 represents the technique procedure. During the process, the normal temperature and pressure are maintained inside the furnace in the presence of inert gas; however, the external environment's temperature is 22–25 °C, as shown in Fig. 11.7. A diverse range of fuel combinations can be consumed to alter powder received as a product after the reaction.

11.5 Application of metal oxide nanomaterials

11.5.1 Catalysis

When involving catalysts in the reaction, metal nanoparticles are electropositive elements and produce attraction for catalysts because of large surface area to unit volume ratio compared to catalyst use in industry of very large amount [25, 26]. There is a certain limitation in the implying of nanoparticles as a catalyst in the large scale industry. The problems are determining nanoparticles for media of reaction, low recycling, and the capability to composite. To a rectify these problems, the movement of metal nanoparticles are not allowed, and they will need support to move [27]. One of the simplest methods is to use porous material with a proper pore identity; by doing this, the pore division by exterior and interior allows the particular molecule to contact nanoparticles [28]. The best thing about the material having pore is the protection and adjoining of Nanoparticle, therefore

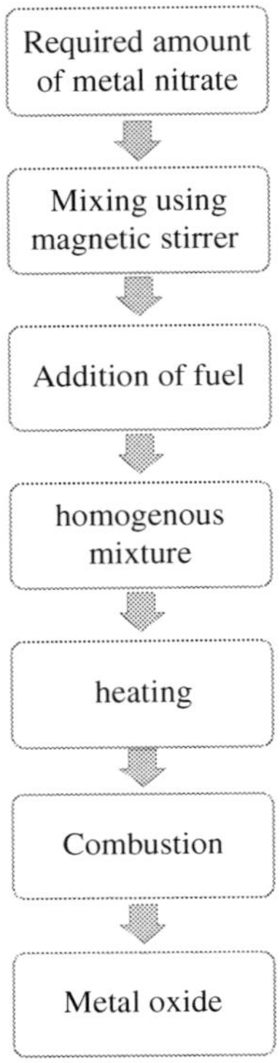

Fig. 11.7 Systematic representation of solution combustion technique.

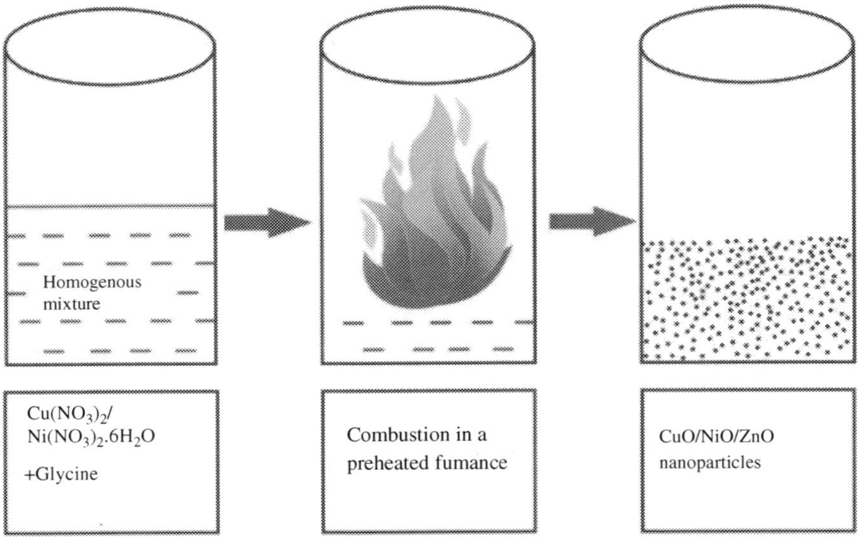

Fig. 11.8 Systematic synthesis of nanoparticles using the combustion technique [24].

Table 11.2 Reactions of metal oxides composite.

Metal oxide composite	Nanoparticles	Weight (%)	Size (nm)
ZIF-8	Pt	3.4	2.5 ± 4.1
ZIF-8	Au	5	4.2 ± 2.6
MIL-101-NH2(Al)	PTA/Pt	–	–
MOF-5	Pd	0.19	–

the rate of recovery from a large amount of solution and protection from the cluster of particles. In this topic, the supporting of metals of periodic group 1A for diversified catalysts by the medium of original porosity of metal oxides in contact with the probability to adapt their size of pore, texture, and structure [29]. In Table 11.2, the reaction is being evaluated in metal oxides that use metal nanoparticles as a catalyst, emphasizing the characteristic. The metal oxide and Nanoparticle, the nanoparticles' physical characteristics, the Nanoparticle's percentile weight in the metal oxides of the metal oxide and the Nanoparticle, the nanoparticle size, and the weight percentage of the Nanoparticle in the metal oxide as shown in Fig. 11.9. The experimental studies are based on oxidation, hydrogenation, C–C coupling, and H_2 production reactions.

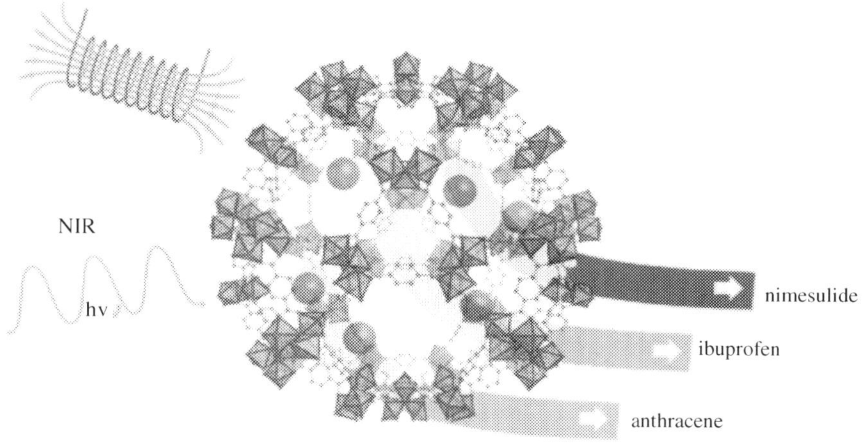

Fig. 11.9 Systematic reactions catalyze metal oxides elaboration [30].

11.5.2 Sensing

Variety of application use in our daily life such as in medical science like diagnostic of any disease. Chemical detection is for safety, quality control of edible stuff such as food and drinks, environmental protection, detection of microcells, and molecules by spending analysis [29]. The physical and chemical properties of typical bacteriophage change when interacting with analytic. The large range of metal nanoparticles is used to detect components as their physical properties as they are optical, electrochemical, or photoelectrochemical and modulate the 14 molecules when interacting with them [31, 32]. Nowadays for the support of chemical probes, as preconcentrators and the filter of molecules, the metal oxides are the main component; for the porous material, metal oxides react in a parallel way to other materials having pores, the element contains mesoporous silica or zeolites. There is a large possibility of metal oxides facilitating the pore size adaption and the chemical study of enhancing the adoption of particular molecules of elements [33].

11.5.3 Gas sensors

In metal oxide nanoparticle application, the gas actual and measured value close to the room temperature is reliable and prior to the application. Many advantages are being observed, like operation parameters to analyzed in polarization effect, modulation in wavelength and fluctuation of light intensity, etc. In particular, the implementation, selection, and easiness of the modulated signal is important in monitoring and exploiting the potential for betterment in the sensing performance used by the plasmatic resonance. For a long time, when semiconductors technology is involved in computer systems, the metal oxides semiconductors were largely used as a switch for sensing; one of the advantages of using metal oxides is the fast switching response and handle the diverse gas amount for a long time. Metal oxide semiconductor uses large power to perform switching for sensing and make limited selections. They were taking large surface area to volume ratio, with specific areas involved with high specific reactivity. Nano metal oxides are used to sense gas. [34, 35]. Fig. 11.10 shows the sensor of gas using the apparatus involving metal oxides contain the wire or layers of heating to suitable derived temperature for working, the conduction of electrodes for measuring the resistance in the wire, and the film of sensing, which change the value of resistance when exposing for sensing.

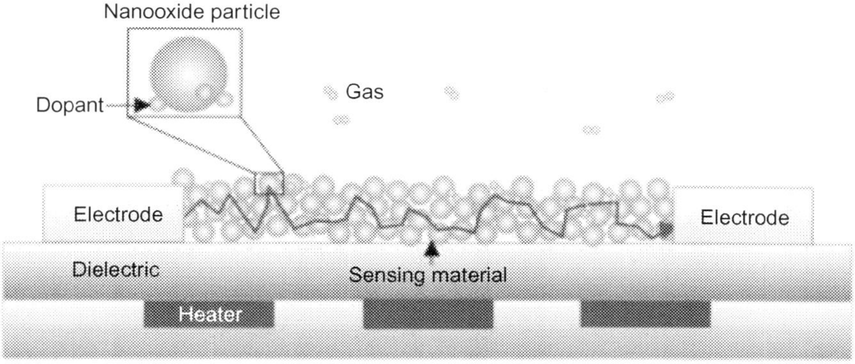

Fig. 11.10 Gas sensing device [36].

11.5.4 Batteries

In concern to the environmental issue, rechargeable batteries get significant importance for the last few years. In electrical storage and distribution, the lithium ion batteries (LIBs) play a major role in embedded cars, cycles, and electromechanical devices because of their large amount of energy and one large turn of the cycle. In the application of metal oxide nanoparticles, the performance of LIBs is increase by allowing prominence to make electrodes for batteries with the help of Nano metal oxides that are utilized in many years for making of LIBs electrodes many materials of TMOs is used. When it comes to the chemical operation TMOs, work on two principles: (a) intercalation/DE intercalation and (b) the 4 reaction displacements for anode material in batteries. Nano TMOs are used because they have a large capacity than the graphite used on a large industrial scale (372 mA hg^{-1}) [37]. Nowadays the only development and focus on the abstraction of many cationic oxides for the extensive amounts in individual rock-salt. For reversible electrochemical storage, these materials are considered significant. The random pseudo-binary systematic arrangement for Cu and Mg atoms was shown in Fig. 11.11.

11.6 Conclusions

Metal oxides nanomaterials are considered as the most relevant nanomaterials for their diverse range of applications in multiple fields. The unique characteristics possessed by these nanomaterials are their changing behaviors and properties. These metal oxides change properties with the change in their

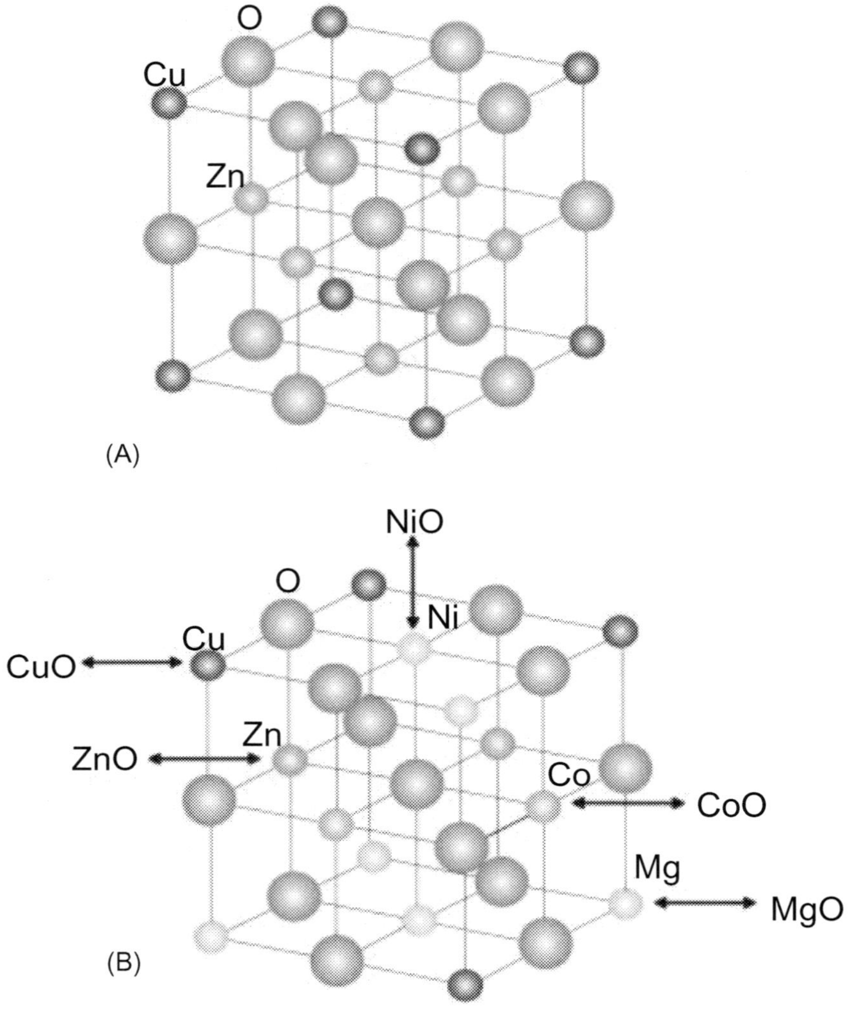

Fig. 11.11 The systematic arrangement of random pseudo-binary (A); Cu and Mg atoms on Cu sublattice (B) [37].

size and shape. The change in size affects the number of corners and edges in the nanomaterial causing it to change its properties. This feature widens the horizon of this application in the field of information technology and chemical synthesis. Metal oxides are consumed for the preparation of different chemicals, batteries, and other gas sensing devices. The method suitable for the synthesis of these nanomaterials is the bottom-up procedure. However, multiple techniques can also be applied to the extraction of nanomaterials.

References

[1] A.K. Srivastava (Ed.), Oxide Nanostructures: Growth, Microstructures, and Properties, CRC Press, 2014.

[2] R. Taylor, S. Coulombe, T. Otanicar, P. Phelan, A. Gunawan, W. Lv, H. Tyagi, Small particles, big impacts: a review of the diverse applications of nanofluids, J. Appl. Phys. 113 (1) (2013) 1, https://doi.org/10.1063/1.4754271.

[3] V. Bansal, P. Poddar, A. Ahmad, M. Sastry, Room-temperature biosynthesis of ferroelectric barium titanate nanoparticles, J. Am. Chem. Soc. 128 (36) (2006) 11958–11963, https://doi.org/10.1021/ja063011m.

[4] W.T. Liu, Nanoparticles and their biological and environmental applications, J. Biosci. Bioeng. 102 (1) (2006) 1–7, https://doi.org/10.1263/jbb.102.1.

[5] C.P. Poole Jr., F.J. Owens, Introduction to Nanotechnology, John Wiley & Sons, 2003.

[6] C. Lang, D. Schüler, D. Faivre, Synthesis of magnetite nanoparticles for bio-and nanotechnology: genetic engineering and biomimetics of bacterial magnetosomes, Macromol. Biosci. 7 (2) (2007) 144–151, https://doi.org/10.1002/mabi.200600235.

[7] Y.W. Jun, J.W. Seo, J. Cheon, Nanoscaling laws of magnetic nanoparticles and their applicabilities in biomedical sciences, Acc. Chem. Res. 41 (2) (2008) 179–189, https://doi.org/10.1021/ar700121f.

[8] M.T. Klem, D.A. Resnick, K. Gilmore, M. Young, Y.U. Idzerda, T. Douglas, Synthetic control over magnetic moment and exchange bias in all-oxide materials encapsulated within a spherical protein cage, J. Am. Chem. Soc. 129 (1) (2007) 197–201, https://doi.org/10.1021/ja0667561.

[9] B.J. Scott, G. Wirnsberger, G.D. Stocky, Chem. Mater. 13 (2001) 3140.

[10] I. Yoffre, Adv. Phys. 50 (2001) 1.

[11] Y.D. Glinka, S.H. Lin, L.P. Hwang, Y.T. Chen, N.H. Tolk, Size effect in self-trapped exciton photoluminescence from SiO 2-based nanoscale materials, Phys. Rev. B 64 (8) (2001), 085421.

[12] L.K. Pan, C.Q. Sun, Coordination imperfection enhanced electron-phonon interaction, J. Appl. Phys. 95 (7) (2004) 3819–3821.

[13] M. Iwamoto, T. Abe, Y. Tachibana, Control of bandgap of iron oxide through its encapsulation into SiO_2-based mesoporous materials, J. Mol. Catal. A Chem. 155 (1–2) (2000) 143–153.

[14] N. Serpone, D. Lawless, R. Khairutdinov, Size effects on the photophysical properties of colloidal anatase TiO_2 particles: size quantization versus direct transitions in this indirect semiconductor? J. Phys. Chem. 99 (45) (1995) 16646–16654.

[15] S. Monticone, R. Tufeu, A.V. Kanaev, E. Scolan, C. Sanchez, Quantum size effect in TiO_2 nanoparticles: does it exist? Appl. Surf. Sci. 162 (2000) 565–570.

[16] G. Reiss, A. Hütten, Applications beyond data storage, Nat. Mater. 4 (10) (2005) 725–726.

[17] W. Wu, Q. He, C. Jiang, Magnetic iron oxide nanoparticles: synthesis and surface functionalization strategies, Nanoscale Res. Lett. 3 (11) (2008) 397.

[18] M. Tanaka, T. Kageyama, H. Sone, S. Yoshida, D. Okamoto, T. Watanabe, Synthesis of Lithium metal oxide nanoparticles by induction thermal plasmas, Nano 6 (4) (2016) 60.

[19] S. Machmudah, H. Kanda, S. Okubayashi, M. Goto, Formation of PVP hollow fibers by electrospinning in one-step process at sub and supercritical CO_2, Chem. Eng. Process. Process Intensif. 77 (2014) 1–6.

[20] J. Quirós, K. Boltes, R. Rosal, Bioactive applications for electrospun fibers, Polym. Rev. 56 (4) (2016) 631–667.

[21] H. Ozawa, S. Machmudah, H. Kanda, M. Goto, Electrospinning of poly (vinyl pyrrolidone) fibers containing metal oxide nanoparticles under dense CO_2, Res. Chem. Intermed. 44 (4) (2018) 2215–2230.

[22] R.A. Yetter, G.A. Risha, S.F. Son, Metal particle combustion and nanotechnology, Proc. Combust. Inst. 32 (2) (2009) 1819–1838.
[23] S.T. Aruna, A.S. Mukasyan, Combustion synthesis and nanomaterials, Curr. Opin. Solid State Mater. Sci. 12 (3–4) (2008) 44–50.
[24] A.J. Christy, M. Umadevi, S. Sagadevan, Solution combustion synthesis of metal oxide nanoparticles for membrane technology, in: Metal Oxide Powder Technologies, Elsevier, 2020, pp. 333–349.
[25] R. Grosse, R. Burmeister, B. Boddenberg, A. Gedeon, J. Fraissard, Xenon-129 NMR of silver-exchanged X- and Y-type zeolites, J. Phys. Chem. 95 (6) (1991) 2443–2447.
[26] J. Liu, D.M. Strachan, P.K. Thallapally, Enhanced noble gas adsorption in Ag@ MOF-74Ni, Chem. Commun. 50 (4) (2014) 466–468.
[27] J.A. Kent (Ed.), Kent and Riegel's Handbook of Industrial Chemistry and Biotechnology, 2007, pp. 1378–1385.
[28] L. Lloyd, Hydrogenation catalysts, in: Handbook of Industrial Catalysts, Springer, Boston, MA, 2011, pp. 73–117.
[29] R.M. Rioux, H. Song, J.D. Hoefelmeyer, P. Yang, G.A. Somorjai, J. Phys. Chem. B 109 (2005) 2192.
[30] P. Falcaro, R. Ricco, A. Yazdi, I. Imaz, S. Furukawa, D. Maspoch, C.J. Doonan, Application of metal and metal oxide nanoparticles@ MOFs, Coord. Chem. Rev. 307 (2016) 237–254.
[31] Y. Huang, Z. Lin, R. Cao, Palladium nanoparticles encapsulated in a metal–organic framework as efficient heterogeneous catalysts for direct C2 arylation of indoles, Chem. Eur. J. 17 (45) (2011) 12706–12712.
[32] S. Joo, R.B. Brown, Chemical sensors with integrated electronics, Chem. Rev. 108 (2) (2008) 638–651.
[33] F. Xiao, J. Song, H. Gao, X. Zan, R. Xu, H. Duan, Coating graphene paper with 2D-assembly of electrocatalytic nanoparticles: a modular approach toward high-performance flexible electrodes, ACS Nano 6 (1) (2012) 100–110.
[34] M. Segev-Bar, H. Haick, Flexible sensors based on nanoparticles, ACS Nano 7 (10) (2013) 8366–8378.
[35] X. Liu, J. Zhang, S. Wu, D. Yang, P. Liu, H. Zhang, H. Zhao, Single crystal α-Fe_2O_3 with exposed {104} facets for high performance gas sensor applications, RSC Adv. 2 (15) (2012) 6178–6184.
[36] M.S. Chavali, M.P. Nikolova, Metal oxide nanoparticles and their applications in nanotechnology, SN Appl. Sci. 1 (6) (2019) 607.
[37] X. Liu, C. Chen, Y. Zhao, B. Jia, A review on the synthesis of manganese oxide nanomaterials and their applications on lithium-ion batteries, J. Nanomater. 2013 (2013), https://doi.org/10.1155/2013/736375.

CHAPTER 12

Composite nanomaterials

Abbreviations

Al_2O_3	alumina
CPCM	composite phase change material
FT-IR	Fourier transform infrared
HA	hydroxyapatite
LFA	laser flash apparatus
MgO	magnesium oxide
PCM	phase change material
SCM	scanning electron microscope
SDs	sodium dodecyl sulfate
SHS	self-propagating high-temperature synthesis
TB	titanium *n*-butoxide
TEOS	tetraethyl orthosilicate
TGA	thermogravimetric
TiO_2	titanium oxide
VSM	vibrating sample magnetometer
ZrO_2	zirconia

12.1 Introduction

The composite nanomaterials are the novel materials of the present generation. These materials are generally formed by amalgamating two or more different nanoscale materials to control and develop new and upgraded structures and features. The term nanocomposites contained diverse materials ranging from 3D metal matrix composites, 2D lamellar composites, and 1D to 0D nanowires shells, representing differentiations in layered and nano mixed materials. The properties of these nanomaterials are not only dependent upon the materials used for its preparation along with this; it also depends upon interface features and morphology.

Hint statement

Nanocomposites have materials which are of 1D, 2D, and 3D. This differentiates in between layered and nano mixed materials.

Synthetic Engineering Materials and Nanotechnology
https://doi.org/10.1016/B978-0-12-824001-4.00014-4

The nanocomposites have advanced features, such as fouling resistance, abrasion resistance, and electronic ion transportation. These features make the nanocomposite significant and essential for industrial applications contrary to other nanoparticles. The industrial products using nanoparticles will have lower price finished goods, as the nanoparticle is price effective for consuming as a raw material. It is estimated that the annual production of nanomaterials products will increase up to 5 million tons approximately. The nanoparticle s prepared from the poly families are considered more essential and have widespread usage. The factors of manufacturing procedures and morphology can be manipulated for obtaining diverse features nanoparticle [1].

The implication of composite nanomaterials is wide; it ranges from the motor vehicle industry to the electronic industry. The motor vehicle industry is currently the biggest consumer of these nanomaterials; however, it is estimated that the electronic industry is rapidly researching new technologies and innovative techniques to produce and implement nanofiber-based composites for practical usage electronic industry. The multifunctional composites are the composites prepared from an amalgamation of two or more base materials. These composites have drawn great attention to their synergy and upgraded properties contrary to their base materials [2]. In the multifunctional domain nanomaterials, the significant level of effort is pivotal on noble metal-based systems by restricting noble metal with organic or inorganic compounds to produce and synthesize the required nanomaterial with the desired multifunction. Furthermore, the procedure is under observation for stacking noble metals on metal oxides of semiconductors; for instance, TiO_2 with the high conducting nanomaterials functions as a buffer or surfactant. It is observed that due to the quantum confinement at the nanoscale, the chemical, biological and physical features of nanomaterials differ from their base materials from which they are prepared. Though these properties are consistent and fundamental for the implication of these nanomaterials, these properties are unpredictable and often contrast from the general properties of base materials.

By the utilization of traditional methods, the synthesis of nanoparticles from solid parent material is difficult. This is because the solid material will have a lesser surface area and stronger atomic bonding that restricts reaction initiation. Each synthesis technique has limitations in the preparation of nanomaterial depended upon the state of the parent material. The application of these synthesis techniques enables nanomaterials' production, which is homogenous and narrow particle size distribution. Table 12.1 represents adequate procedures for the synthesis of nanomaterials possessing multifunction.

Table 12.1 Application of different composite nanomaterials [3].

Type of phase	Mode of application	Applications	Properties
NP (single phase)	Combination of metal oxide and polymer matrix	Electronics, aerospace, and surface wetting	Thermally stable with a larger surface area
NF (single phase)	Metal oxides	Effective energy, high flux, ion-transportation	Higher-strength and stiffness
NF (two-phase)	Combination of metal oxide and polymer matrix	Thermal and electric stability with multifunctionality	Coatings, thermal interfaces
NP (two-phase)	Metal oxides	Resistant	Textiles, specie identification

12.2 Methods of producing composite nanomaterials

12.2.1 Biological method

Multiple biological methods can be utilized to synthesize composite Nanomaterials; this chapter will specifically explain nanomaterial's biological synthesis method using the synthesis of TiO_2 and $NiFe_2O_4$. The synthesis of TiO_2 and $NiFe_2O_4$ cores is segmented into two portions; in the first step of the procedure, the monocrystalline nickel ferrites ($NiFe_2O_4$) by utilizing the micelle methods [4, 5]. By usage of the first step, two microemulsion systems are prepared. The first microemulsion constitutes of an oil-phase microemulsion consisting of surfactant diso-octylsulphoccinate (AOT) and isooctane. Whereas, the second microemulsion consisting of aqueous phase emulsion constitutes of similar materials of isooctane and AOT along with reactant salts that are hydrated nickel chloride and hydrate iron chloride.

Hint statement

Oil-base microemulsions are consisting of surfactants such as; diso-octylsulphoccinate and isooctane.

The chemicals proportion in the first microemulsion system was; 2.4 mL H_2O + 30% concentrated NH_4OH + 66 mL of 0.50 M AOT-isooctane 10 min sonicated. On the contrary, the second microemulsion system constitutes of; 66 mL of AOT-iso_octane sonicated for 10 min + 0.275 g of $FeCl_2$ Æ$_4H_2O$ + 0.164 g $NiCl_2$ Æ$_6H_2O$ dissolved in 8 mL of H_2O. THE

NH_4OH was the precipitating agent for the first microemulsion system. The microemulsion systems were exposed to 75 min of frequent mechanical stirring. The precipitation of metal hydroxides was carried in the reverse micelles water pool and was oxidize to ferrite. Whereas, the $NiFe_2O_4$ was precipitated as per the following reaction:

$$NiCl_2 \cdot 6H_2O + 2FeCl_2 \cdot 4H_2O + 6NH_4OH + \frac{1}{2}O_2 \longrightarrow NiFe_2O_4 \downarrow + 6NH_4Cl + 17H_2O \quad (12.1)$$

The monocrystalline of nickel ferrite was formed under the microreactor. The acidic titanium salt solution in an aqueous form was introduced in the nickel ferrite microemulsion product, associating the $TiCl_4$ hydrolysis. As explained in the following equation:

$$TiCl_4 + H_2O \rightarrow TiOCl_2 + 2HCl \quad (12.2)$$

The coating of nickel ferrite nanoparticles with anatase polymorphs of titania was also performed. The synthesis of multiple polymers of titania depends on the amount of HCL acid used and the titanium salt concentration. When the reaction of precipitation is carried out in a highly acidic medium, the brookite phase is formed [6]. However, the kinetics of precipitations is the objective of the medium acidity and the titanium concertation. In the case of low concentration of titanium and acidic concentration, the kinetics of anatase TiO_2 is significantly rapid, whereas, contrary to the anatase phase, the precipitation is significantly lower in the brookite phase in this condition shown in Fig. 12.1. According to Palmero [7], if the concentration of titanium is 0.15 mol/dm^3, the primary phase of brokite, which is 70%–80%, will achieve the acidic level of 2–6.

Hint statement
The HCL acid utilized in titanium salt concentration, synthesis the polymer of titania.

12.2.2 Chemical methods

The chemical synthesis methodology was applied to the synthesis of tetraethoxysilane (TEOS), and titanium *n*-butoxide (TB) was utilized as the source of titanium and silica, respectively. The steps involved in the procedure of titania–silica nanoparticles are mentioned pictorially represented in Fig. 12.2.

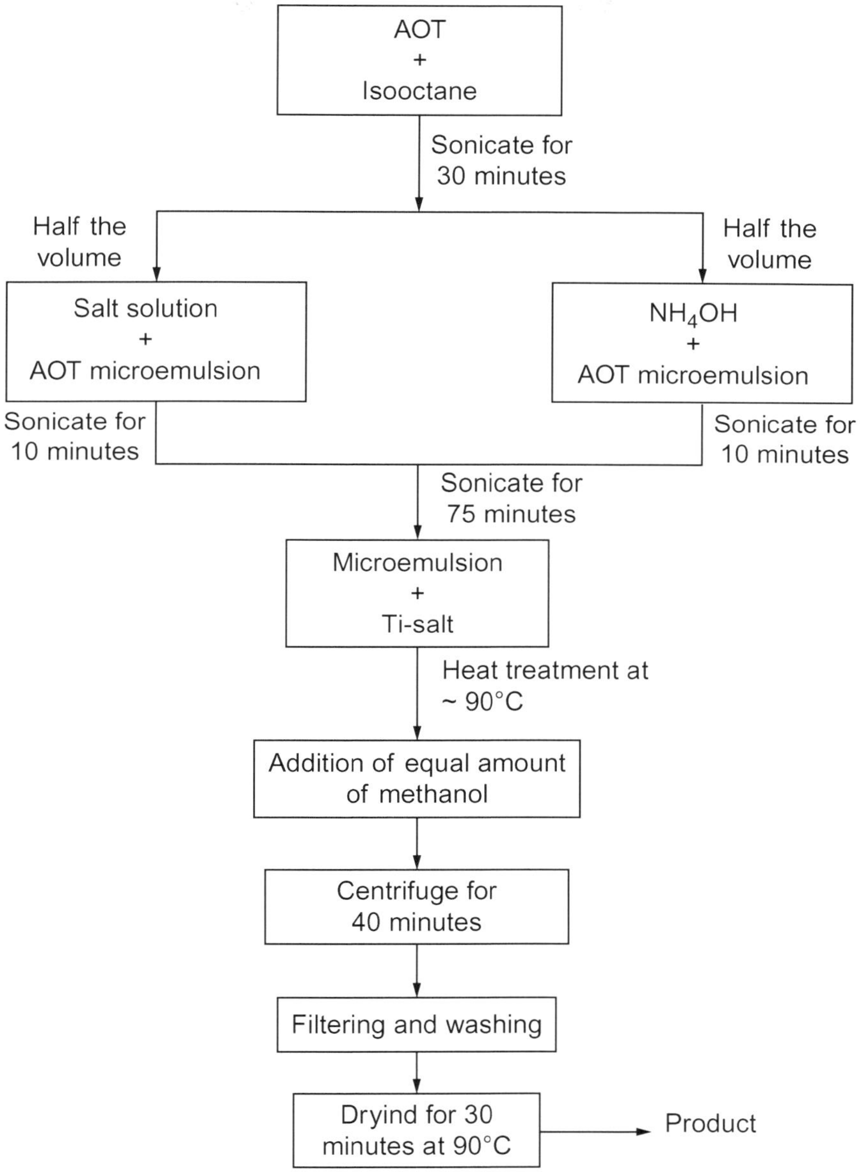

Fig. 12.1 Systematic procedure for the synthesis of TiO_2.

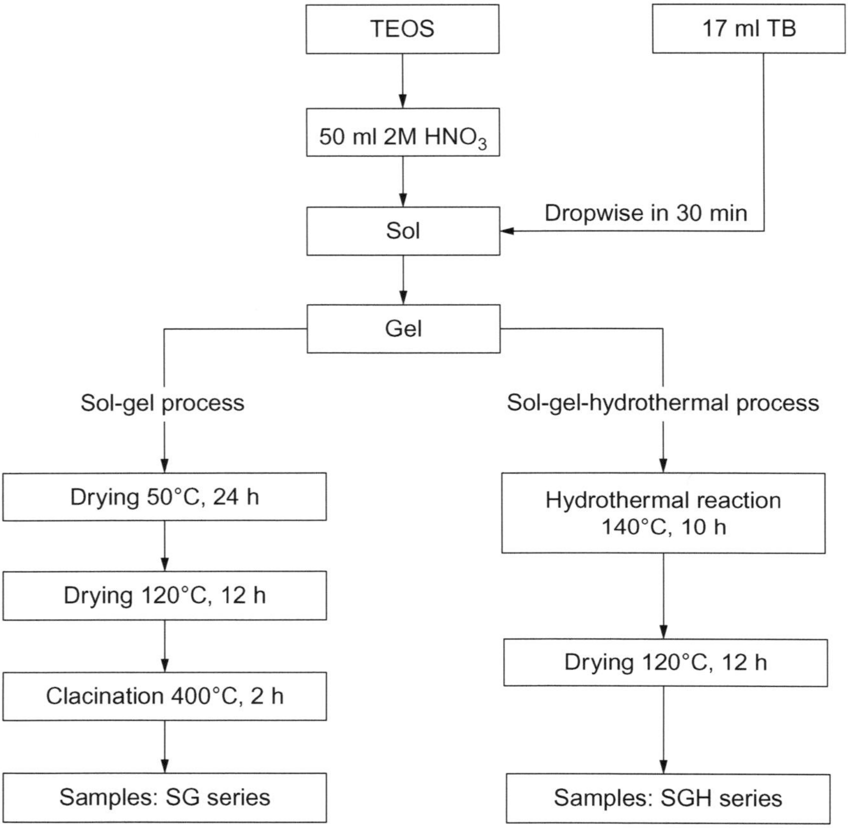

Fig. 12.2 The synthesis procedure of titania nanoparticles.

The solution was prepared in the first step by using TEOS + 55 mL of $2MHNO_3$ solution at 50 °C along with TB added dropwise approximately 17 mL under 0.5 h magnetic stirring. For the gel formation, the solution was stirred continuously for a period of 1 h. The two different processes involved the treatment of prepared gel, the Sol–gel process, and the Sol–gel-hydrothermal process. In the sol–gel process, the gel was treated in the oven for 24 h at the temperature of 50 °C, and then for drying, it was heated at the temperature of 120 °C. In the final step, the synthesized amorphous form was reduced for 2 h in the air at 400 °C. The final product obtained is the titania-silica composite nanomaterial and is signified as SG-series. In the second process of the sol–gel–hydrothermal process, the gel is matured for 0.5 h at 20 °C. Then in the second step, the gel is processed through a hydrothermal reaction in the Teflon-lined autoclave for the time period of 10 h at the temperature of 140 °C; after this, the obtained hydrothermal product is dried at the temperature of 120 °C, and the titania-silica composites can be obtained after the process of drying. The obtained product is denoted as the SGH-series. According to Eq. (12.3):

$$SiO_2\% = \frac{W_{SiO_2}}{W_{SiO_2} + W_{TiO_2}} = 0 \tag{12.3}$$

where W stands for the weight of TiO_2 or SiO_2 in the elements, these elements were thereof, construct as SG, SG0, SG1–SG7, and continues for the SG series. The samples were for further synthesis using different temperatures for heating 400, 600, 800, and 1000 °C for a period of 2 h in the air. The X-ray diffraction resolute the crystalline phase and phase of the transformation of titania particles where; Cu-Kα, 100, Mega Ampere, XRD, 40 kV, and D/max = 2500. The initial fraction of the rutile phase was calculated using the following Eq. (12.4):

$$Rutile\% = 1/\left[1 + 0.8\left(\frac{I_A}{I_R}\right)\right] \tag{12.4}$$

where I_A and I_R are integration intensity of X-ray (101), a reflection of rutile, and reflection of anatase (110). The Scherrer formulae were utilized for calculating the crystalline gram size.

$$L = K\lambda/(\beta \cos\theta) \tag{12.5}$$

In this equation, λ represents the wavelength of the X-ray radiation that is; Cu-Kα = 0.15406 nm. Whereas K is a constant considered as 0.89, which is the half-maximum height of the line width, and the diffracting angle is represented by θ.

Example 12.1

Calculate the gram size using Scherrer formulae, where

$$B = 0.5° \lambda = 0.154\,\text{nm}\, 2\theta = 27° \theta = 13.5° \text{Cos}\,(13.5) = 0.972$$

Solution:

$$360° = 2 \times 3.142 0.5 = 0.5 \times 2 \times 3.142/(360) = 0.00873 t = 16.3\,\text{nm}$$

Example 12.2

In case of $t = 2\,\text{mm}$ calculate the reflection by using the similar data from 12.1.

Solution:

$$\begin{aligned} B &= 0.9\lambda/(t \cos\theta) = 1.386/2 \times 10^{-3} \times 0.972 \\ &= 7.12 \times 10^{-4}\,\text{radians}7.12 \times 10^{-4}\,\text{rads} = 7.12 \times 10^{-4} \times 360/(6.284) \\ &= 0.047^{\circ} \end{aligned}$$

12.2.3 Combustion method

Self-Propagating High-Temperature Synthesis (SHS) includes the powder mixtures that require ignition as the activation energy to initiate an exothermic chemical reaction. This reaction produces temperature, which ranges from 1000 to 3000 °C; adiabatic conditions are compulsory for the reaction's initiation. By the usage of this technique nanomaterials, composite is extracted in the amorphous form. Furthermore, by the usage of this technique single and multidimensional nanomaterials can be synthesized. The technique is widely consumed in the extraction of different types and compositions of nanocomposites, as presented in Table 12.2.

Table 12.2 Synthesis of composite nanoparticles and their composition [8].

Synthesis category	Composition	Type of composition
Mechanochemical	$MgO/MgTiO_3$/HA ZrO_2; Al_2O_3/TiB_2 Sic/Nbc	Oxide/oxide Oxide/non-oxide Non-oxide/non-oxide
Vapor phase	SiO_2/ZrO_2; V_2O_5/TiO_2 Sic/ZrC; Sic/Si_3N_4	Oxide/oxide Non-oxide/non-oxide
SHS	Mullite/TiB_2; SiC/Al_2O_3 TiN/Si_3N_4; $MoSi_2$/ Si_3N_4 Si_3N_4-TiN-SiC; ZrC-SiC-ZrSi-ZrB_2	Oxide/oxide Oxide/non-oxide Non-oxide/non-oxide
Sol–gel	Mullite/TiO_2; ZrO_2/Mullite Mullite/SiCAIN/BN; SiC/Al_2O_3 Mullite/SiCAIN/BN	Oxide/oxide Oxide/non-oxide Non-oxide/non-oxide

The chemical reaction that occurs during the synthesis is mentioned in Eq. (12.6):

$$wTi\,(s) + xB\,(s) + yC\,(s) + zCu\,(s) \rightarrow TiB_2(s) + mTiC\,(s) + zCu\,(s) \tag{12.6}$$

By using the solution of Si and Ti, the nanomaterial composite of TiN/ Si_3N_4 can be produced from the combustion procedure [9]. The components are rapidly deformed in a while the reaction due to nitrides' property to produce higher exothermic energy released in the form of heat; thus, deforming the elements involved in the reaction. The nitrogen infiltration is hindered at the combustion forefront by the external medium; this leads to an incomplete chemical reaction. For the prevention of this hindrance while the chemical reaction, the inert diluent generally of TiN of Si_3N_4 is initiated in the initial solution.

Another prevention technique that can be implemented in the procedure is the usage of compounds for the initial reaction instead of individual elements. For instance, the mixture of $TiSi_2$–SiC is consumed to synthesize TiN-SiC-Si_3N_4 under the 130 MPa pressure of nitrogen [10]. For this process, multiple raw materials form the titanium silicide category have been tested [11]. The relative experiments and thermodynamics studies have reflected that, for this purpose, at the significantly lower nitrogen pressure, almost 5 Mpa, Ti_5Si_3 is the most suitable raw material. In the study of [12], bi, tri, and tetra phasic amorphous forms are synthesized, including ZrC/ ZrB_2/SiC, SiC/ZrC, and ZrC/ZrSi/ZrB_2. These amorphous elements were obtained by utilizing the combustion method consuming solution of NaCl, $ZrSiO_4$, Mg, B, and C as a raw material. The Mg was used for the reduction of $ZrSiO_4$, and NaCl functions as a diluent that controls the size of the particle and composition of amorphous nanomaterials. It was concluded that with the increase in NaCl, the particle combustion temperature and size decreases [12].

12.2.4 Mechanochemical synthesis

The mechanochemical synthesis of composite nanomaterial is performed in a controlled environment and utilizes high energy milling techniques. It is used to synthesize composites of complex compositions, and intermetallic compounds reinforced the process [13]. The composite nanomaterials are synthesized and consumed to prepare non-oxide, oxide, and mixed materials [14]. It is suggested by the fracture mechanism theory that the least possibility for the synthesis of the smallest size particle ranges from 5 to 100 nm [15]. This secures this technique's significant position, as larger size particles contain smaller size grains; however, it is almost impossible to synthesize this smaller size range particle from the milling methods. The localize damage also initiated the localize phase transformations [16]. The procedures also have its drawbacks as it fails to produce the finest grain nanomaterials and is also potent to the mixture with the elements pre-existing in grinding equipment in repeated or frequent usage of the apparatus.

12.3 Techniques

12.3.1 Microwave induced technique

The samples were induced in the quartz tube and were treated by microwave plasma. For the activation energy, a spark is required; for this purpose, the electronically conductive materials must be placed inside the reaction vessel. In the case of the non-conductive samples, a small conductive piece of metal, approximately 2.2 cm, shall be placed in the quartz tube for initiating the reaction along with at least 2 g of the required sample. In the procedure exemplified in this chapter involves microwave induction synthesis. The modest vacuumed conditions were subjected to the tube by using a vacuum pump, which was sealed, and the standard apparatus for the microwave induction that is household microwave was utilized with the power supply of 700 W and 2.45 GHz. The period varied from sample to sample; it generally ranges in between 30 and 180 s. An optical pyrometer was used in the technique for the measurement of enhanced temperature and time of heating inside the oven.

As pictorially presented in Fig. 12.3, the temperature plot expands inside the oven as heating time increases; it was measured on the sample's surface where nanoparticles are present by using a radiation pyrometer. For the generation of significant heat in the oven, microwave radiation can be utilized. Maghemite is considered an adequate element for this, as it reflects the significantly higher abortion of microwaves [17]. In the magnetic and electric field, the compressed magnetic amorphous form can reach up to the temperature of 600 °C within a time duration of 60 s and generally remains constant for a longer period at the temperature of 650 °C. The temperature inside the quartz tube can be raised up to 800 °C within 60 s, and in 180 s, it reaches up to 900 °C. The plasma present in the vacuum quartz tube is initiated by microwave radiations emitted from the oven. The plasma can be identified as bright and shiny light because of the remaining gas in the tube. After this step, the tube was cooled at room temperature. The sample

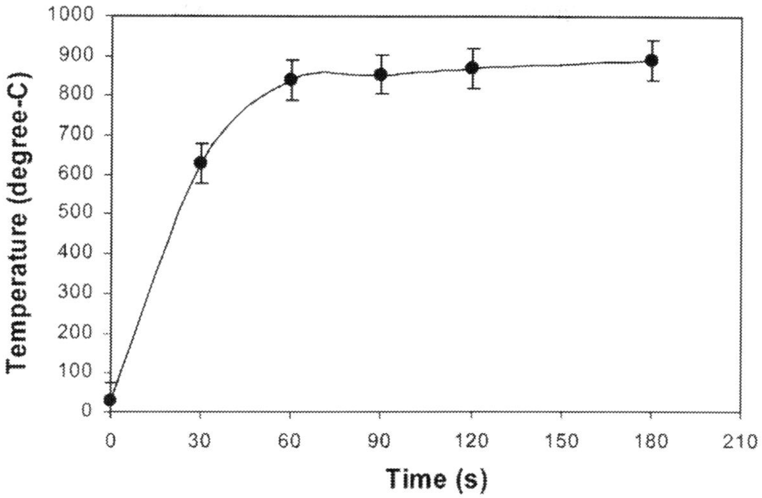

Fig. 12.3 The shift in temperature on the functioning of microwave plasma heating.

can be collected from the tube by utilizing ethanol; the collected material is then sonicated for 10 min. This enables all the particles to distribute and dissolve in the prior solvent synthesis. For observing the size and morphology of obtained nanoparticle, the usage of a transmission electron microscope with a power supply of 200KV is adequate. Furthermore, the diffractometer can be consumed for observing the X-ray diffraction of nanoparticle using radiation of Cu-Kα. With the scan timing of 2 s/step, step size to be 0.02, obtained composite nanoparticles can be scanned for $2\theta = 10°–80°$. The thermos scientific spectrometer can be utilized for identifying Fourier transform infrared (FT-IR) using a KBr sample with a concentration of less than 5%. The sample can also be processed with the Thermogravimetric (TGA) analysis at the heating rate of 100–700 °C. At room temperature with the 15 kOe, the Vibrating Sample Magnetometer (VSM) can be consumed to identify the composite nanomaterial synthesized magnetic properties.

12.3.2 Solution evaporation technique

The materials consumed in the solution evaporation technique use the following materials: to prepare eutectic salts of binary carbonates, 97% pure concentrated anhydrous lithium, and 99% pure concentrated anhydrous potassium. The salt is the ratio of 38 mol% K_2CO_3: 62, mol. % Li_2CO_3, this is a collection of original Phase Change Material (PCM). The tabular representation of the thermophysical properties of eutectic salts is presented. The melting point statistics, temperature, and enthalpy were adapted from

[18, 19]. For the preparation of real microstructures and geometric parameters of four different kinds, multiple additives are used: graphene, Single and Multi-Walled Carbon Nanotube, and C60 prepare Composite Phase Change Material (CPCM), these additives are amalgamated with PCM.

The four nanomaterials Structure and geometric parameters are shown in Table 12.3. The general nanomaterial characteristics and structural properties are studied using a Scanning Electron Microscope (SCM), as shown in Fig. 12.4. For creating the nanomaterials dispersivity in the CPCM, 99% concentrated Sodium Dodecyl Sulfate (SDs) can be used as a surface-active agent.

The nanomaterials of CPCM were prepared using the evaporation technique. The step-wise procedure for the preparation of nanomaterial is mentioned in this paragraph. In the first step, the moisture is removed from the chemicals; for this, using a vacuum oven at the temperature of 150 °C, Li_2CO_3 and K_2CO_3 are heated for 3 h. Fig. 12.5 shows and prepares the eutectic salts of binary compounds, 267.05 mf K_2CO_3, and Li_2CO_3 was taken 232.95 mg. In the second step of the procedure, the nanomaterials are weigh in a fixed ratio ranging from 0.1% to 2.5% of the eutectic salts. Then in the 1:1 mass ratio to nanomaterial of SDS is weigh to prepare the uniform nanomaterial solution; a sample prepared from the mixture of nanomaterials and SDS along with deionized water is vibrated ultrasonically for 3 h. The CPCM particles in the solid form can be gained after evaporating the solution. The removal of water moisture and bicarbonates from the synthesized sample, Muffle furnace is used, the particles are heated, and the moisture is removed from the sample. In the last step, the sample is triturated, and composite nanomaterial is obtained. By repeating the same procedure, 20 different kinds of composite nanomaterials can be obtained with different proportions of 0.1%, 0.5%. 1.0%, 1.5%, and 2.5%.

Table 12.3 Structure and geometric parameters for the four nanomaterials [20].

Nanomaterials	MWCNT	SWCNT	Graphene	C_{60}
Structure/shape	Short cylinder	Long cylinder	Square sheet	Spherical
Diameter/thickness	10–50 nm	5–20 nm	10–20 nm	500 nm to 2 μm
Length/width	500 nm to 1 μm	1–5 μm	1–5 μm	–
Specific surface area	647.62 m^2/g	1781.48 m^2/g	2013.33 m^2/g	28.24 m^2/g

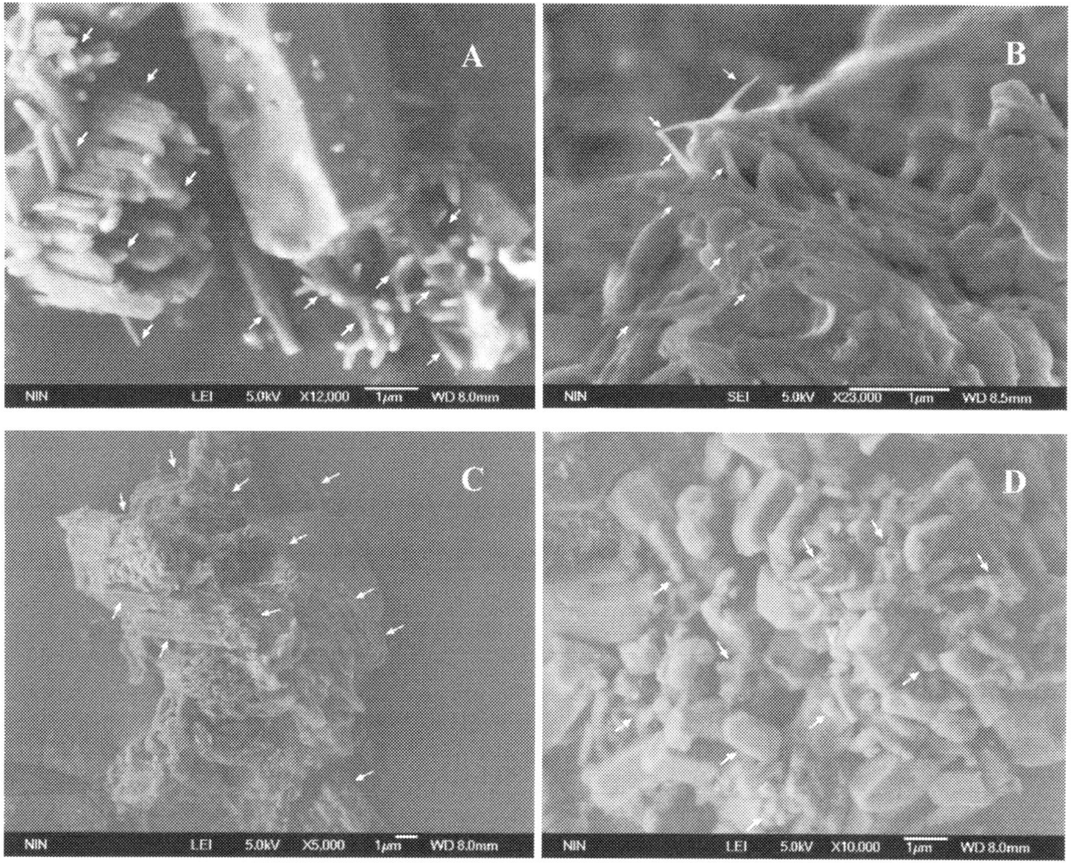

Fig. 12.4 Microstructures of nanomaterials: (A) MWCNT; (B) SWCNT; (C) graphene; and (D) C_{60} [20].

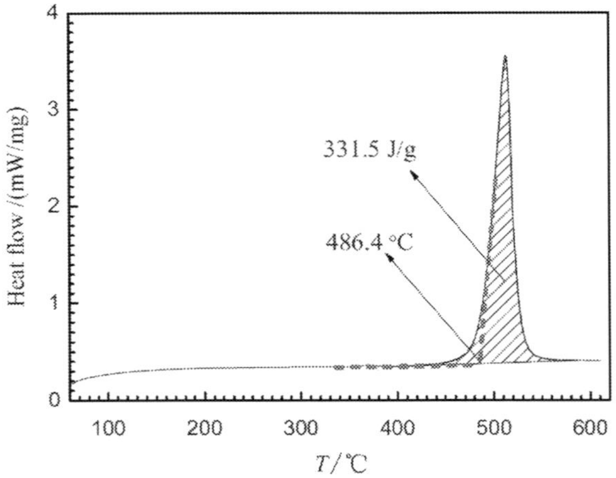

Fig. 12.5 Thermograms of binary carbonate eutectic salts [20].

The thermal properties of the extracted sample were observed using (DSC); the accuracy of the DSC was 3%. The nanomaterial's thermal conductivity was identified using the Laser Flash Apparatus (LFA). The nanomaterials thermal conductivity can be obtained by utilizing the following equation:

$$k = \alpha \times \rho \times C_\rho \tag{12.7}$$

In this equation, thermal conductivity is represented by K; whereas, thermal diffusivity by $\mathrm{Wm}^{-1}\,\mathrm{K}^{-1}$; $\acute{\alpha}$ and density by $\mathrm{m}^2\,\mathrm{s}^{-1}$; ρ. The LFA has 3% of the rated thermal diffusivity testing accuracy. Therefore, the estimated accuracy for density is to be 3% [19]. The uncertainty in thermal conductivity be obtained by using error transfer formulae:

$$E_k = \sqrt{E_\alpha^2} + E_\rho^2 + E_{C_\rho}^2 = \sqrt{0.03^2} + 0.03^2 + 0.03^2 = 5.2\% \tag{12.8}$$

The further validation of the obtained results, the values of melting point, specific heat, and melting enthalpy were verified with the original values extracted from the published papers. The present obtained values of melting enthalpy and melting point are 331.5 L gL and 486.4 °C, respectively. It's pictorially represented in Fig. 12.7, and these values are complying with the published values: 488 °C and $342\,\mathrm{J\,g}^{-1}$ [18].

Fig. 12.6 represents the contrasting values in between the obtained values of nanomaterial and published values initially.

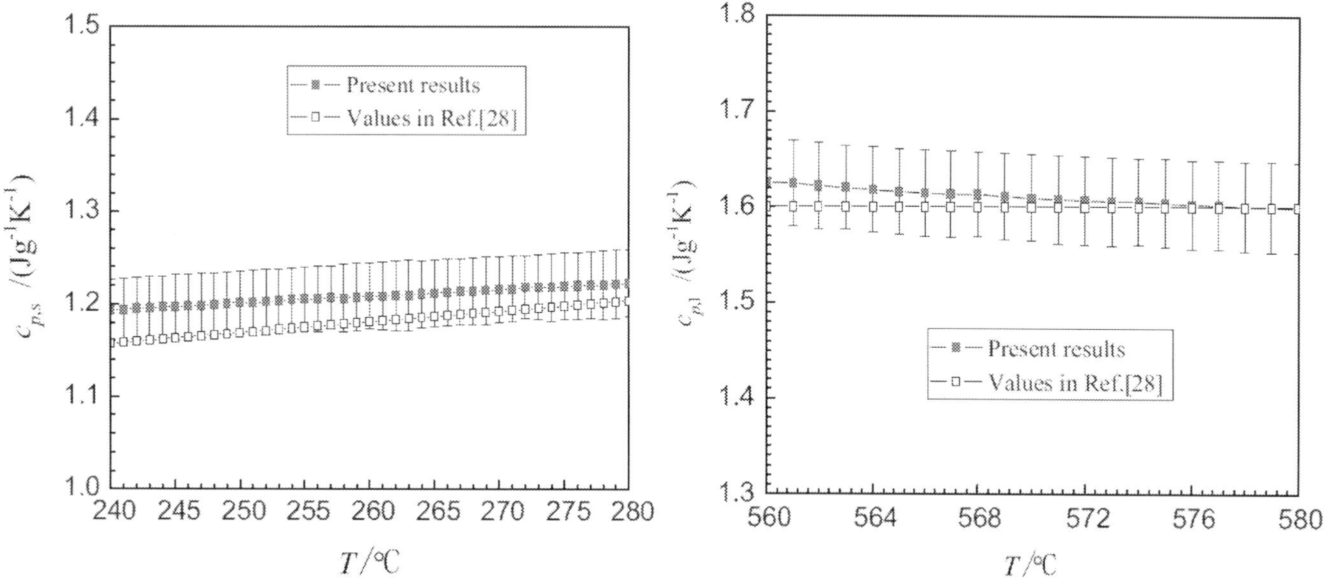

Fig. 12.6 Specific heat for binary carbonate eutectic salt comparison [20].

It can be concluded from the above descriptions that the values obtained during the procedures are nearly complying with the published values initially; this includes the meting point, specific heat, and melting enthalpy. This reflects the validity and significance of this technique. It showcases that composite nanomaterials' uniform production prepared from different materials can have similar properties and features by using this technique.

12.4 Properties of composite nanomaterials

The fractured result in ceramics results from ceramics' physical property to make ceramics suitable for the application of engineering, and many methods are applied that are not associating the ductile property of metal or embedded the ceramics into the form of a matrix. Table 12.4 shows the ceramic's nanoparticle composition, and the changes in the physical

Table 12.4 Ceramic matrix nanocomposites and their properties [21].

Matrix/ reinforcements	Properties
Si_3N_4/SiC	Improved strength and toughness
$MoSi_2/ZrO_2$	–
B_4C/TiB_2	–
Al_2O_3/SiC	–
MgO/SiC	–
Mullite/SiC	–
Al_2O_3/ZrO	–
Al_2O_3/Mo, Al_2O_3/W	–
$Al_2\ O_3\ NdAlO_3$	Improved photoluminescence

Table 12.5 Properties of Al_2O_3/SiC nano and microcomposites.

Properties/material	$Al_2O_3/SiCp$ composite	$Al_2O_3/SiCp$ nanocomposite
Vickers hardness (GPa)	–	22
Young's modulus (GPa)	–	383
Fracture strength (MPa)	106–283	549–646
Fracture toughness ($MPam^{1/2}$)	2.4–6.0	4.6–5.5

properties of ceramics are also investigated and compared to the monotheistic materials. Table 12.5 shows the comparison result of the Al_2O_3 and SiC system's physical properties, along with it a micro combination of the eternal part [20, 22, 23].

Table 12.5 enlightened the proper arrangement of Co and Ni Nano combination, or there are CNTs in the micro combination structure of these ceramic in Al_2O_3 and Fe_2O_3. The doping of metals leads to an improvement in the physical properties like electrical and thermally conductive properties. The field properties are magnetic, electronic, and light properties and the creation of nanomaterials and interesting salient characteristics because of extremely tiny components (Fig. 12.7).

In Table 12.6 shows many new properties are coming forward, which are due to the nanostructure combination of metal. The best example is α-Fe/$Fe_{23}C_6$/Fe_3B. Table 12.7 shows the calculated value of hardness (GPa), the ingot prototype, and the ribbon made from this system [24]. The alloy made by Branagan [25] is the hardness values of Vickers and were got to be 10.3 and 11 GPa in the scenario of solid-state. A different ribbon form indicates the hardness increased as the temperature involved is up, gives a high value of 16.2, and temperature of 973 K. Temperature involves very high compared to any steel and alloys descending to 10.5 GPa at the temperature of 1123. The result is used in comparing the type of ingot (8 and 6.6 GPa at 873 and 973 K, respectively).

The external micro part has fewer advantages than the metal Nano combination when evaluated by the Al/SiC system [26–28]. Fig. 12.8 illustrates the graphical statistical plots of the numerical content of SiC with the graphical plots of Vickers's hardness. The modulus of shear and Young's is presented as the content of SiC. The straight line is increasing in the graph of hardness with the growth of the volumetric reaction of the harder state

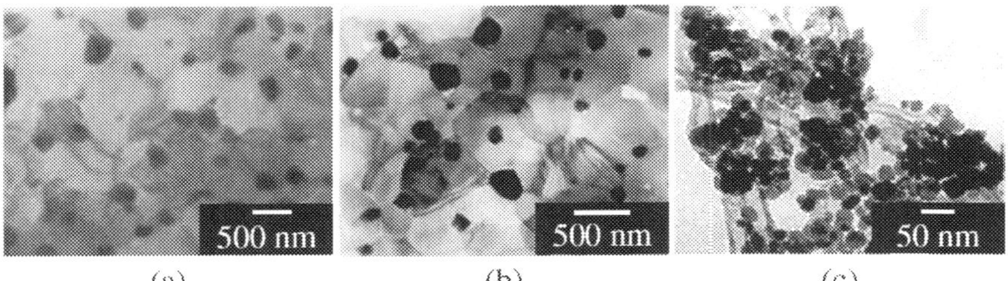

Fig. 12.7 Composite nanomaterials: (A) Al_2O_3/10 wt; (B) Al_2O_3 5% volume Ni; and (C) Fe_3O_4/CNTs nanocomposites [21].

Table 12.6 Metal nanocomposites and their properties.

Matrix/reinforcement	Properties
Ag/Au	Improvement in catalytic activity
Ni/PSZ and Ni/YSZ	Improved hardness and strength
Cu/Nb	Improved microhardness
Al/AlN	Higher compression resistance and low strain rate
Al/SiC	Improved hardness and elastic moduli
CNT/Sb and CNT/$SnSb_{0.5}$	Improvements in Li + intercalation properties
α-Fe/$Fe_{23}C_6$/Fe_3B	Drastic improvement in hardness
Cu/Al_2O_3	Improved microhardness
CNT/Fe_3O_4	Improved electrical conductivity

Table 12.7 Hardness values (GPa) of the ingot and ribbon samples prepared from the Fe/$Fe_{23}C_6$/Fe_3B nanocomposite.

Sample	Ingot	Ribbon
As-solidified	10.3	11.0
600 °C	8.0	11.0
650 °C	–	15.6
700 °C	6.6	16.2
750 °C	–	12.2
800 °C	6.5	12.0
850 °C	–	10.5

SiC till the highest range of 2.6 GPa for the data having 10 volume % SiC. The young modulus expansion and the expanding of the SiC component, the nanoparticle production, involve brittle associated with Al's matrix.

Hint statement

The structural property of nanomaterials varies, it can improve tensile strength, strain at break, stiffness, and tensile stress.

The structural property of composite nanomaterials will be elaborated in this section. The mechanical properties and morphology of nanomaterials are summarized in the tabular format.

In Table 12.8 shows the matrix nanocomposites and their properties for the polymer. By using the dimension as a tool, the raw materials utilized in the production of composite nanomaterials can be distinguished. For instance, the nanometer scale represents the raw material structure to be three dimensional; it is termed as is dimensional nanoparticle [29]; this

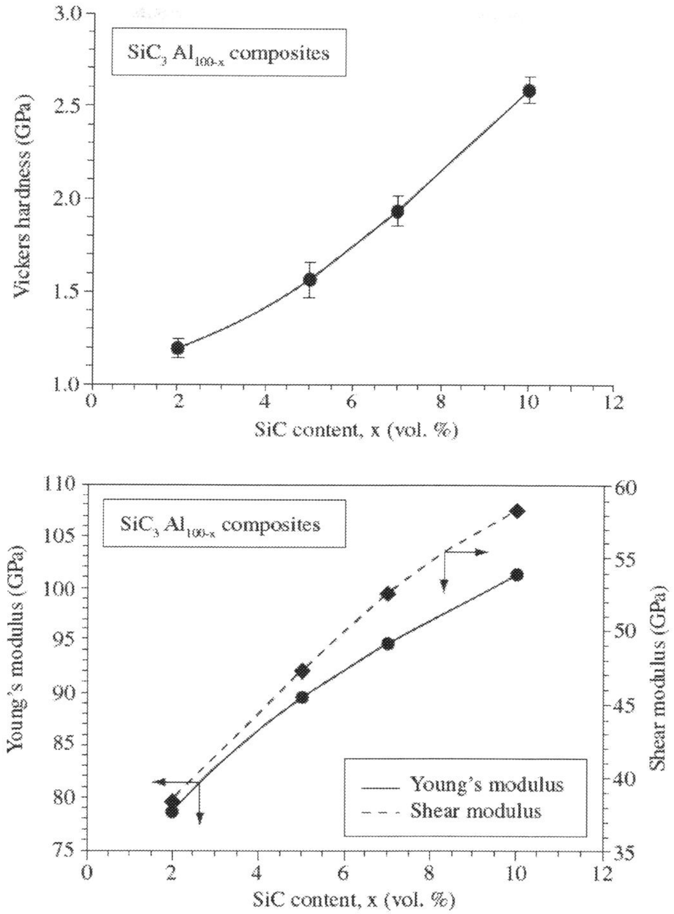

Fig. 12.8 Relation between SiC content, *x*, and Vicker's hardness [21].

includes semiconductor nanoclusters, metal particles, and silica. The second kind can be distinguished as created in nanotubes or whiskers; it consists of a 2D structure with one larger than another, forming an elongated oval shape. The significantly researched carbon nanotubes and cellulose whiskers are inclusive in this group. Its one-dimensional structure range identifies the type 3 of reinforcement in nanometer [30–32].

This group is termed as the polymer-layered nanocomposites; in this group, the nanotube is only a few meters thick and thousands of meters in length [33–35]. The polymer intercalation inside the layered host gallery is the process that is utilized to synthesize these products. Many crystalline

Table 12.8 Polymer-matrix nanocomposites and their properties.

Matrix/reinforcement	Properties
Polypropylene/ montmorillonite	Improved tensile strength, strain at break, stiffness, Young's modulus, and tensile stress
Nylon-6/layered-silicates	Improved storage modulus, tensile modulus, HDT, tensile stress, and reduced flammability
Polylactide/layered-silicates	Improved bending modulus, bending strength, distortion at the break, storage modulus, gas barrier properties, and biodegradability
Epoxy/layered-silicates	Improved tensile strength and modulus
Polyimide/ montmorillonite	Improved tensile strength, elongation at break, and gas barrier properties
Polystyrene/layered-silicates	Improved tensile stress and reduced flammability
Polyethylene oxide/ layered-silicates	Improved ionic conductivity

hosts, either synthesized or natural, are used to intercalate a polymer under specific conditions. These crystalline hosts can be; layered silicate, graphite, saponite, metal chalcogenides, and double-layered hydroxides. The layer was silicated, and the clay-based nanocomposites are widely used due to their significant intercalation with the chemistry field [36, 37]. As a property of the volume percentage of TiO_2 in the polyester, the quasi-static fracture variation can be seen. The addition of TiO2 significantly impacts the fracture toughness. The toughness level increases at the loading of 1%, 2%, and 3% of volume; however, it was observed that the toughness reduced at 4% of volume. These variations in the nanocomposites can be explained pictorially in Figs. 12.9 and 12.10.

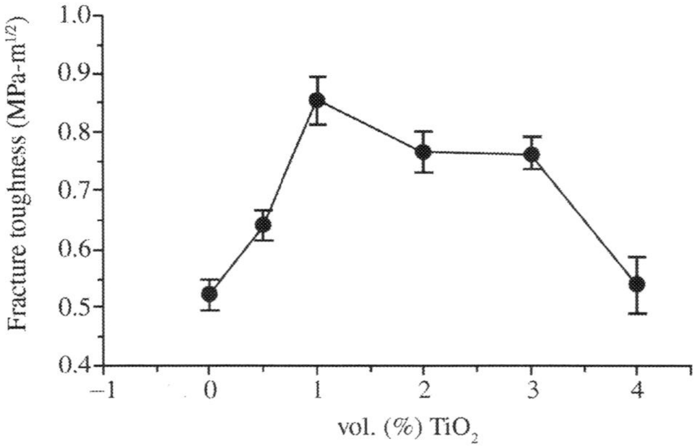

Fig. 12.9 Quasi-static fracture toughness variations [21].

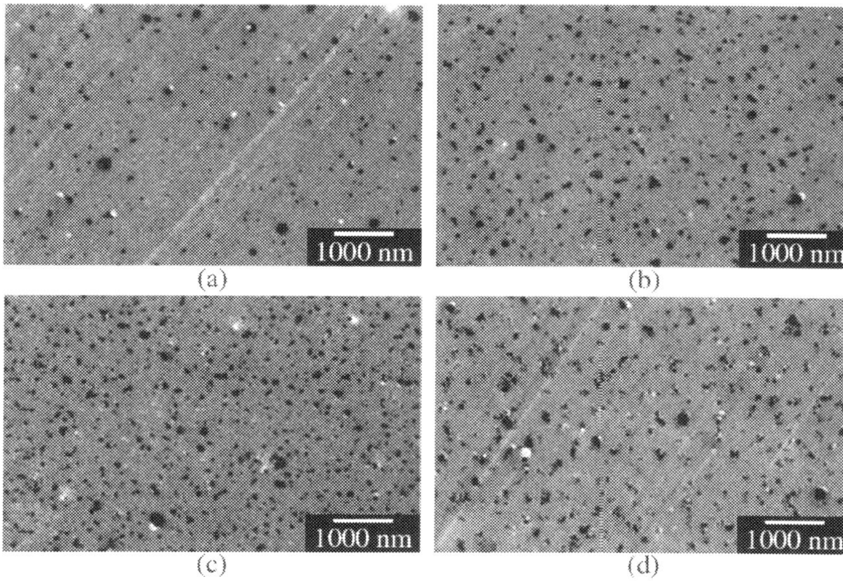

Fig. 12.10 TiO_2 nanoparticles dispersion: (A) 1% TiO_2 nanoparticles; (B) 2% TiO_2 nanoparticles; (C) 3% TiO_2 nanoparticles; and (D) 4% TiO_2 nanoparticles [21].

12.5 Conclusion

The composites nanomaterials are considered to be the most significant form of nanomaterials. These nanomaterials are produced by using a mixture of two or more elements. Thereof, these metals reflect their own individual properties independent of their parent material; this secures its significance, as because of this feature, multifunctional nanocomposites can be prepared which possess a wide range of properties, which is also dependent upon their structure and morphology. Different methods are explained in the chapter above for the synthesis of these nanomaterials; however, the chemical synthesis of these nanomaterials is considered effective and efficient. The technique that can be adequate applied for the synthesis is the microwave-induced technique. Nanocomposites are being widely consumed in the preparation of different motor vehicle pars. In contrast, the electronic domain is extensively performing researches for the effective implication of these nanomaterials in the field. These nanomaterials are also largely consumed in the ceramic industry.

References

[1] M. Akbulut, P. Ginart, M.E. Gindy, C. Theriault, K.H. Chin, W. Soboyejo, R.K. Prud'Homme, Generic method of preparing multifunctional fluorescent nanoparticles using flash nanoprecipitation, Adv. Funct. Mater. 19 (5) (2009) 718–725.

[2] C.S. Wu, C.T. Wu, Y.S. Yang, F.H. Ko, An enzymatic kinetics investigation into the significantly enhanced activity of functionalized gold nanoparticles, Chem. Commun. 42 (2008) 5327–5329.

[3] R. Sahay, V.J. Reddy, S. Ramakrishna, Synthesis and applications of multifunctional composite nanomaterials, Int. J. Mech. Mater. Eng. 9 (1) (2014) 25.

[4] R.D.K. Misra, S. Gubbala, A. Kale, W.F. Egelhoff Jr., A comparison of the magnetic characteristics of nanocrystalline nickel, zinc, and manganese ferrites synthesized by reverse micelle technique, Mater. Sci. Eng. B 111 (2–3) (2004) 164–174.

[5] S. Gubbala, H. Nathani, K. Koizol, R.D.K. Misra, J. Phys. B 348 (2004) 317.

[6] A. Mills, S. Le Hunte, An overview of semiconductor photocatalysis, J. Photochem. Photobiol. A Chem. 108 (1) (1997) 1–35.

[7] P. Palmero, Structural ceramic nanocomposites: a review of properties and powders' synthesis methods, Nano 5 (2) (2015) 656–696.

[8] A. Kay, M. Grätzel, Low cost photovoltaic modules based on dye sensitized nanocrystalline titanium dioxide and carbon powder, Sol. Energy Mater. Sol. Cells 44 (1) (1996) 99–117.

[9] H. Wanbao, Z. Baolin, Z. Hanrui, L. Wenlan, Combustion synthesis of Si_3N_4–TiN composite powders, Ceram. Int. 30 (8) (2004) 2211–2214.

[10] J.C. Han, G.Q. Chen, S.Y. Du, J.V. Wood, Synthesis of Si_3N_4–TiN–SiC composites by combustion reaction under high nitrogen pressures, J. Eur. Ceram. Soc. 20 (7) (2000) 927–932.

[11] K.V. Manukyan, S.L. Kharatyan, G. Blugan, J. Kuebler, Combustion synthesis and compaction of Si_3N_4–TiN composite powder, Ceram. Int. 33 (3) (2007) 379–383.

[12] H.Y. Ryu, H.H. Nersisyan, J.H. Lee, Preparation of zirconium-based ceramic and composite fine-grained powders, Int. J. Refract. Met. Hard Mater. 30 (1) (2012) 133–138.

[13] S.E. Aghili, M.H. Enayati, F. Karimzadeh, Synthesis of (Fe, Cr) 3Al–Al_2O_3 nanocomposite through mechanochemical combustion reaction induced by ball milling of Cr, Al and Fe_2O_3 powders, Adv. Powder Technol. 25 (1) (2014) 408–414.

[14] E.M. Sharifi, F. Karimzadeh, M.H. Enayati, Preparation of Al_2O_3–TiB_2 nanocomposite powder by mechanochemical reaction between Al, B_2O_3 and Ti, Adv. Powder Technol. 22 (4) (2011) 526–531.

[15] Z. Zhang, X. Du, J. Wang, W. Wang, Y. Wang, Z. Fu, Synthesis and structural evolution of B4C–SiC nanocomposite powders by mechanochemical processing and subsequent heat treatment, Powder Technol. 254 (2014) 131–136.

[16] O. Torabi, S. Naghibi, M.H. Golabgir, H. Tajizadegan, A. Jamshidi, Mechanochemical synthesis of NbC–NbB_2 nanocomposite from the Mg/B_2O_3/Nb/C powder mixtures, Ceram. Int. 41 (4) (2015) 5362–5369.

[17] T.H. Hsieh, K.S. Ho, X. Bi, Y.K. Han, Z.L. Chen, C.H. Hsu, Y.C. Chang, Eur. Polym. J. 45 (2009) 613–620.

[18] M.M. Kenisarin, High-temperature phase change materials for thermal energy storage, Renew. Sust. Energ. Rev. 14 (3) (2010) 955–970.

[19] N. Araki, M. Matsuura, A. Makino, T. Hirata, Y. Kato, Measurement of thermophysical properties of molten salts: mixtures of alkaline carbonate salts, Int. J. Thermophys. 9 (6) (1988) 1071–1080.

[20] Y.B. Tao, C.H. Lin, Y.L. He, Preparation and thermal properties characterization of carbonate salt/carbon nanomaterial composite phase change material, Energy Convers. Manag. 97 (2015) 103–110.
[21] P.H.C. Camargo, K.G. Satyanarayana, F. Wypych, Nanocomposites: synthesis, structure, properties and new application opportunities, Mater. Res. 12 (1) (2009) 1–39.
[22] C.C. Anya, Microstructural nature of strengthening and toughening in Al_2O_3-SiC (p) nanocomposites, J. Mater. Sci. 34 (22) (1999) 5557–5567.
[23] L.A. Timms, C.B. Ponton, M. Strangwood, Processing of Al_2O_3/SiC nanocomposites—part 2: green body formation and sintering, J. Eur. Ceram. Soc. 22 (9–10) (2002) 1569–1586.
[24] L.P. Ferroni, G. Pezzotti, T. Isshiki, H.J. Kleebe, Determination of amorphous interfacial phases in Al_2O_3/SiC nanocomposites by computer-aided high-resolution electron microscopy, Acta Mater. 49 (11) (2001) 2109–2113.
[25] D.J. Branagan, Y. Tang, Developing extreme hardness (>15 GPa) in iron based nanocomposites, Compos. A: Appl. Sci. Manuf. 33 (6) (2002) 855–859.
[26] D.J. Branagan, in: D.E. Alman, J.W. Newkirk (Eds.), Powder Metallurgy, Particulate Materials for Industrial Applications, TMS Publication, St. Louis, 2000.
[27] H. Liu, L. Wang, A. Wang, T. Lou, B. Ding, Z. Hu, Study of SiC/Al nanocomposites under high pressure, Nanostruct. Mater. 9 (1–8) (1997) 225–228.
[28] B. Venkataraman, G. Sundararajan, The sliding wear behaviour of Al-SiC particulate composites—I. Macrobehaviour, Acta Mater. 44 (2) (1996) 451–460.
[29] M.S. El-Eskandarany, Mechanical solid state mixing for synthesizing of SiCp/Al nanocomposites, J. Alloys Compd. 279 (2) (1998) 263–271.
[30] N. Herron, D.L. Thorn, Nanoparticles: uses and relationships to molecular cluster compounds, Adv. Mater. 10 (15) (1998) 1173–1184.
[31] V. Favier, J.Y. Cavaille, G.R. Canova, S.C. Shrivastava, Mechanical percolation in cellulose whisker nanocomposites, Polym. Eng. Sci. 37 (10) (1997) 1732–1739.
[32] L. Chazeau, J.Y. Cavaille, G. Canova, R. Dendievel, B. Boutherin, Viscoelastic properties of plasticized PVC reinforced with cellulose whiskers, J. Appl. Polym. Sci. 71 (11) (1999) 1797–1808.
[33] M. Ogawa, K. Kuroda, Preparation of inorganic–organic nanocomposites through intercalation of organoammonium ions into layered silicates, Bull. Chem. Soc. Jpn. 70 (11) (1997) 2593–2618.
[34] D. Schmidt, D. Shah, E.P. Giannelis, New advances in polymer/layered silicate nanocomposites, Curr. Opinion Solid State Mater. Sci. 6 (3) (2002) 205–212.
[35] M. Alexandre, P. Dubois, Polymer-layered silicate nanocomposites: preparation, properties and uses of a new class of materials, Mater. Sci. Eng. R. Rep. 28 (1–2) (2000) 1–63.
[36] S.S. Ray, M. Okamoto, Polymer/layered silicate nanocomposites: a review from preparation to processing, Prog. Polym. Sci. 28 (11) (2003) 1539–1641.
[37] L.C. Stearns, J. Zhao, M.P. Harmer, Processing and microstructure development in Al_2O_3-SiC' nanocomposites', J. Eur. Ceram. Soc. 10 (6) (1992) 473–477.

CHAPTER 13

Membrane-derived nanomaterials

Abbreviations

ILCS	impregnated layer combustion synthesis
ISC	impregnated support combustion
MLV	multilamellar vesicles
MSN	silica nanocapsules
PCTDE	track-etched polycarbonate
PLGA	poly lactic-*co*-glycolic acid
PVA	polyvinyl alcohol
RBC	red blood cell
SDS-PAGE	sodium dodecyl sulfate polyacrylamide gel electrophoresis
SEM	scanning electron microscopy
SHS	high temperature synthesis
TEM	transmission electron microscopy
VCS	volume combustion synthesis
WBC	white blood cell

13.1 Introduction

Nanomaterials have significant implications in diverse fields for the betterment of humanity [1, 2]. The nanomaterials are being synthesized from a diverse range of raw materials, and every nanomaterial has a different property. Therefore, they are consumed for different purposes. The membrane-based nanomaterials extracted by utilizing a membrane as a nanomaterial are considered the most significant nanoparticle. This is because these nanomaterials are generally consumed for the treatment of cancer and other life-threatening diseases. Moreover, these nanomaterials have specialized properties to cater to the complex internal environment of living beings. These nanomaterials are considered to be chemically resistant and can have a higher tolerance level to many chemicals found in living organisms [3, 4].

Several techniques and synthesis processes are derived from producing nanomaterials that can comply with the biological and chemical environment [5]. This is considered difficult as living bodies do not accept any particle that does not match the pre-existing organelles in the living body [6]. Moreover, it is significantly difficult for researchers to synthesize a

Synthetic Engineering Materials and Nanotechnology
https://doi.org/10.1016/B978-0-12-824001-4.00005-3

nanomaterial, which is artificially replicable to body organelles due to their sensitivity and specificity. A new nanomaterial is being synthesized by the researchers, which is derived from different membranes; these nanomaterials possess biologically replicable features and have a similar sensitivity. This is why these nanomaterials are ranked as the most significant among other nanomaterials derived from different nonliving raw materials [7]. Extensive research and synthesis techniques are being involved in developing this field; however, it has fewer synthesis techniques than other nanomaterials [8]. The membrane-derived nanomaterials function because they are derived from cells. The human body cells are the most basic and essential functional unit; all of the basic living organisms' basic functions are dependent upon cells. These cells collectively form membranes, which is the thin layer of cells, generally covering an organ or tissue. These membranes, which are solely formed from cells, are utilized for the synthesis of membrane-derived nanomaterials. This is why the derived nanomaterials have similar properties because they are prepared to begin the cells' core bulk.

Hint statement

Every cell in the human body is designed to perform different task.

As every cell has a specialized task to perform in the body. For instance, the red blood cells (RBCs) are responsible for the transportation of oxygen and circulation in the body. As the cells control the functionality of the membranes are providing differentiating features to the cells. This bio-macromolecules presented in the lipid bilayer are controlling cellular compartmentalization. Therefore, the nanomaterials derived from the membrane-based RBCs will have similar RBCs. Fig. 13.1 represents the membrane-derived materials and their applications. The individual membrane components generally utilized are proteins and glycans.

13.2 Significance of membrane-derived nanoparticle

The raw materials utilized for the extraction of nanomaterials include different sorts of membranes and membrane-bounded organelles. The recent development in this field regards the usage of cellular components as raw material to synthesize nanoparticles. The development will provide significant convenience in the synthesis and artificial replication of nanomaterials, as they are directly produced by the cells; they do not require artificial replication and additional features to be suitable for the living environment. Therefore, the nanomaterials can be sorted into two major categories,

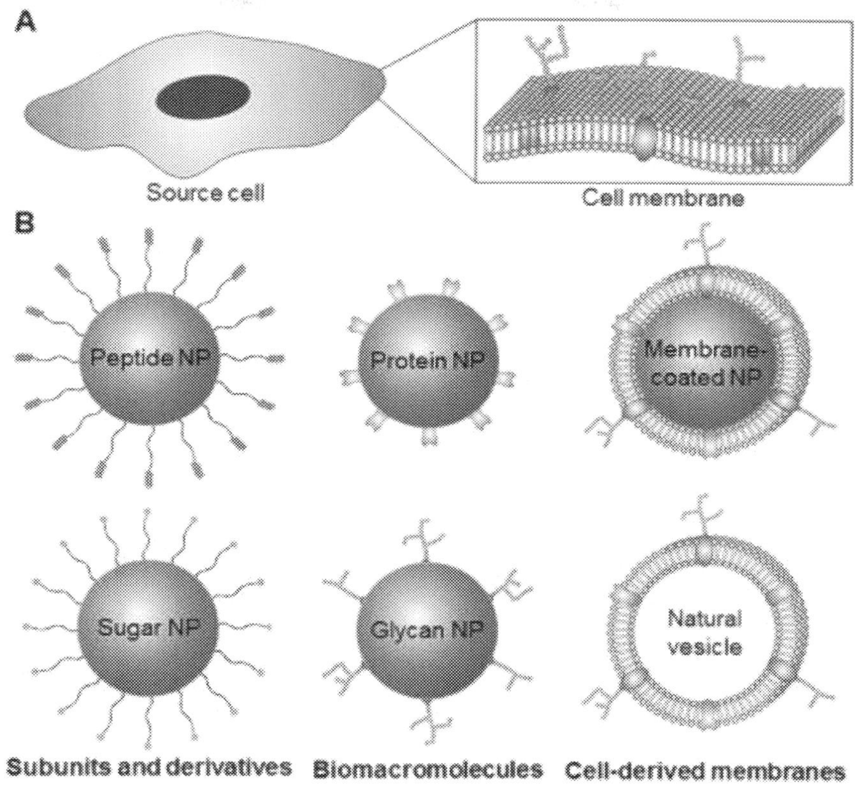

Fig. 13.1 Functionalization of nanomaterials: (A) membrane composition deriving cellular properties and (B) multiple cell-membrane based strategies [9].

the first category is naturally derived nanomaterials, and the second is of nanomaterials obtained from a membrane-bounded raw material. These developments will provide nanomaterials with more complying naturally occurring organelles characteristics and require lesser effort to prepare.

> **Hint statement**
>
> Different membrane bounded organelles are utilized as a raw material for the synthesis of membrane bounded nanoparticles.

Moreover, there is a large range of these nanomaterials as per the cell utilized for the preparation. The natural analog nanomaterial of liposome is a structured vesicular membrane-derived nanomaterial consumed for the preparation and synthesis of different chemicals was shown in Fig. 13.2.

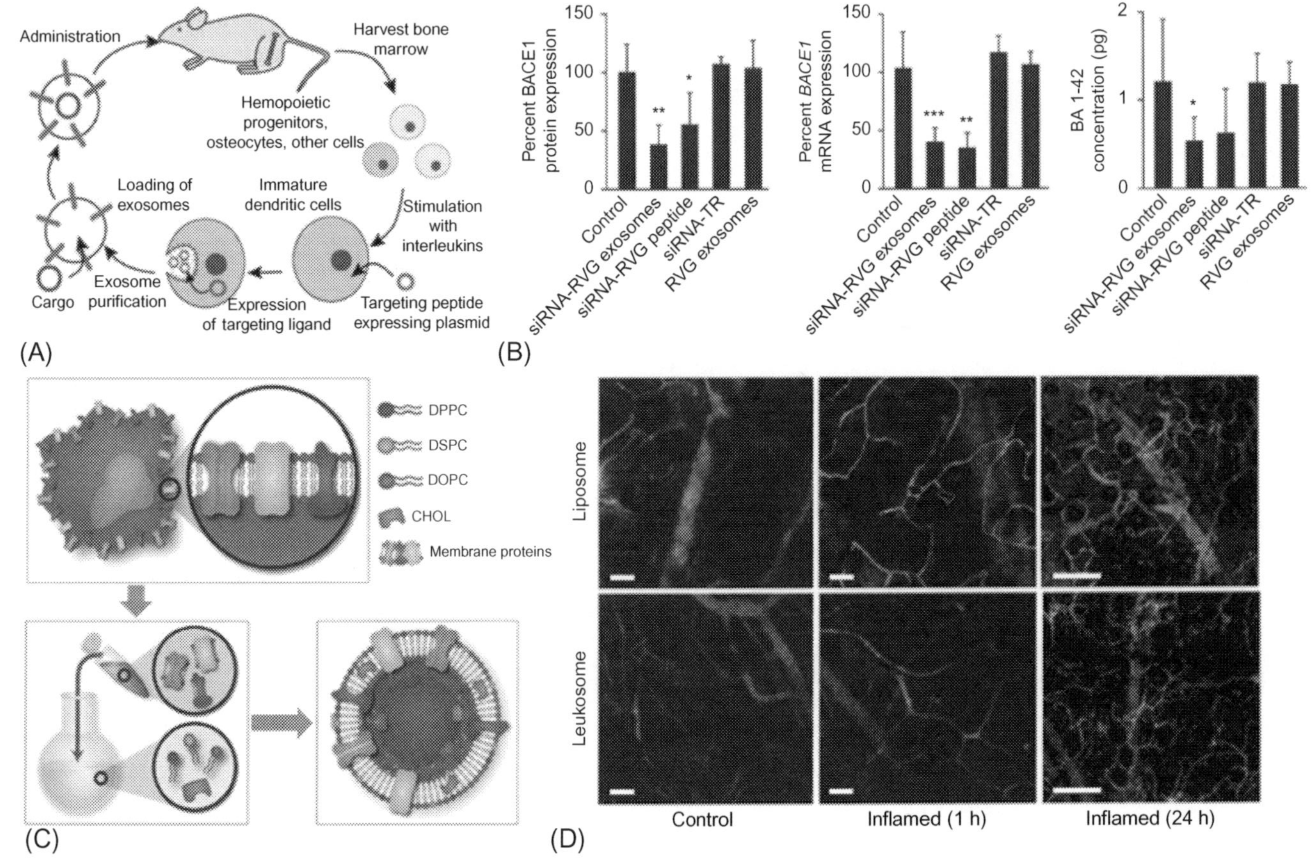

Fig. 13.2 Natural membrane vesicles: (A) workflow of membrane fabrication; (B) statistical data of BACE1 protein and mRNA; (C) fabrication of leukocytes; and (D) microscopic image of inflamed tissue [9].

These nanomaterials can be extracted from larger molecular-sized membranes or synthesized from naturally occurring substances [10]. For instance, the naturally occurring vesicle nanomaterial is the exosomes found in the mammal's body [10]. These sorts of nanomaterials are largely consumed in cancer treatment and diagnosis [11, 12]. The nanomaterial's products, such as the exosomes, are considered the most relevant discovery and widely consumed in the bimolecular and meditational research. A large number of clinical investigations and researches are conducted based on membrane-derived nanomaterials. The most adequate and essential features of these nanomaterials are that they are pre-equipped with the necessary elements and functioning required to exist and survive in the living environment. Moreover, they are specifically designed to perform biological functions accurately [13].

At the initial stage, when the procedural discovery of the first membrane-bounded nanomaterials occurs, extensive attention was attracted to these nanomaterials. More experiments were performed for the improved synthesis of these nanomaterials. It was observed that doxorubicin and polymeric cores could be utilized for more convenience; moreover, the polymer tethers can be used for the penetration of different elements in the outermost layer of the membrane that will enable the synthesis of nanomaterials having varied properties [14, 15], these vary properties prepare nanomaterials that can diagnose and treat tumor [16–18]. Other nanomaterials can be synthesized by utilizing similar techniques, which are applicable to reducing the toxicity and functions in the replacement of drugs [19]. The nanomaterials derived from RBC have biocompatibility features and non-immunogenic nature that enables them to administer the additional circulations [20]. Along with RBC, other cells are utilized as raw material, and each cell utilization provides a unique sort of nanomaterial shown in Fig. 13.3A. The low immunogenic nature of cancer cells attracts similar cells towards itself; at this point, the nanoparticle with injected drugs can be utilized in the placement of drugs to treat cancer [21–23]. Similarly, the nanomaterials extracted from White Blood Cell (WBC) membrane can be active and establish a defense mechanism against alien particles [24] and on the abnormal clusters of cells made from similar cells, that is, cancerous or tumorous cells [25]. The membrane-derived nanomaterials are reported to be adaptable for multiple diseases targeting devices and treatment options. These nanomaterials filled with cytotoxic drugs can reduce restenosis and pathogen clearance in the living body [26]. They can function as the loaded antibiotic for the treatment of multiple diseases [27, 28].

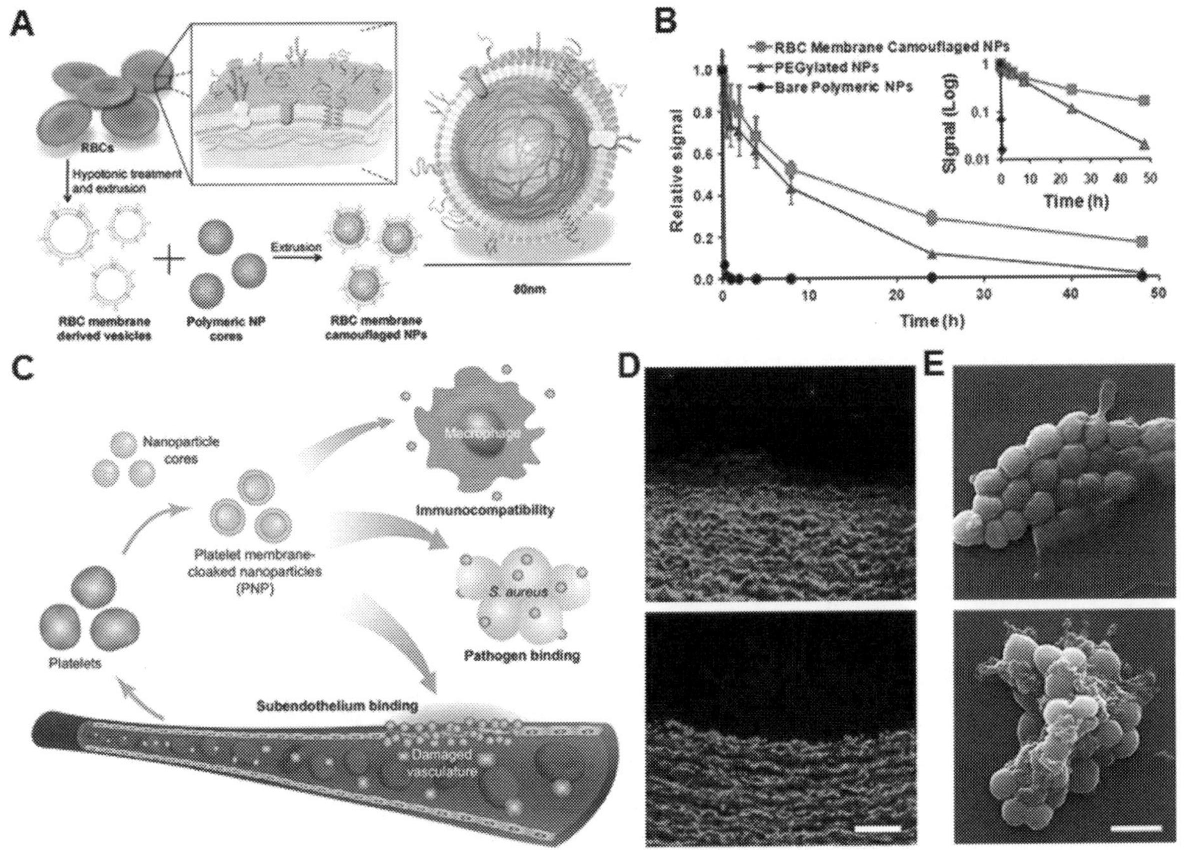

Fig. 13.3 Nanoparticles that are membrane coated: (A) RBC membrane coated; (B) luorescent nanoparticle demonstrating increased circulation; (C) membrane coated platelet; and (D) bonding of membrane coated platelet; € Methicillin-resistant *Staphylococcus aureus* (*top*) platelet membrane coated nanoparticles (*bottom*) [9].

13.3 Methods

13.3.1 Biological method

The biological synthesis of membrane-derived nanomaterials is based on the three major steps. As shown in Fig. 13.4, the procedure includes extraction of the membrane, nanoparticle preparation, and nanomaterials fusion. For the adequate production of these nanomaterials, every step is considered to be critical.

13.3.1.1 Extraction of membrane

Cell membranes are considered significant for their capabilities of performing bio-functions. These membranes are asymmetrical and have a protein base with purposeful external proteins [29]. The denaturation reduces the membrane-oriented proteins; the plasma membrane extraction shall be performed critically, including the procedural steps of purifying the membrane and cell lysis. According to the presence of a nucleus in the cell, the procedure is altered.

The nanomaterial extraction from nucleus free cells is scarce and has a higher intensity to perform different functions; for instance, the nucleus free cells obtained from the human body can perform additional cargo delivery functions. The membrane obtained from RBCs and platelets falls in this category as they are responsible for delivering growth factors in all the injured parts of the body. The CD47 protein is present in the RBC cells for enabling it to perform its functions adequately; the presence of these membranous proteins increases the circulation of RBC cells up to 120 days [30]. On the contrary, the CD44 and P-selection proteins are present in platelets that functions under the pathological condition [31]. The membranes' extraction forms the cells, and platelets require blood fraction isolations and methods based on centrifugation. In the second step, the obtained membrane is either treated with the hypotonic or freeze–thaw process. The centrifugation process is performed again to obtain the soluble protein, which is sieved from polycarbonate membranes with porous cell walls, enabling the extraction of pure nanomaterials for the fusion process. For the maintenance of biological activities in the nanomaterials, they are stored at the temperature of 4 °C [18, 32].

13.3.1.2 Core nanoparticles

The inner cores of nanomaterials generally consist of synthetic materials, which include gelatin, poly-lactic-*co*-glycolic acid (PLGA), caprolactone,

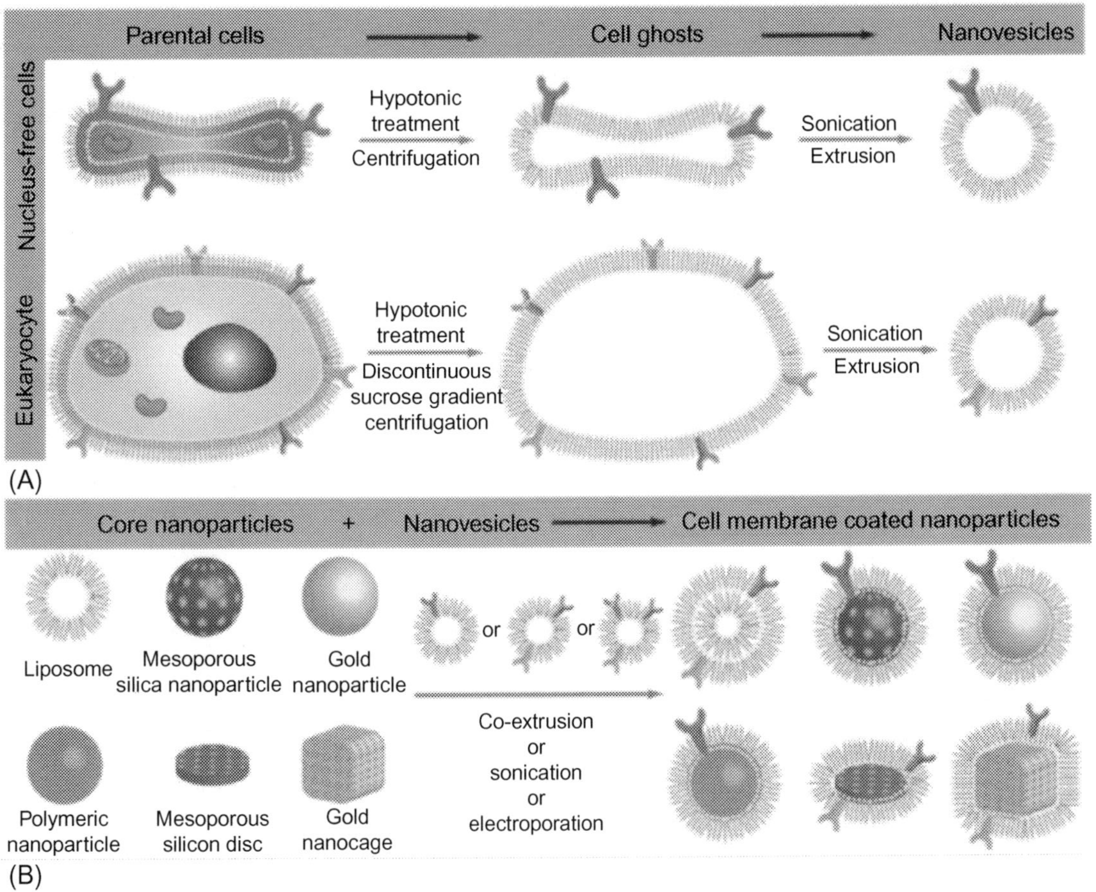

Fig. 13.4 Preparation of cell membrane coated nanoparticle: (A) intact cell extraction and (B) nanoparticles being utilized as inner core.

mesoporous silica nanocapsules (MSNs), gold nanoparticles, and liposomes [33–38]. The selection of nanomaterial varied according to the usage and purpose of the synthesis of nanomaterials. PLGA and PCL are the widely consumed nanomaterials generally utilized in the replacement of drugs. The PLGA secures its significant position due to its higher biocompatibility and biodegradable nature [39]. Therefore, nanomaterials' production can be consumed, having varied size range [40, 41]. PLGA containing the terminal carboxyl group is generally considered adequate for the preparation of these nanomaterials. A negative charge is generated by the carboxyl group, which repels the outer leaflet of the membrane. This enables the accurate topological arrangement of cell membranes in nanoparticles [42].

In contrast, the effect of size and amount of drug is considered to be negligible in creating any significant effect on the efficiency of membrane nanomaterials [43, 44]. The PCL derived nanomaterials have lower glass transition property along with higher biodegradability and biocompatibility [45]. On the contrary of PLGA, the PCLs are more hydrophobic, having slow degradation, and therefore, it is considered adequate.

13.3.1.3 Process of fusion

The extracted nanomaterials required the fusion process to properly eliminate any other foreign particles and amalgamate the nanomaterial properly [23, 36, 37]. The fusion process shall be performed with critical care as it should not denature proteins or cause leakage in the loaded drugs in the membrane of nanoparticle [46–48]. The sequential extrusion of samples was performed using different size pores, and this step is termed as membrane extrusion technique. Despite the adequacy of this technique, its industrial-scale implementation is difficult. Several self-recognized proteins in the nanoparticles are derived from the membrane; therefore, the process of sonication is considered significant for nanoparticles' industrial-scale preparation. The sonication process's effectiveness and avoidance of protein maturation are possible when the power, frequency, and duration of the process is enhanced. For the adequate surface coverage of the cell membrane, the ratio between the membrane and nanomaterial is required to be balanced [49]. The recent development in this technique includes the microfluidic electroporation method; in this method, the voltage, velocity, and duration are required to be increased [20].

Example 13.1

Ge has a crystal structure (diamond) as Si. A lattice constant of $a = 5.64\,\text{A} = 0.564\,\text{nm}$. Calculate the atomic density (atoms/cm^3).

Solution:

Cubic unit cell's volume: $V_u = a^3$ ($a = 0.564 \times 10^{-7}$ cm).

Atoms in the cubic unit cell: $N_u = 8 \times 1/8 + 6 \times 1/2 + 4 = 8$.

(Eight on the corners, shared with 8 neighbors +6 on the faces, each one shared with a nearest neighbor +4 in the interior.) See fig. 1.4 Pierret, SDF.

Atomic density: Nu $V_u = 8/a^3/8/$ $(0.564 \times 10^{-7)3} = 4.46 \times 10^{22}$ atoms/cm.

13.3.2 Chemical methods

The nanomaterials preparation from liposome techniques is considered the most adequate and convenient nanomaterial preparation technique [38, 50]. In the nanomaterial extraction procedures, the multilamellar vesicles (MLVs) are sieved with force from nanoporous channels to prepare unilamellar liposome vesicles; the procedure is termed as the nanoporous membrane extrusion, as shown in Fig. 13.5. During the procedure's conduction, the natural or synthetic lipids, which may include; cholesterol and 1,2-dioleoysln-glycero-3-phosphocholine, utilizing the evaporator or gas chamber are dried.

Hint statement

In the chemical synthesis of nanoparticle derived membranes, the natural lipids are utilized which includes; cholesterol and 1,2-dioleoysln-glycero-3-phosphocholine.

In the second step, the dried lipid films are hydrated using an aqueous solution. Due to this rehydration process, the molecules of phospholipids present in lipid films arrange themselves in MLVs. The MLVs created during the dispersion in aqueous solution ranges from 0.5 to 10 μm in size. For increasing the efficiency of obtained MLVs, they are generally processed through 10 freeze or thaw cycles, which sustains the rearrangement of lipid molecules in the bilayers of lipid. Using the double-stacked nanoporous membranes, the formed MLVs can be extrusion; however, the membrane's pore size should range between 100 and 200 nm [52]. While conduction the procedures, the prepared MLVs are forcefully sieved from the membrane having a smaller size of pores, which cause

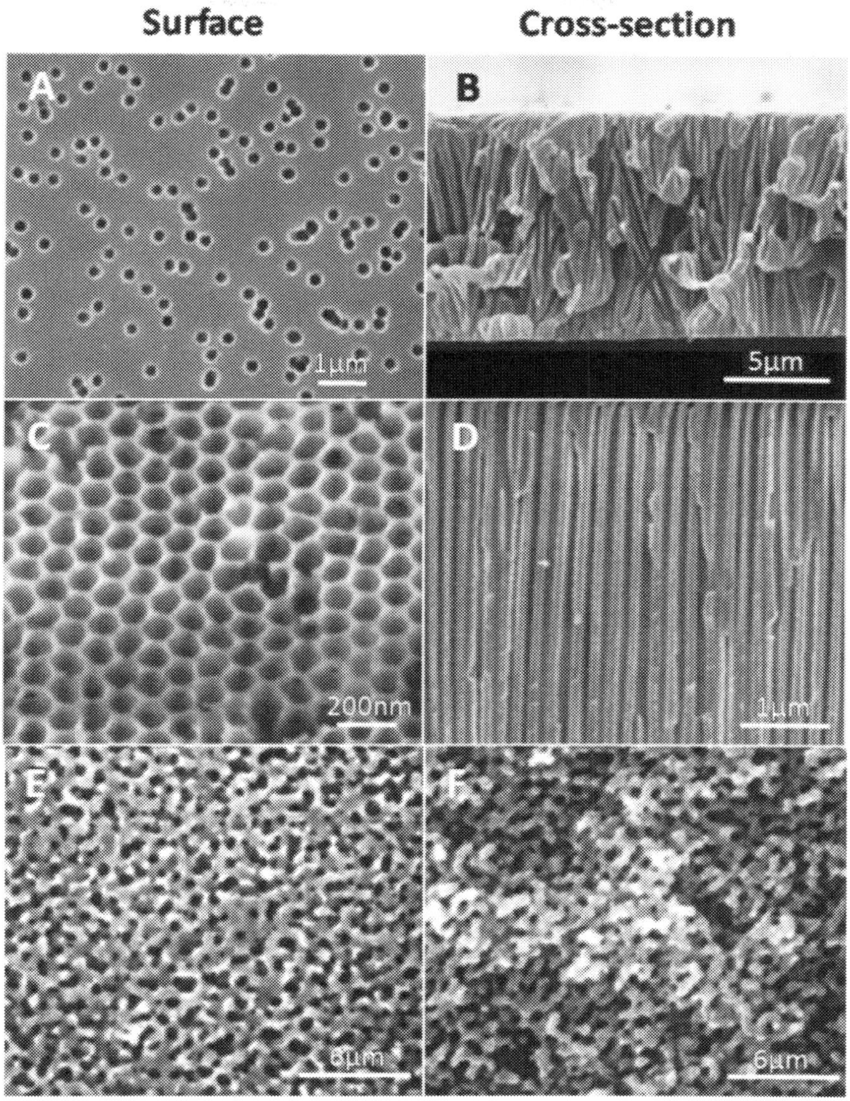

Fig. 13.5 Scanning electron microscopy (SEM) images of (A and B) a track-etch membrane (PCTE), (C and D) an AAO membrane, and (E and F) an SPG membrane [51].

the lipid membrane of MLV to rupture and creates the single layer lipid bilayer inside the channel. The formed unilamellar liposomes traveled through the channels due to the higher pressure and are dumped at the nanoporous channel [53]. This process in general extraction procedures is repeated 5–10 times in order to obtain the nanoparticle that

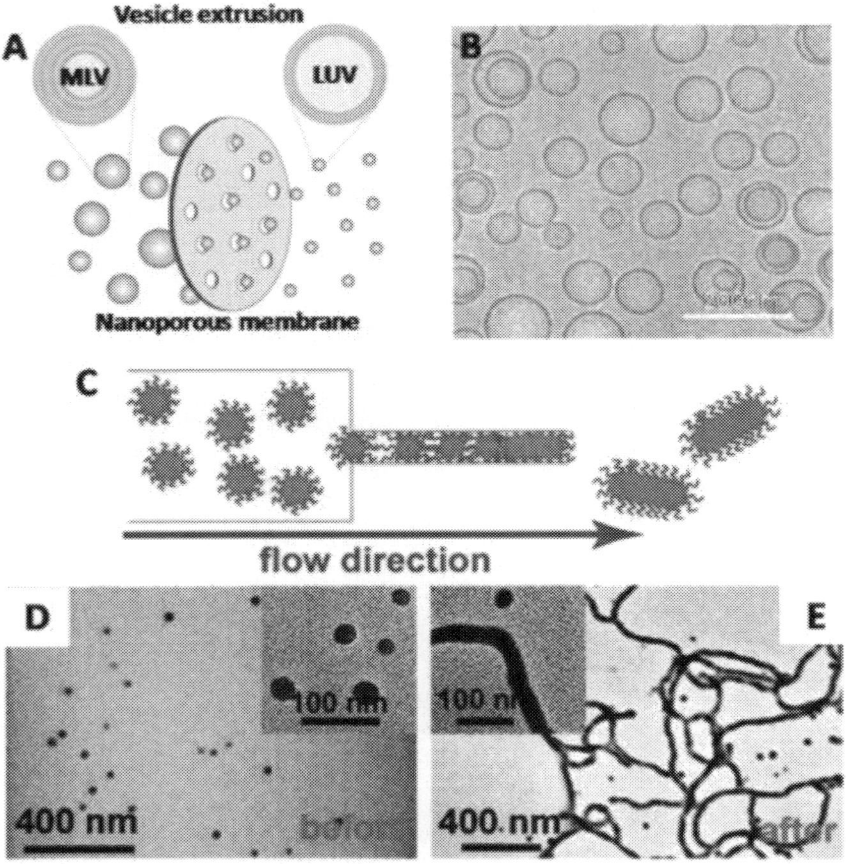

Fig. 13.6 (A) Extrusion process; (B) images of TEM; (C and D) schematic illustration of vesicle extrusion [51].

has the desired size replicating of that membrane, as shown in Fig. 13.6. The critical conditions required for adequate synthesis by this process is the high temperature [54]. The higher temperature causes the nanomaterials to form a gel-liquid. The nanoparticle is then entered into the liquid state, which provides more strength and flexibility to the particles for their consumption. The membrane most adequately used for nanoparticle extrusion through vesicle application is Track-etched polycarbonate (PCTE) [55].

13.4 Techniques

13.4.1 Electrospinning

Nanomaterials extraction is using the electrospinning technique. It's based on three essential apparatus, as reflected in Fig. 13.7. The power supply of high voltage ranging from 0 to 40 kV, a syringe which works as a container filled with melted with a needle or polymer solution and various constructs grounded collector that can be either flat or drum type. The metallic adapter equipped with a needle or a whole needle consisting of metal is supplied with high voltage, the needle is filled with the melt of a polymer solution. The droplets are formed in the cone shape due to the continuous push of melt or polymer solution form the syringe, and the high voltage electric field overpowers the surface tension, as reflected in Fig. 13.7, shown an ultrafine nanoparticle is obtained from the tip of the needle and is collected at the ground collector. The nanoparticle is processed through electrostatic repulsion and Coulombic force, due to which it receives its elongated shape. The needleless apparatus is also consumed to prepare nanomaterials using the electrospinning technique [57, 58].

Hint statement

The apparatus of electrospinning utilizes metal-based needles to maintain the consistency of voltage supplied and liquid stored in it.

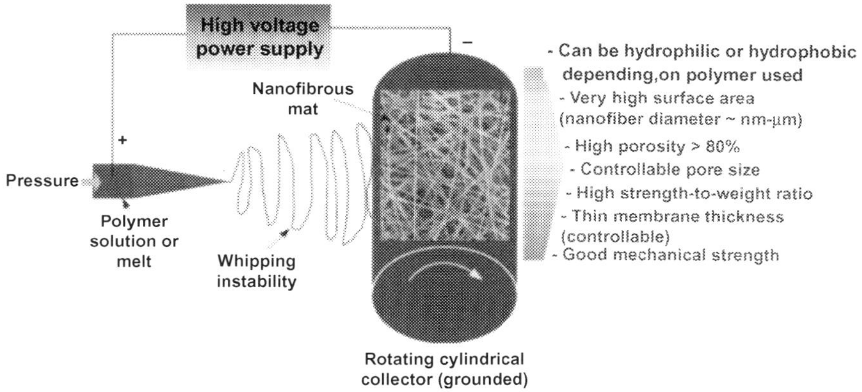

Fig. 13.7 Electrospinning apparatus [56].

Although obtaining nanomaterial procedures was from electrospinning, and it is considered a convenient procedure, it requires multiple optimization parameters. The advantage of using this technique is that the nanomaterials prepared from this technique can be prepared variedly with a diverse range of properties by controlling the environmental, procedural, material, and post-processing parameters [56]. The most potent parameters in causing the diversification are operational parameters that may include; tip-to-collector, applied voltage, tip size, drum type, and set-up configuration. Whereas, molecular weight, viscosity, type of polymer, concentration, and solution conductivity are counted as material parameters. The chamber heat the temperature is considered as environmental parameters. The drying conditions, hot pressing, and heat treatments are the post-procedural parameters that significantly impact nanomaterials' features. The conditions mentioned above must be adequately achieved, and these conditions can be manipulated to achieve authentically and desires nanomaterial, as shown in Fig. 13.8.

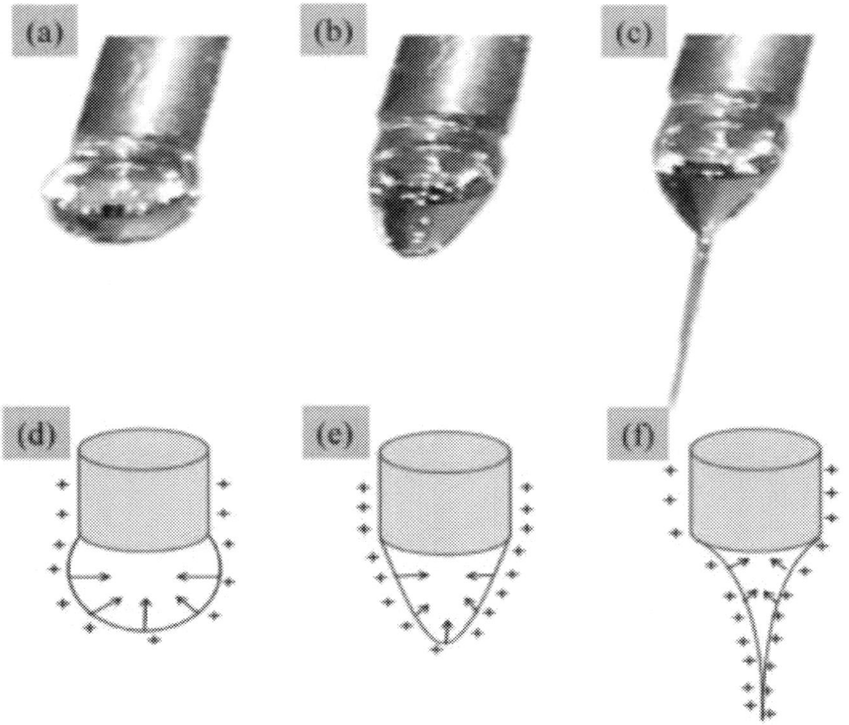

Fig. 13.8 Deformation of PVP after electrospinning [59].

Example 13.2

Silicon (Si) has a diamond crystal structure. Answer the following questions about Si. (Assume a lattice spacing of $a=5.42$ A.) Compute the density of Si atoms per cm^2 on {100} planes.

Solution:

Suppose, there are 5 atoms on the face, but the 4 on the corners are shared between 4 adjacent unit cells, so the total number is $N=4\times 1/4+1=2$ per face of a cell.

The density per unit area is $N_S=2/a^2=2/(5.42\times 10^{-8}\ cm)^2=6.81\times 1014\ cm^{-2}$ N.

13.4.2 Combustion synthesis

Among the bottom-up approaches, the c7ombustion synthesis is the most extensively consumed technique for the extraction of nanomaterials due to convenient, time-effective, and economical procedure. These processes required specialized laboratory equipment, including beakers and hot plates, to conduct the process [60]. This technique's utilization is considered significant in the provision of a wide range of nanomaterials obtained from metal oxides, non-metal oxides, ceramics, and alloys with pure and homogeny nanoparticles [61]. In the traditional combustion synthesis method, the amorphous form of two elements is pressed together and is heated uniformly until they react with each other [62]. Fig. 13.9 shows the combustion synthesis in which solid materials are utilized is categorized into two different sorts based on the ignition, namely, high-temperature synthesis (SHS) and volume combustion synthesis (VCS) [64].

A reactive liquid solution is prepared in the SCS model. Generally, the water-soluble precursors, including; metal nitrate and fossil fuels, are combined homogeneously to prepare the solution. The water solubility of metal nitrates makes them adequate to be used in the procedure as metal nitrates; moreover, they are effective oxidizers. Whereas on the contrary, fossil fuels are considered on the availability of carboxylic and amine groups in the fuel and its ability to perform as a reducing agent. As per Ref. [65], the fuels can be; hydrazine, glycine, glucose, urea, and polyvinyl alcohol (PVA). The relevancy and efficiency of a fueling agent can be observed by using the following equation:

$$\text{Fueling agent}=\frac{\Sigma\text{ valencies of oxidizing and reducing agents of reactants}}{(-1)\Sigma\text{ valencies of oxidizing and reducing agent in fuel}} \tag{13.1}$$

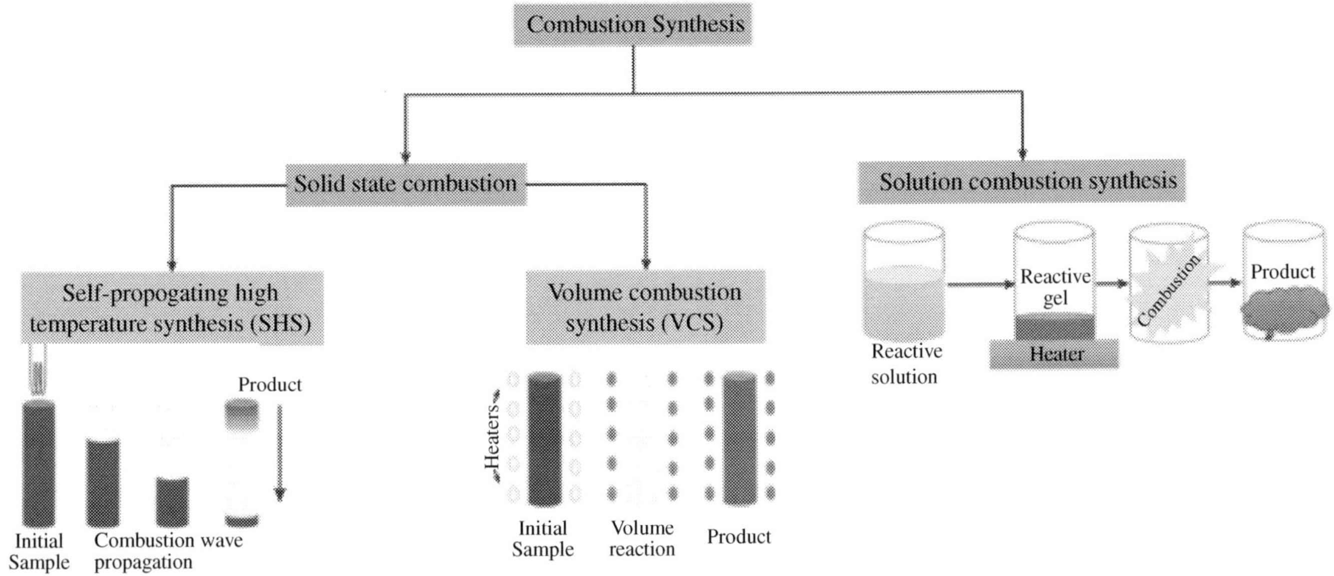

Fig. 13.9 Combustion synthesis classification [63].

These studies also concluded that the amount of fuel use has a significant impact on synthesized nanomaterial properties. According to [66], the energy required by the combustion synthesis reaction can be obtained by utilizing NH_3, Al_2O_3, and HNO_3; these fuels undergo decomposition reaction and release a huge amount of energy that be utilized in the process. The equation below represents the chemical reaction when Al_2O_3 is used as an oxidizer/fuel agent.

$$2Al(NO_3)_3 + 3CO(NH_2)_2 \rightarrow AL_2O_3 + 6H_2O + 6N_2 + 3CO_2 \quad (13.2)$$

Hence, they are considered to be the most appropriate combination to be used as fuel. The amount of energy released by the exothermic reaction is enough to crystalize and purified the synthesized nanomaterials, and any additional procedure of thermal treatment and purification can be exempted [66]. The combustion reaction between the metal nitrates and glycine is represented by the stoichiometric Eq. [65]. In the equation, ν is the metal valence, and the metal oxidizer to fuel ratio is mentioned by φ:

$$\begin{aligned} &M^{\nu}(10/9\varphi)NH_2CH_2COOH + \frac{10}{4}(\varphi - 1)O_2 \\ &M^{\nu}O_{\nu/2}(S) + \left(\frac{20}{9}\varphi\right)CO_2(g) \\ &+ \frac{25}{9}\varphi H_2O(g) + \left(5\varphi + \frac{9}{9}\right)N_2(g) \end{aligned} \quad (13.3)$$

The maximum temperate is attained, which in reaction maintains the size of the particles, surface area, number of gaseous particles, and porosity. The temperature after the combustion reaction is decreased by the additional gaseous fumes emitting from the reaction; this is because; these gaseous molecules absorb energy emitting from the chemical reaction. Furthermore, these fumes cause channels to be aligned, creating pores in nanomaterials, due to which the nanoparticle obtained haves large pores, reduced aggregation, and smaller particle sizes. The significance of a single step combustion technique lies based on the following key steps:

- The homogenous mixture of reactants is possible at the molecular level due to the liquid phase.
- The level of crystallinity and purity is higher due to the increased temperature of the reaction.
- High temperatures and gaseous emissions sustain the smaller size and increased pores.
- The requirement of external energy is minimized because of the exothermic reaction.

Although the process is considered significant and has numerous benefits and convenience in nanomaterials synthesis, the reaction also possesses its side effects in the pore' uncontrollable size form and agglomeration observation in the synthesized product. These limitations can be reduced by controlling the combustion temperature of the reaction and chilling of amorphous nanomaterials prior to the combustion. For the adequate pre colling of the amorphous nanoparticles, several techniques are being evolved and used; among them, the Impregnated Layer Combustion Synthesis (ILCS) is considered sufficient, as shown in Fig. 13.10. In this technique, a substrate is reacted with the amorphous nanoparticle, which functions as initial energy that initiates the reaction similarly to the performed in the combustion reaction technique [65]. Based on the nature of the substrate used, the ILCS process is further divided into two categories. The usage of the inert substrate leads to the process of Impregnated Support Combustion (ISC); whereas, the active substrates lead to Impregnated Active Layer Combustion Synthesis (IALC) [65].

The higher surface areas that support catalyst $>200\,m^2/g$ utilizes the ISC method, in which the reaction of combustion takes place in the heterogeneous environment [63]. This reaction technique is not considered adequate for the chemical reactions that require lower exothermic reactions as additional fuel will be required because the reaction will not control the wave of self-sustained combustion. On the contrary, the IALC is used to provide additional energy required for the combustion in the reduced exothermic systems [63].

13.5 Properties

13.5.1 Physicochemical properties

For the observation of characteristics possessed by the nanoparticles synthesized from the membrane, the instrument of Transmission Electron Microscopy (TEM) is considered to be significant [36]. The apparatus enables us to elaborate on the membrane's inner structure, as shown in Fig. 13.11. The first part of Fig. 13.11A represents the internal structure of the synthesized nanoparticle. It is visible by TEM usage that the nanoparticle is off white and has cylindrical shapes; the gray circles represent the presence of a cell membrane. In comparison, the nanoparticles' inner structure is spherical with a sharp color that contrasts, and the white color patches represent the vesicles' hollow cells. The additional membrane coating increases the applicability of nanoparticles enriching it with potential and colloidal

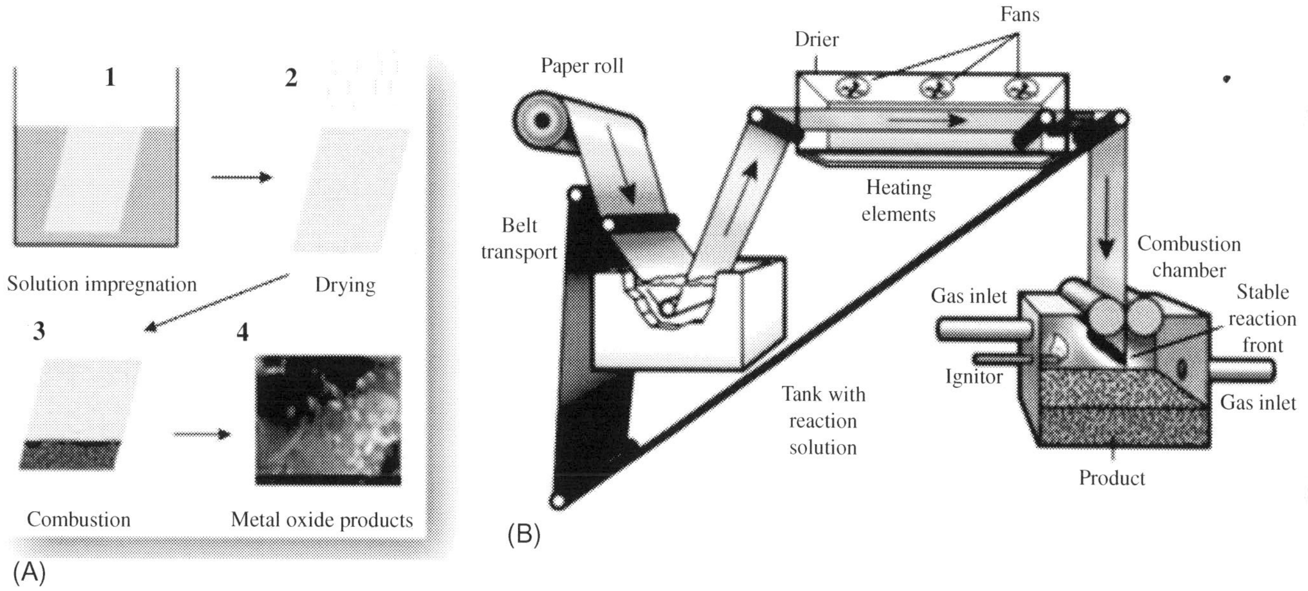

Fig. 13.10 (A) Sequential steps involve ILCS synthesis and (B) apparatus for nano-powders synthesis [63].

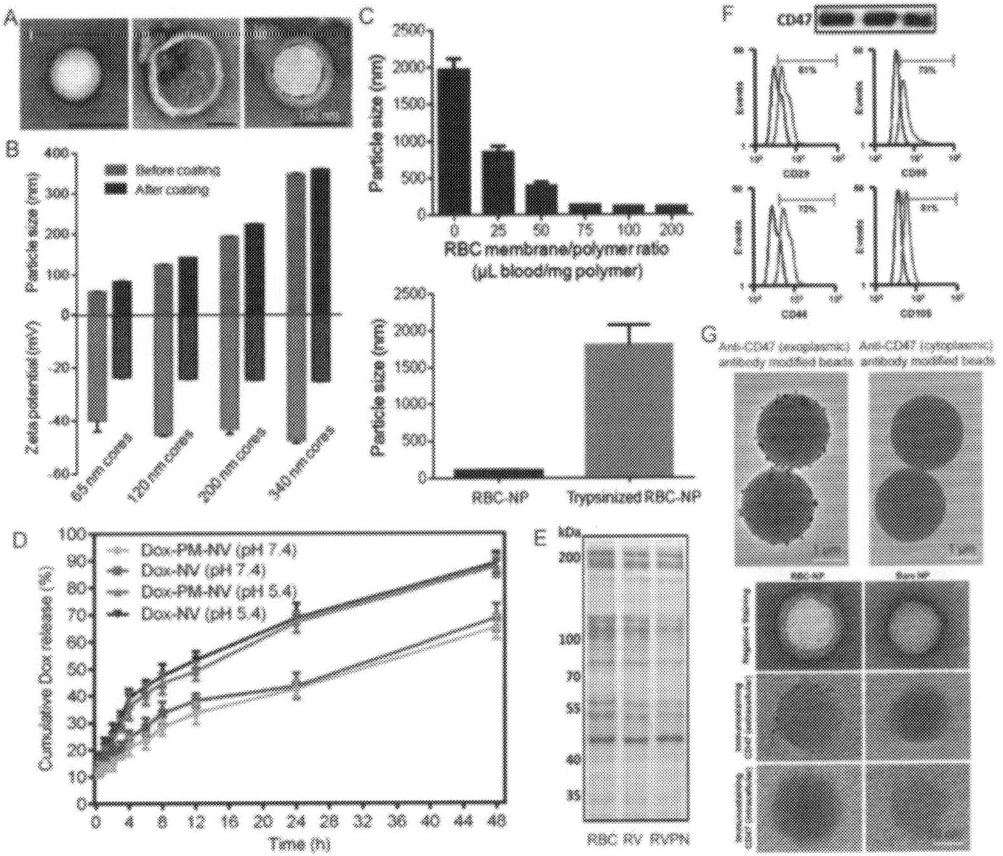

Fig. 13.11 Characterization of coated nanomaterials: (A) TEM image of nanoparticle; (B) difference in size and potential of nanoparticle; (C) cell membrane to polymer ratio; (D) dox loaded vesicles; € SDS-PAGE results; (F) Western blot experiment; and (G) TEM image of nanoparticle collide with antibodies.

stability. The surfactant usage stabilizes the nanoparticles would be aggregate by the electrostatic field repulsion if stored for a longer period. On the contrary, if the nanoparticle is synthesized from zwitterionic protein-membrane. It will reflect resistance to hypertonic treatments. The diffusion barrier was in the form of membrane coating that can be induced to prevent the spillage of drugs injected in the nanoparticles. These features enable them to be implemented in multiple fields generally related to medication and treatment of diseases [47].

13.5.2 Biological properties

The biological functioning and compliance are significant to be observed prior to the usage of nanoparticles. These nanoparticles are widely consumed in medicine for treating living beings; therefore, their characteristics and functionality shall replicate the originally occurring substances. The protein expression of cells is observed using sodium dodecyl sulfate-polyacrylamide gel electrophoresis (SDS-PAGE) along with the Western Bolt technique [67]. The SDS-PAGE is consumed for contrasting the properties of synthesizing nanoparticle and the properties of the base membrane, as reflected in Fig. 13.11; the western bolt is consumed to identify membrane coated nanoparticles [35]. The biological functionality of nanoparticles is evaluated based on protein profile and protein markers as the parameters. Moreover, the symmetry of nanoparticles is considered significant as they derive the desired functions of the nanoparticles. Antibodies and gold conjugated antibodies are used as a standard benchmark to evaluate the nanoparticles' features reflected in Fig. 13.11.

13.6 Conclusions

The membrane-derived nanoparticles are considered significant because they can be used in the replacement of drugs. The replacement is possible because they replicate biological features found in living organisms. This is why this domain has received significant attention in previous years and is counted as an adequate medical research area. The preparation and synthesis techniques and procedures are being evolved, and more convenient and efficient method procedures are being used. The methods involved in the synthesis of these nanoparticles involve; the synthesis by membrane extraction, which is a biological synthesis method and is considered the most relevant synthesis method among all. The chemical synthesis method is by vesicle extraction. The techniques involved in the synthesis are the electrospinning and combustion techniques, among which the combustion technique is considered to be more effective in the synthesis of these nanoparticles' membrane. In combustion technique processes, an exothermic reaction occurs in which excessive temperature and gaseous fumes are released. The membrane-derived nanomaterials are widely consumed in the medical fields and have various features depending upon their parent cell. The nanoparticle's ability to replicate as a biological material is considered an adequate replacement for drugs.

References

[1] R.H. Fang, L. Zhang, Nanoparticle-based modulation of the immune system, Annu. Rev. Chem. Biomol. Eng. 7 (2016) 305–326.
[2] A.Z. Wang, R. Langer, O.C. Farokhzad, Nanoparticle delivery of cancer drugs, Annu. Rev. Med. 63 (2012) 185–198.
[3] R.H. Fang, A.V. Kroll, L. Zhang, Nanoparticle-based manipulation of antigen-presenting cells for cancer immunotherapy, Small 11 (41) (2015) 5483–5496.
[4] J.C. Harris, M.A. Scully, E.S. Day, Cancer cell membrane-coated nanoparticles for cancer management, Cancer 11 (12) (2019) 1836.
[5] D. Dehaini, R.H. Fang, L. Zhang, Biomimetic strategies for targeted nanoparticle delivery, Bioeng. Transl. Med. 1 (1) (2016) 30–46.
[6] F. Alexis, E. Pridgen, L.K. Molnar, O.C. Farokhzad, Factors affecting the clearance and biodistribution of polymeric nanoparticles, Mol. Pharm. 5 (4) (2008) 505–515.
[7] C.M.J. Hu, R.H. Fang, L. Zhang, Erythrocyte-inspired delivery systems, Adv. Healthc. Mater. 1 (5) (2012) 537–547.
[8] R.H. Fang, C.M.J. Hu, L. Zhang, Nanoparticles disguised as red blood cells to evade the immune system, Expert Opin. Biol. Ther. 12 (4) (2012) 385–389.
[9] R.H. Fang, Y. Jiang, J.C. Fang, L. Zhang, Cell membrane-derived nanomaterials for biomedical applications, Biomaterials 128 (2017) 69–83.
[10] M. Colombo, G. Raposo, C. Théry, Biogenesis, secretion, and intercellular interactions of exosomes and other extracellular vesicles, Annu. Rev. Cell Dev. Biol. 30 (2014) 255–289.
[11] W. Xu, Z. Yang, N. Lu, From pathogenesis to clinical application: insights into exosomes as transfer vectors in cancer, J. Exp. Clin. Cancer Res. 35 (1) (2016) 1–12.
[12] X. Zhang, Z. Pei, J. Chen, C. Ji, J. Xu, X. Zhang, J. Wang, Exosomes for immunoregulation and therapeutic intervention in cancer, J. Cancer 7 (9) (2016) 1081.
[13] S.E. Andaloussi, I. Mäger, X.O. Breakefield, M.J. Wood, Extracellular vesicles: biology and emerging therapeutic opportunities, Nat. Rev. Drug Discov. 12 (5) (2013) 347–357.
[14] S. Aryal, C.M.J. Hu, R.H. Fang, D. Dehaini, C. Carpenter, D.E. Zhang, L. Zhang, Erythrocyte membrane-cloaked polymeric nanoparticles for controlled drug loading and release, Nanomedicine 8 (8) (2013) 1271–1280.
[15] B.T. Luk, R.H. Fang, C.M.J. Hu, J.A. Copp, S. Thamphiwatana, D. Dehaini, L. Zhang, Safe and immunocompatible nanocarriers cloaked in RBC membranes for drug delivery to treat solid tumors, Theranostics 6 (7) (2016) 1004.
[16] A.V. Kroll, R.H. Fang, L. Zhang, Biointerfacing and applications of cell membrane-coated nanoparticles, Bioconjug. Chem. 28 (1) (2017) 23–32.

[17] R.H. Fang, C.M.J. Hu, K.N. Chen, B.T. Luk, C.W. Carpenter, W. Gao, L. Zhang, Lipid-insertion enables targeting functionalization of erythrocyte membrane-cloaked nanoparticles, Nanoscale 5 (19) (2013) 8884–8888.
[18] J. Su, H. Sun, Q. Meng, Q. Yin, S. Tang, P. Zhang, Y. Li, Long circulation red-blood-cell-mimetic nanoparticles with peptide-enhanced tumor penetration for simultaneously inhibiting growth and lung metastasis of breast cancer, Adv. Funct. Mater. 26 (8) (2016) 1243–1252.
[19] H. Sun, J. Su, Q. Meng, Q. Yin, L. Chen, W. Gu, Y. Li, Cancer cell membrane-coated gold nanocages with hyperthermia-triggered drug release and homotypic target inhibit growth and metastasis of breast cancer, Adv. Funct. Mater. 27 (3) (2017) 1604300.
[20] L. Rao, B. Cai, L.L. Bu, Q.Q. Liao, S.S. Guo, X.Z. Zhao, W. Liu, Microfluidic electroporation-facilitated synthesis of erythrocyte membrane-coated magnetic nanoparticles for enhanced imaging-guided cancer therapy, ACS Nano 11 (4) (2017) 3496–3505.
[21] J.Y. Zhu, D.W. Zheng, M.K. Zhang, W.Y. Yu, W.X. Qiu, J.J. Hu, X.Z. Zhang, Preferential cancer cell self-recognition and tumor self-targeting by coating nanoparticles with homotypic cancer cell membranes, Nano Lett. 16 (9) (2016) 5895–5901.
[22] D. Zhang, J. Zhang, Surface engineering of nanomaterials with phospholipid-polyethylene glycol-derived functional conjugates for molecular imaging and targeted therapy, Biomaterials 230 (2020) 119646.
[23] R.H. Fang, C.M.J. Hu, B.T. Luk, W. Gao, J.A. Copp, Y. Tai, L. Zhang, Cancer cell membrane-coated nanoparticles for anticancer vaccination and drug delivery, Nano Lett. 14 (4) (2014) 2181–2188.
[24] A. Parodi, N. Quattrocchi, A.L. van de Ven, C. Chiappini, M. Evangelopoulos, J.O. Martinez, L. Isenhart, Biomimetic functionalization with leukocyte membranes imparts cell like functions to synthetic particles, Nat. Nanotechnol. 8 (1) (2013) 61.
[25] C. Gao, Z. Lin, B. Jurado-Sánchez, X. Lin, Z. Wu, Q. He, Stem cell membrane-coated nanogels for highly efficient in vivo tumor targeted drug delivery, Small 12 (30) (2016) 4056–4062.
[26] M.J. Hu, R.H. Fang, K.C. Wang, B.T. Luk, S. Thamphiwatana, D. Dehaini, P. Nguyen, P. Angsantikul, et al., Nanoparticle biointerfacing by platelet membrane cloaking, Nature 526 (7571) (2015) 118–121.
[27] Q. Hu, W. Sun, C. Qian, C. Wang, H.N. Bomba, Z. Gu, Nanomedicine: anticancer platelet-mimicking nanovehicles, Adv. Mater. 27 (44) (2015) 7014.
[28] Q. Hu, C. Qian, W. Sun, J. Wang, Z. Chen, H.N. Bomba, Z. Gu, Engineered nanoplatelets for enhanced treatment of multiple myeloma and thrombus, Adv. Mater. 28 (43) (2016) 9573–9580.
[29] K. Simons, W.L. Vaz, Model systems, lipid rafts, and cell membranes, Annu. Rev. Biophys. Biomol. Struct. 33 (2004) 269–295.
[30] N.G. Sosale, K.R. Spinler, C. Alvey, D.E. Discher, Macrophage engulfment of a cell or nanoparticle is regulated by unavoidable opsonization, a species-specific 'Marker of Self'CD47, and target physical properties, Curr. Opin. Immunol. 35 (2015) 107–112.
[31] D.L. Sprague, B.D. Elzey, S.A. Crist, T.J. Waldschmidt, R.J. Jensen, T.L. Ratliff, Platelet-mediated modulation of adaptive immunity: unique delivery of CD154 signal by platelet-derived membrane vesicles, Blood 111 (10) (2008) 5028–5036.
[32] J.M. Pawelek, K.B. Low, D. Bermudes, Bacteria as tumour-targeting vectors, Lancet Oncol. 4 (9) (2003) 548–556.
[33] K. Neumann, A. Lilienkampf, M. Bradley, Responsive polymeric nanoparticles for controlled drug delivery, Polym. Int. 66 (12) (2017) 1756–1764.
[34] W. Gao, C.M.J. Hu, R.H. Fang, B.T. Luk, J. Su, L. Zhang, Surface functionalization of gold nanoparticles with red blood cell membranes, Adv. Mater. 25 (26) (2013) 3549–3553.

[35] J. Liu, H.J. Li, Y.L. Luo, C.F. Xu, X.J. Du, J.Z. Du, J. Wang, Enhanced primary tumor penetration facilitates nanoparticle draining into lymph nodes after systemic injection for tumor metastasis inhibition, ACS Nano 13 (8) (2019) 8648–8658.
[36] C.M.J. Hu, L. Zhang, S. Aryal, C. Cheung, R.H. Fang, L. Zhang, Erythrocyte membrane-camouflaged polymeric nanoparticles as a biomimetic delivery platform, Proc. Natl. Acad. Sci. 108 (27) (2011) 10980–10985.
[37] A. Narain, S. Asawa, V. Chhabria, Y. Patil-Sen, Cell membrane coated nanoparticles: next-generation therapeutics, Nanomedicine 12 (21) (2017) 2677–2692.
[38] F. Szoka Jr., D. Papahadjopoulos, Comparative properties and methods of preparation of lipid vesicles (liposomes), Annu. Rev. Biophys. Bioeng. 9 (1) (1980) 467–508.
[39] R.L. McCall, R.W. Sirianni, PLGA nanoparticles formed by single-or double-emulsion with vitamin E-TPGS, J. Vis. Exp. 82 (2013), e51015.
[40] D. Nie, Z. Dai, J. Li, Y. Yang, Z. Xi, J. Wang, R. Wang, Cancer-cell-membrane-coated nanoparticles with a yolk–shell structure augment cancer chemotherapy, Nano Lett. 20 (2) (2019) 936–946.
[41] L. Xing, Q. Shi, K. Zheng, M. Shen, J. Ma, F. Li, L. Du, Ultrasound-mediated microbubble destruction (UMMD) facilitates the delivery of CA19-9 targeted and paclitaxel loaded mPEG-PLGA-PLL nanoparticles in pancreatic cancer, Theranostics 6 (10) (2016) 1573.
[42] B.T. Luk, C.M.J. Hu, R.H. Fang, D. Dehaini, C. Carpenter, W. Gao, L. Zhang, Interfacial interactions between natural RBC membranes and synthetic polymeric nanoparticles, Nanoscale 6 (5) (2014) 2730–2737.
[43] S.D. Li, L. Huang, Pharmacokinetics and biodistribution of nanoparticles, Mol. Pharm. 5 (4) (2008) 496–504.
[44] B.T. Luk, L. Zhang, Cell membrane-camouflaged nanoparticles for drug delivery, J. Control. Release 220 (2015) 600–607.
[45] V.R. Sinha, K. Bansal, R. Kaushik, R. Kumria, A. Trehan, Poly-ϵ-caprolactone microspheres and nanospheres: an overview, Int. J. Pharm. 278 (1) (2004) 1–23.
[46] M. Xuan, J. Shao, L. Dai, J. Li, Q. He, Macrophage cell membrane camouflaged au nanoshells for in vivo prolonged circulation life and enhanced cancer photothermal therapy, ACS Appl. Mater. Interfaces 8 (15) (2016) 9610–9618.
[47] J.G. Piao, L. Wang, F. Gao, Y.Z. You, Y. Xiong, L. Yang, Erythrocyte membrane is an alternative coating to polyethylene glycol for prolonging the circulation lifetime of gold nanocages for photothermal therapy, ACS Nano 8 (10) (2014) 10414–10425.
[48] C. Li, X.Q. Yang, J. An, K. Cheng, X.L. Hou, X.S. Zhang, Y.D. Zhao, Red blood cell membrane-enveloped O_2 self-supplementing biomimetic nanoparticles for tumor imaging-guided enhanced sonodynamic therapy, Theranostics 10 (2) (2020) 867.
[49] C.M.J. Hu, R.H. Fang, B.T. Luk, K.N. Chen, C. Carpenter, W. Gao, L. Zhang, 'Marker-of-self'functionalization of nanoscale particles through a top-down cellular membrane coating approach, Nanoscale 5 (7) (2013) 2664–2668.
[50] A. Samad, Y. Sultana, M. Aqil, Liposomal drug delivery systems: an update review, Curr. Drug Deliv. 4 (4) (2007) 297–305.
[51] P. Guo, J. Huang, Y. Zhao, C.R. Martin, R.N. Zare, M.A. Moses, Nanomaterial preparation by extrusion through nanoporous membranes, Small 14 (18) (2018) 1703493.
[52] F. Szoka, F. Olson, T. Heath, W. Vail, E. Mayhew, D. Papahadjopoulos, Preparation of unilamellar liposomes of intermediate size (0.1–0.2 μm) by a combination of reverse phase evaporation and extrusion through polycarbonate membranes, Biochim. Biophys. Acta Biomembr. 601 (1980) 559–571.
[53] S. Farrell, K.K. Sirkar, Controlled release of liposomes, J. Membr. Sci. 127 (2) (1997) 223–227.
[54] R.C. MacDonald, R.I. MacDonald, B.P.M. Menco, K. Takeshita, N.K. Subbarao, L.R. Hu, Small-volume extrusion apparatus for preparation of large, unilamellar vesicles, Biochim. Biophys. Acta Biomembr. 1061 (2) (1991) 297–303.

[55] F. Olson, C.A. Hunt, F.C. Szoka, W.J. Vail, D. Papahadjopoulos, Preparation of liposomes of defined size distribution by extrusion through polycarbonate membranes, Biochim. Biophys. Acta Biomembr. 557 (1) (1979) 9–23.
[56] L.D. Tijing, J.S. Choi, S. Lee, S.H. Kim, H.K. Shon, Recent progress of membrane distillation using electrospun nanofibrous membrane, J. Membr. Sci. 453 (2014) 435–462.
[57] O.O. Dosunmu, G.G. Chase, W. Kataphinan, D.H. Reneker, Electrospinning of polymer nanofibres from multiple jets on a porous tubular surface, Nanotechnology 17 (4) (2006) 1123.
[58] A.L. Yarin, E. Zussman, Upward needleless electrospinning of multiple nanofibers, Polymer 45 (9) (2004) 2977–2980.
[59] B. Bhushan (Ed.), Encyclopedia of Nanotechnology (No. 544.1), Springer, Dordrecht, Netherlands, 2012.
[60] W. Wen, J.M. Wu, J.P. Tu, A novel solution combustion synthesis of cobalt oxide nanoparticles as negative-electrode materials for lithium ion batteries, J. Alloys Compd. 513 (2012) 592–596.
[61] A. Varma, A.S. Mukasyan, A.S. Rogachev, K.V. Manukyan, Solution combustion synthesis of nanoscale materials, Chem. Rev. 116 (23) (2016) 14493–14586.
[62] F. Maglia, U. Anselmi-Tamburini, G. Spinolo, Z.A. Munir, Zirconia-based metastable solid solutions through self-propagating high-temperature synthesis: synthesis, characterization, and mechanistic investigations, J. Am. Ceram. Soc. 83 (8) (2000) 1935–1941.
[63] A. Ashok, A. Kumar, F. Tarlochan, Preparation of nanoparticles via cellulose-assisted combustion synthesis, Int. J. Self-Propag. High-Temp. Synth. 27 (3) (2018) 141–153.
[64] B. Akgün, H.E. Çamurlu, Y. Topkaya, N. Sevinç, Mechanochemical and volume combustion synthesis of ZrB2, Int. J. Refract. Met. Hard Mater. 29 (5) (2011) 601–607.
[65] R.K. Lenka, T. Mahata, P.K. Sinha, A.K. Tyagi, Combustion synthesis of gadolinia-doped ceria using glycine and urea fuels, J. Alloys Compd. 466 (1–2) (2008) 326–329.
[66] N. Kasapoğlu, A. Baykal, Y. Köseoğlu, M.S. Toprak, Microwave-assisted combustion synthesis of $CoFe_2O_4$ with urea, and its magnetic characterization, Scr. Mater. 57 (5) (2007) 441–444.
[67] T. Sun, Y.S. Zhang, B. Pang, D.C. Hyun, M. Yang, Y. Xia, Engineered nanoparticles for drug delivery in cancer therapy, Angew. Chem. Int. Ed. 53 (46) (2014) 12320–12364.

CHAPTER 14

Nanomaterial-based coatings

Abbreviations

CO_2 carbon dioxide
Cu copper
CVD chemical vapor deposition
MA mechanical alloying
MPa megaPascal
PTW potassium titanate whisker
TiO_2 titanium dioxide
TTIP titanium tetraisopropoxide
ZnO zinc oxide

14.1 Introduction

The transformation of nanostructured materials in the form of coatings enhance the characteristics, i.e., chemical characteristics, the resistance toward oxidation and wear, flame retardants, electrical features, mechanical characteristics, surface features, and permeability. For instance, borides, oxides, and transition metals carbides were amalgamated into a thermoset resin for manufacturing a callous coating by employing plasma assistance techniques. Adding up the layered structure material, i.e., Nano clay and double-layered oxides, helps increase the coating's permeability. The thermal resistance could be enhanced by using carbonaceous materials, i.e., nanotube, graphene-based materials, black carbon, and carbon fiber [1]. Few of the photocatalytic additives, i.e., ZnO and TiO_2, were included for creating coating to improve its photocatalytic tendency [2].

Hint statement

The coating through nanocomposites is not only effective in preserving the quality of the materials but they play greater role as multipronged nanofillers.

Synthetic Engineering Materials and Nanotechnology
https://doi.org/10.1016/B978-0-12-824001-4.00017-X

Treating surfaces can facilitate superhydrophobic coating for a myriad of applications. These coatings have comprised the water-repellent agents, and their basis has been laid on perfluorinated compounds or saline compounds [3]. Similarly, nanoparticles of silver oxides could be used in the coating to increase the quality of its antibacterial activity. Moreover, self-healings and anticorrosive coatings are used as nano-filler to manufacture nanocontainers. It carries nanosized volume and comprises active constituents that are a healing agent, loaded nanocontainer, and inhibitor-loaded nanocontainer. These nanocontainers can avert the immediate contact of these active agents and the matrix of coating. Nanomaterials are also found as well-disseminated in the coating matrix when unfettering the healing agents amid critical settings. These may be any mechanical rapture, light change, and variation in pH [4]. The halloysite clay nanotubes are also part of nanocontainers [5, 6], supramolecular nanocontainers [7], nano-silica [8], zirconia nanospheres [7], and nanoceria [9, 10].

For the organic matrix or the polymer-based nanocomposite, thermoplastics and elastomers have been used to manufacture nanocomposite coatings, i.e., epoxy [11], chitosan [12], polypyrrole [13], polymers containing reactive trimethylsilyl [14], polystyrene [15], polyurethane [16], chitosan [12], polyethylene glycol [17], polyvinylidene fluoride, polyaniline [18], polyamic acid and polyimide [19], rubber-modified polybenzoxazine [20], pullulan, fluoroacrylic polymer, ethylene tetrafluoroethylene, poly (*N*-vinyl carbazole), polycarbonate, fluorinated polysiloxane, polyester, polyacrylic, polyvinyl alcohol, polydimethylsiloxane, polyamide, polyacrylate, and ultraviolet-curable polymers [21]. Moreover, in the inorganic matrix, metal matrix composites comprised the second phase that has been remain under investigation [22].

Hint statement

Coating the nanomaterials increase the characteristics i.e., chemical characteristics, the resistance toward oxidation and wear, flame retardants, electrical features, mechanical characteristics, surface features, and permeability.

Various methods and techniques are adopted to process the molecular disposition on a specific substrate for synthesizing nanomaterial coatings. Some manufacturing nanomaterial-based coatings are sol–gel methods, chemical vapor deposition (CVD), physical vapor deposition, cold spray, etc. Simultaneously, the techniques included layer-by-layer assembly, self-assembly [23], and atomic layer deposition [24]. Since, a plethora of methods are available to synthesize the nanomaterials' coatings, and they are selected based on their applications. These applications include antiwear, anticorrosion, self-cleaning, antibacterial, and superhydrophobic and electronics. The coating through nanocomposites does not merely work to preserve the quality of the materials but also play a greater role as multipronged nanofillers.

Some examples of preserving the materials by nanomaterial-based coatings include comfortable and clean coatings for the buildings, making substrates to protect, and energy and water conservative apparatus [1].

Example 14.1

Calculate the free volume/unit cell if the Chromium has a BCC structure, and its atomic radius is 0.1249 nm.

Solution:

Atomic radius of chromium, $r = 0.1249$ nm.

If "a" is the BCC unit cell edge length, then the relation between "a" and "r" is

$$a = \frac{4}{\sqrt{3}} r = \frac{4}{\sqrt{3}} \times 0.1249$$

The volume of the unit cell, $V = a^3 = (0.28845)^3$ nm^3.
$= 0.024$ nm^3.

Number of atoms in BCC unit cell = 2.

Hence the volume of atoms in unit cell:

$$v = \frac{4}{3} \pi r^3 \times 2 = 0.01633 \text{ nm}^3$$

Free volume/unit cell = $V - v = 0.00767$ nm^3.

Hint statement

The usage of nanoparticles of silver oxides increases the quality of its antibacterial activity.

14.2 Methods of manufacturing nanomaterial-based coatings

14.2.1 Sol–gel method

This method is regarded as an appropriate method to acquire enhanced quality films that are characterized by micron thickness. The effectiveness of this method is considered equivalent to the techniques of physical deposition.

However, the method contains some limitations, such as the tensile stress tendency that may be developed during drying, despite its effectiveness. This can lead to fissures in the films. In the inorganic matrix, the sol–gel method is used for inorganic nanofillers, i.e., I/I coatings of I/I coatings. When there is an organic matrix, the sol–gel method is considered an appropriate way to formulate inorganic nanophases in the organic matrix phase. The metal oxides of aluminum, titanium, silicon, and zirconium are used to create nanocomposite coatings. Various types of oligomers, including organic constituents with lower molecular mass, are known as the organic phase precursors. Amid the predefined settings, organic molecules and silanes could create coatings that carry silica nanoparticles or nanophases of silica. If the coupling agent does not include an agent, the inorganic and organic phases could be associated covalently [21]. Fig. 14.1 provides the process of creating coatings via the sol–gel method.

Hint statement

These coatings are useful in preserving the building materials, making substrates to protect and energy and water conservative apparatus.

14.2.2 Cold spray method

In contrast to the conventional thermal spray, cold spraying facilitates the production of coatings while having the least temperatures compared to the point of melting of the materials that are melted. The cold spray method is processed at a lower temperature to avert the corrosion and decay along with the transitional variations amid the process. The coatings that are manufactured are characterized by the lower porosity (280 MPa) as well as stronger adhesion (>70 MPa). The method is employed to manufacture the coatings of nanocomposite features with the metallic matrix, i.e., Cu, aluminum, carbon monoxide or alloy matrix, and nanofillers. These nanofillers are diamond, boride, nitrite, and carbide, etc. [26]. In order to formulate the powders of nanocomposites through the cold spray method, the technique of mechanical alloying (MA) can be employed with the powders of metallic matrix and nanoparticles [27]. Fig. 14.2 presents the process of cold- spray method to create nanomaterial-based coatings.

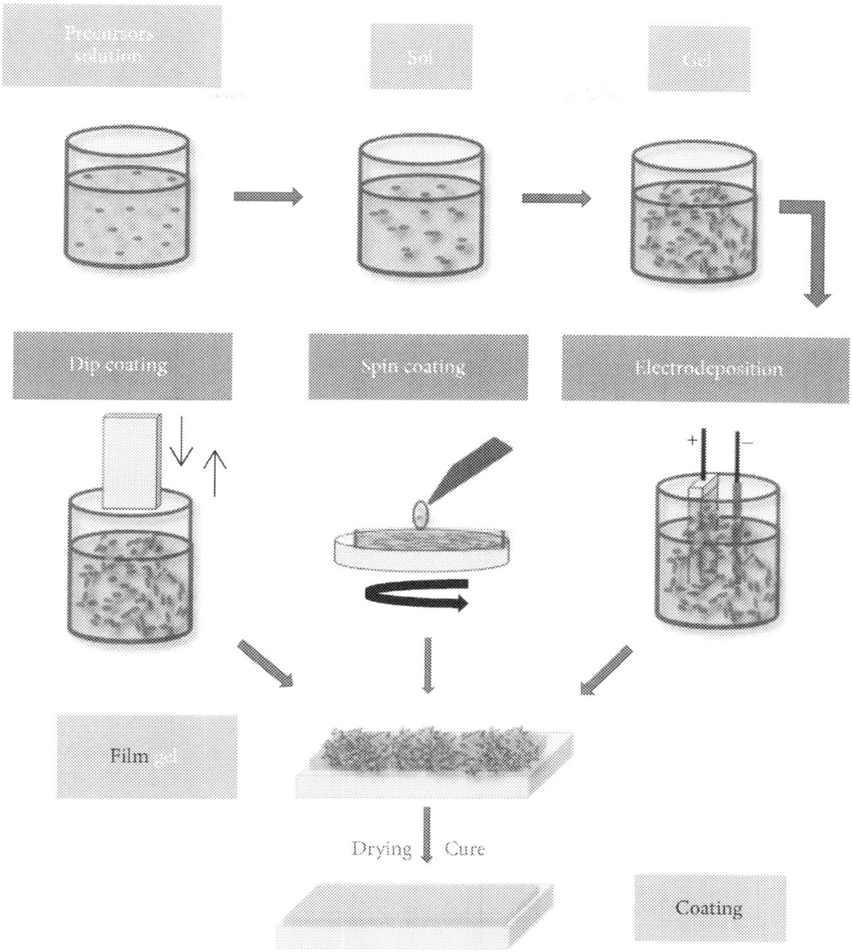

Fig. 14.1 Sol–gel method of manufacturing nanomaterial-based coatings [25].

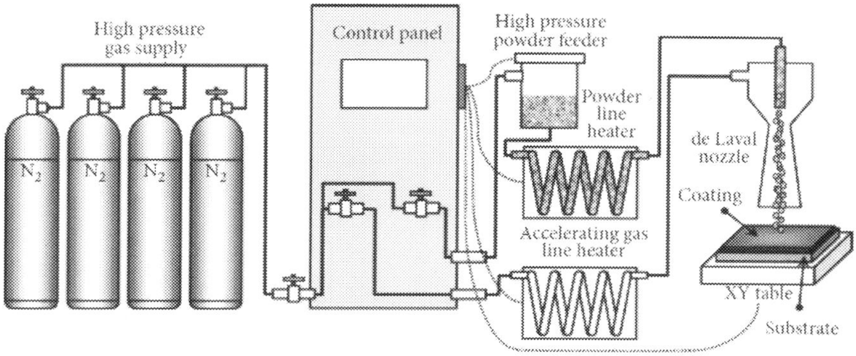

Fig. 14.2 Cold spray method [28].

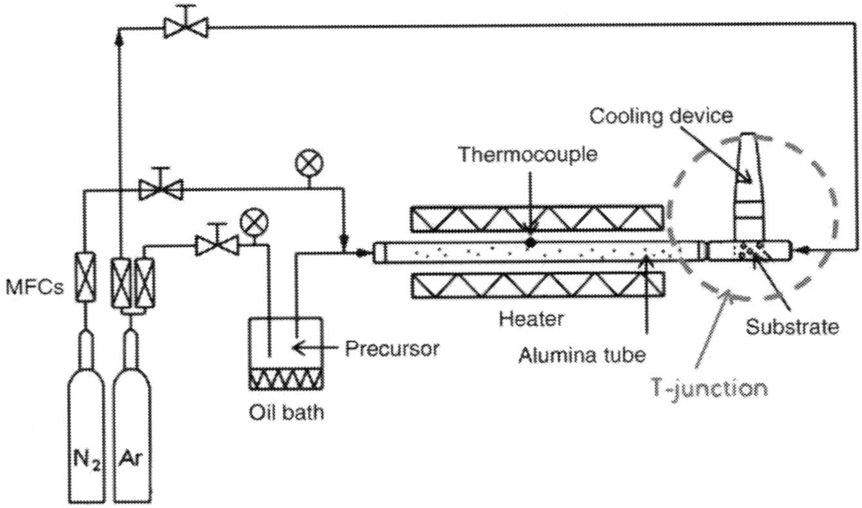

Fig. 14.3 CVD method [29].

14.2.3 Chemical vapor deposition method

The method of CVD is used to manufacture the coatings of I/I nanocomposite. Their major constituents are the inorganic matrix and nanofillers. It has enhanced the quality of the coating; the aerosol-assisted CVD method is considered appropriate. While the coatings of O/I are nanocomposite, the CVD method is also used along with the platinum precursors and hexafluoroacetylacetonate. It permits creating the layers of nanocomposite that are made of the organic substrate through a single phase that exhibits both the electric and ionic conductivities. One of the advantages of this method includes the construction of the finest quality films. Fig. 14.3 shows the experimental mechanism for the nanoparticle TiO_2-glass bead coating via the CVD method. The stimulant of the TiO_2 included titanium tetra isopropoxide. It was melted and further went through vaporization at the temperature limit of 90 °C inside a bubbler while using argon gas. The solvent mixture of Titanium tetraisopropoxide (TTIP) that has been volatilized and the combination of nitrogen and argon gases were undergone through the tube of alumina. The temperature was remained constant at the level of 900 °C. Further, alumina's tube was linked with a T-junction, horizontal shape that comprised soda-lime glass beads [30].

Example 14.2

Calculate the maximum kinetic energy of an electron ejected from silver by a 3.13×10^{15} Hz photon.

Solution:

$$KE_{max} = hf - \varnothing.$$

$$KE_{max} = 4.14 \cdot 10 \cdot {}^{15}eVs \cdot 3.13 \cdot 10 \cdot {}^{15}Hz - 4.7eV$$

$$KE_{max} = 8.22eV$$

14.2.4 Supercritical antisolvent (SAS) process

The illustration of this method has been shown in Fig. 14.4 this method's experimental apparatus includes a system that could supply carbon dioxide CO_2, a system for solution delivery availability. This vessel could provide higher pressure with a capacity of 1 L (Parr Instruments, USA). The vessel that carries higher pressure is delved into the water to maintain the temperature level at a constant level amid the experience. Moreover, a metering pump (Model EL-1A, AMERICAN LEWA®, USA) was used to dispense CO_2 in liquid form via a cylinder for the vessel [31]. More importantly, before inserting the pump's head, the temperature for the liquid form of CO_2 was reduced at the lowest levels. It was aimed to deteriorate the cavitation. Once the pump head is left in the vessel, a liquid form of CO_2 was provided through a heating tape. Subsequently, a solution that carried

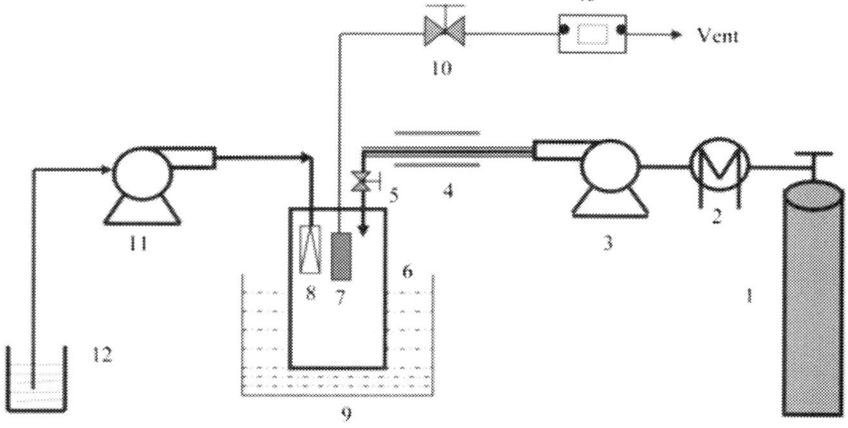

Fig. 14.4 Supercritical antisolvent (SAS) process.

polymer was arranged, and Eudragit was dissolving in acetone. The silica nanoparticles were also dissolved into the polymer solution to obtain an anticipated proportion of polymer. Notably, the 600 nm of silica particles' surface area is not more than 6–20 nm silica; therefore, not a higher amount of polymer was required to manufacture the coatings of 600-nm via nanoparticles. Consequently, 14%–20% of the polymer was employed to coat the silica amounted to 600-nm. Also, an ultrasonicator was employed to breakdown the agglomerates of nanoparticles in the suspension of silica-acetone. Amid the experiments, pressure and heat were retained 8.27 MPa and 305.5 K sequentially. As the steady settings arrive at the higher levels of pressure and heat, the temperature of CO_2 reached the stabilized position that led to the delivery of suspension via a high-pressure pump (Beckman, 110B). The duration of spraying was 20 min, and settling took 30 min. Consequent to the cleaning, the high-pressure vessel gradually lost pressure, and samples were taken to evaluate the features. Fig. 14.4 exhibits the coating process of nanomaterials (Table 14.1).

According to Wang et al. [31], for evaluating the effectiveness of the CO_2 in the process of SAS, the hydrophilic silica nanoparticles and hydrophobic silica were used, and their size was different. These were acquired from Degussa, USA, and Catalysts & Chemicals Ind. Co., Japan. Table 14.2 shows the details of silica particles that were used during the experiment.

Table 14.1 provides the experimental parameters [31].

Experiments	Parameters: Polymer concentration (g/100 mL)	Parameters: The ratio of polymer to nanoparticles (g/g)
16 nm hydrophobic silica coating	0.8	1:2
20 nm hydrophobic silica coating	0.8	1:1
600 nm hydrophobic silica coating	0.4	1:4

Table 14.2 description of silica nanoparticles.

Suppliers	Catalyst and Chemical Industry Company, Japan	Degussa (USA)	
Trade name	COSMO 55	Aerosil ® 90	Aerosil ® 90
Particle size (nm)	600	20	16
Surface property	Hydrophilic	Hydrophilic	Hydrophilic

14.2.5 Layer by layer method

The polyelectrolytes multilayer are the alternating charges that can be placed on the substrate of metal. The corrosion inhibitor can be embedded in between the deposited polyelectrolyte layer, as elaborated in the figure. For the controllable growth of multilayers, the core factors include; pH and strength of ion, that is, the degree of ionization of dissociation of the polyelectrolytes. The thickness of multilayers and deposition of polyelectrolytes is significantly affected by these factors [32]. The addition of pH or additional salt can affect the conformation of the polyelectrolytes. The polymeric chains in the layer are collapsed due to the excessive salt, and the expanded chain is transformed into coil form. The polyelectrolytes strength is also dependent on the effectiveness of these factors; the weak polyelectrolyte is easily affected by the change in pH. The maximum charge density can be achieved when the pH is increased, and ions are completely dissociated; this pH range is the adequate range for change charge density (Fig. 14.5).

14.2.6 Emulsion polymerization

The process of emulsion polymerization is utilized to prepare the capsule, as presented in Fig. 14.6. The biphasic liquid is utilized to prepare the capsule by strong agitation; the liquid can be oil and water [34, 35]. These liquids are sonicated, which creates small droplets, and these droplets are then consumed as the base material to form a capsule. The organic interface is being polymerized to form the shell of the capsule. The polymer formation can be formed by polymerizing immiscible monomers together at the organic interface [36]. The free radical of situ can also be consumed for the conduction of the polymerization process.

14.2.7 Nano container-based synthesis

Contrary to the microcapsules, the tubular nano containers are considered to be more significant as they reflect superior aero and hydrodynamic features that will increase the processability. By utilizing the well-defined nanocontainers, the fine-tuned properties can be extracted in the coating material.

The usage of halloysite nanotubes, hydroxyapatite, and double-layered hydroxides are the adequate elements for the conduction, storage, and transportation of inhibitors present in the self-healing and anticorrosive coatings. The whole process can be achieved in three simpler steps, as shown in Fig. 14.7. The initial step is based on the inclusion of organic/inorganic inhibitors into the Nano container. The nanomaterials absorption is

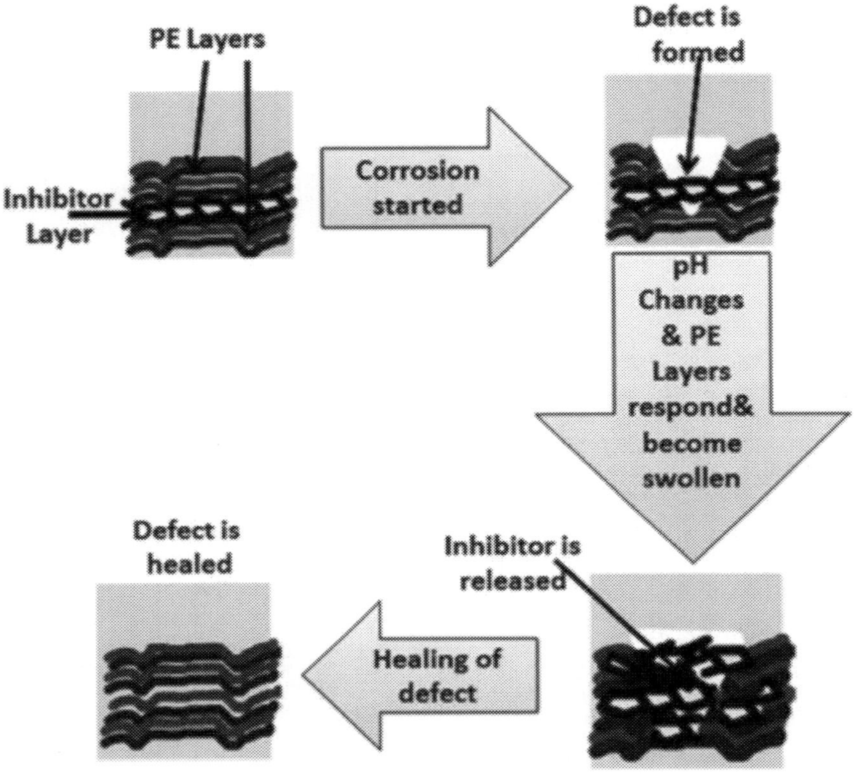

Fig. 14.5 Layer-by-layer deposition process [33].

processed through a porous Nano container structure; the emulsion polymerization is utilized for the encapsulating process. It can also be conducted by utilizing the ion exchange method using counter negative/positive ions in the nanocontainer. The pH-sensitive polyelectrolyte multilayer is used for the coating of filled nanocontainer in the second step. In the final step, relevant polymeric matrix material, either organic or inorganic, is used to dispersion the inhibitor-filled Nano container.

14.3 Techniques of manufacturing nanomaterial-based coatings

14.3.1 Directly solution dip-coating solution

It is one of the conventional techniques for film disposition concerning the exterior of fibrous constituents. Typically, they are arranged during the

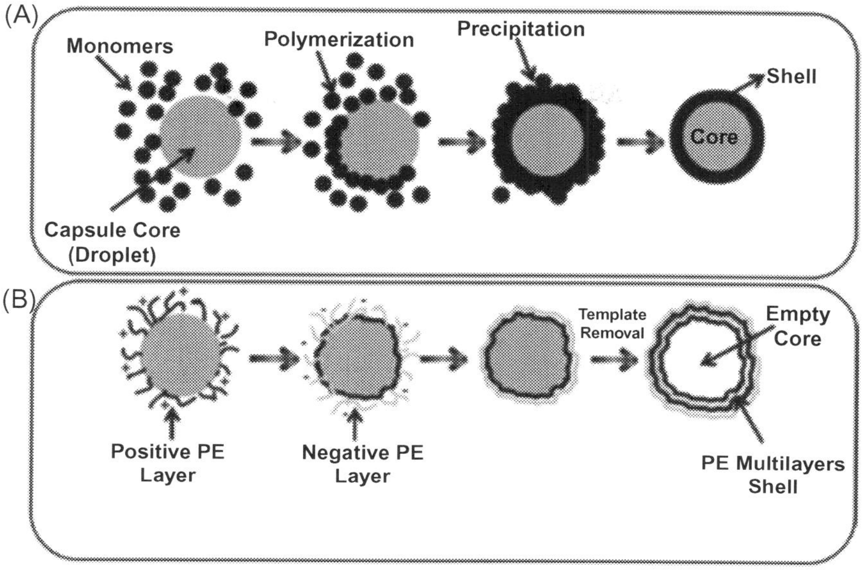

Fig. 14.6 Microcapsule preparation: (A) polymeric shell prepared from emulsion polymerization and (B) layer-by-layer deposition process [33].

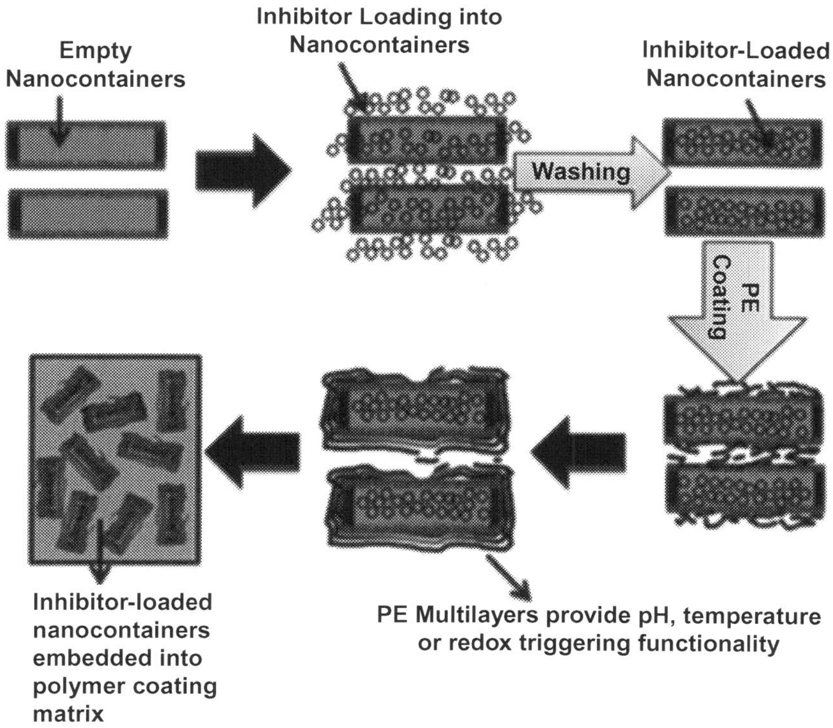

Fig. 14.7 Preparation of inhibitor-filled nano container [33].

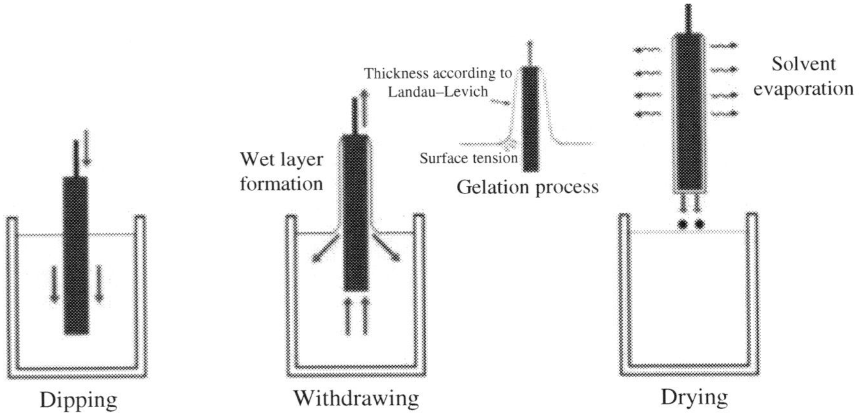

Fig. 14.8 Directly solution dip-coating solution [37].

production of coatings in the industry of textile. In this technique, filaments are drowned into sizing agents' solutions to add up the tenderness and diminish the breaks that are occurred due to weaving. This technique is also useful in implanting the interfacial characteristics of the composites that are fiber-reinforced. These carbon fabric composites are transformed along with the potassium titanate whisker (PTW) and are also manufactured by the solution- dip technique. In ultrasonic processing and to provide heat for the evaporation (Fig. 14.8) [38].

14.3.2 Sol–gel dip-coating

This technique is used in effective textile finishing, and does not involve a higher cost. It is also recognized to have lower consumption of chemicals and improved functions. To enhance the resistance toward fractures of the composites of reinforced fabrics and modify the fiber matrix's bonding conditions, the tetragonal zirconia interfacial layer was initiated by the sol–gel dip-coating for reinforced fiber. Amid the process, the initial precursors comprise metalloid factors and ligands, while the most common precursors are titanium alkoxides and silicon alkoxides. Dip coating is considered a superficial strategy for making film deposition. Notably, composites sol–gel is a finishing method for the fibrous materials, i.e., hybrid SiO_2/HTEOS/CPTS sol, SiO_2/TiO_2-doped cationic coating, and cationic SiO_2/TiO_2 hybrid sol, and F/TiO_2 hybrid sol, for transfer printing [39]. Fig. 14.9 shows the process of sol–gel dip-coating [39].

Moreover, a significant aspect of the process of dip-coating is to evaluate the thickness of the deposited films that can be estimated by using Landau

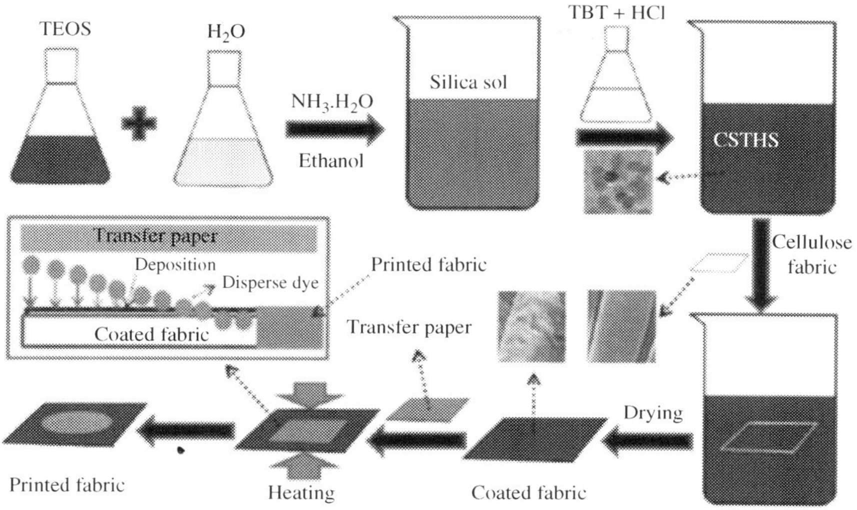

Fig. 14.9 Sol–gel dip-coating process [39].

Levich's theory. The theory predicts the deposited films' thickness while using the equation below:

$$t_1 = 0.944 C_a^{1/6} \left(\frac{nu}{pg} \right)^{1/2} \tag{14.1}$$

C_a refers to the capillary number and given, and ρ shows the thickness, tension at the surface, and the coating solutions' compactness. While U denotes the speed of withdrawal and g reflects the acceleration of gravitation constant. Groenveld devised a model to forecast the viscosity of deposited films:

$$t_1 = j \left(\frac{nu}{pg} \right)^{1/2} \tag{14.2}$$

where J refers to dimensionless flow. For examining the dip-coating process of the solutions that have high viscosity, Guglielmi et al. [40] formulated the following equations:

$$t_1 = t_p \frac{P_p}{c} \tag{14.3}$$

$$t_p = 0.944 C_a^{1/6} \frac{C}{P_p} \left(\frac{nu}{pg} \right)^{1/2} \tag{14.4}$$

$$t_p = J \frac{C}{P_p} \left(\frac{nu}{pg} \right)^{1/2} \tag{14.5}$$

where t_p demonstrates the thickness of heat-treated coatings. P_p reflects the density of heat-treated coating ($g\,cm^{-3}$), while c refers to the number of coating solutions ($g\,cm^{-3}$). Moreover, J should be considered to calculate the thickness of multiple types of coatings [41].

14.3.3 In situ polymerization method

The organic matrices are fabricated in this process by using the nanocomposites coatings, which conducts polymer [42]. Other monomers with initiators can also be used [43]. Whereas, the nanofiller that are utilized can be of metals and metal oxides. The electrodeposition method is consumed for the polymerization, oxidizing agents, and photons [44]. For the organic matrices' methods can be emulsion polymerization or latex emulsions [45] (Fig. 14.10).

14.4 Applications of the nanomaterial-based coatings

There is a myriad application of nanomaterial-based coatings. Self-cleaning applications are one of them. For instance, they are used for the self-cleaning of the glass windows, solar panels, oxidation prevention, stain resistance in the textile, and the antibiofouling agents' surfacing. Also, they are useful in creating hydrophobic surfaces (Table 14.3).

Moreover, the consumption and significance of nanomaterial-based coatings in the paint and lacquers industry are infinite. It is estimated that, on average, consumption of the nanomaterial coating, around 2.6 million tons of paint is produced [41]. The establishment and usage of smart coatings are considered to technological innovation are expected to increase the

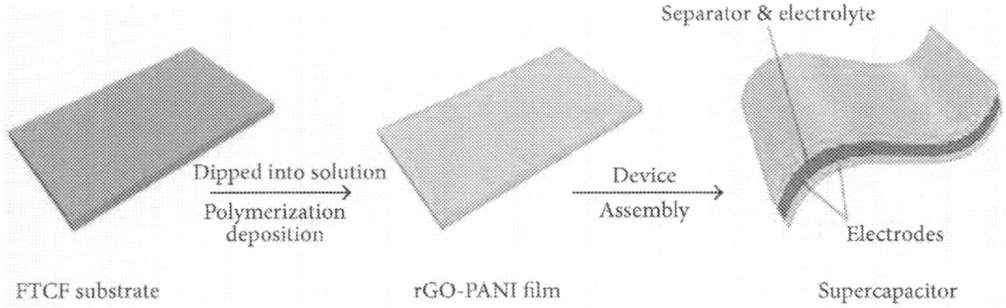

Fig. 14.10 Situ polymerization technique [45].

Table 14.3 The different applications of the nanomaterial-based coatings while characterizing their features.

The nanomaterial-based coatings surface features	Applications
Mechanical properties	Equipment protection and machinery and fragile equipment
Wetting features	Antifouling and self-cleaning of surface
Thermal and chemical features	The oxidation is preventing in the equipment and machinery, resistance toward heat They were used in thermal insulation processing in equipment and infrastructure.
Biological characteristics	For providing biological implants Manufacturing medical instruments and tools and in the dressing of wounds
Electronic and magnetic features	For manufacturing excessively thin dielectrics that are used in the field of transistors and data memory
Catalytical features	For improving the catalytic effectiveness via higher surface-to-volume
Optical features	For manufacturing photo and electrochromatic windows, solar cells, and antireflective screens

turnover to 20% by utilizing smart approaches. The additives prepared from nanomaterial coatings are being widely used to prepare paints and other lacquers, for instance, the usage of barium sulfate and iron oxides as the coloring pigments. They were maintaining the fluidity of lacquers by using powdered silica. The observation and characteristics of nanoscale materials are being studied recently for this technology sector's development and growth. These developments have enhanced nanomaterials' usage in different paint industry regions due to their capability to replicate parent material properties. It is adequate, as a single material can be utilized in fulfilling multiple application requirements. These layers' coatings of the nanomaterials are being consumed as the solution to many authentic problems. It is being used as a coating for anticorrosion. It can also protect against biological and weathering effects. The specialized color range provided by these coatings increased their visual appealing [46]. The nanomaterial coatings are utilized for securing the houses, buildings, and other materials from the external environmental effect. A larger portion of resource conservation depends on these nanomaterials coatings as they increase the product's lifeline and period. It also enhances the durability of the product and increases the replacement times of the objects. Environmental pollution and hazardous materials are significantly reduced by the usage of these nanocoating [47].

14.5 Conclusions

The nanoparticles are used to synthesize liquid material known as coating, and the product obtained from nonmaterial is known as nanomaterial derive coatings. These coatings have larger usage in the manufacturing of lightweight and durable products; the paint industry is one of these coatings' largest consumers. Nanomaterial- based coatings are widely used in the manufacturing industries. The nanocomposites were transformed in coatings that enhance the properties and effectiveness, and wider applicability of the nanocomposites. Creating nanomaterial-based coating includes Supercritical Antisolvent (SAS) Process, Layer by Layer Method, Sol–Gel Method, Cold Spray Method, Chemical Vapor Deposition (CVD) Method, Layer by Layer Method, Emulsion Polymerization, and Nano container-Based Synthesis. On the other hand, directly solution dip-coating solution and sol–gel dip-coating are some of the examples of techniques that are used in the process of creating the nanomaterial-based coatings. The usage of these coatings in different industries has significantly affected the environmental sector as these coatings' usage eliminates heavy and toxic materials.

Acknowledgments

To my loving parents, Mr. Mohammed and Modhi Alarifi, for giving me their full support during my overseas study; to my wife Helen Alarifi and my family who openhandedly provided me with a healthy environment to proceed with my studies here in the United States; to my dear brothers and sisters back home who have supported me during my absence. I want to express my true appreciation to my supervisor, Dr. Ramazan Asmatulu, for his support and for providing me with clear guidance, inspiration, suggestions, criticism, and financial assistance on my writing book journey. I could not have asked for a better mentor, and I owe him much for his support over the years. His leadership consistently challenged me to be a better researcher and provided the right environment for me to achieve my goals.

Engagement Ring

Ibrahim Alarifi

Condolences

My condolences are on the passing of my father, Mohammed Alarifi. I am deeply saddened by the loss that my family has encountered—my condolences. My deepest sympathies go out to my family. I offer you my thoughts, prayers, and well-wishes during this dark time in my life.

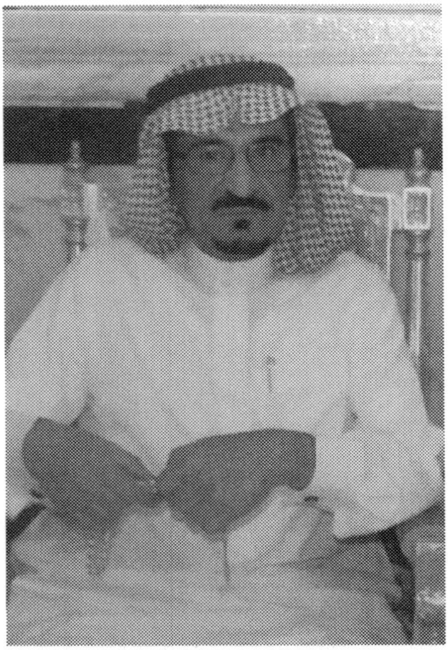

Mohammed Alarifi (Father)

References

[1] P.N. Tri, T.A. Nguyen, S. Rtimi, C.M.O. Plamondon, Nanomaterials-based coatings: an introduction, in: Nanomaterials-Based Coatings, Elsevier, 2019, pp. 1–7.

[2] T. Nardi, S. Rtimi, C. Pulgarin, Y. Leterrier, Antibacterial surfaces based on functionally graded photocatalytic Fe_3O_4@TiO_2 core–shell nanoparticle/epoxy composites, RSC Adv. 5 (127) (2015) 105416–105421, https://doi.org/10.1039/C5RA19298F.

[3] A. Bonnefond, E. González, J.M. Asua, J.R. Leiza, E. Ieva, G. Brinati, C. Pulgarin, Stable photocatalytic paints prepared from hybrid core-shell fluorinated/acrylic/TiO_2 waterborne dispersions, Crystals 6 (10) (2016) 136.

[4] E. Hagtvet, T.J. Evjen, E.A. Nilssen, D.R. Olsen, Assessment of liposome biodistribution by non-invasive optical imaging: a feasibility study in tumour-bearing mice, J. Nanosci. Nanotechnol. 12 (3) (2012) 2912–2918.

[5] D.G. Shchukin, S.V. Lamaka, K.A. Yasakau, M.L. Zheludkevich, M.G.S. Ferreira, H. Möhwald, Active anticorrosion coatings with halloysite nanocontainers, J. Phys. Chem. C 112 (4) (2008) 958–964.

[6] E. Shchukina, D. Grigoriev, T. Sviridova, D. Shchukin, Comparative study of the effect of halloysite nanocontainers on autonomic corrosion protection of polyepoxy coatings on steel by salt-spray tests, Prog. Org. Coat. 108 (2017) 84–89.

[7] A. Chenan, S. Ramya, R.P. George, U.K. Mudali, Hollow mesoporous zirconia nanocontainers for storing and controlled releasing of corrosion inhibitors, Ceram. Int. 40 (7) (2014) 10457–10463.

[8] M. Yeganeh, A. Keyvani, The effect of mesoporous silica nanocontainers incorporation on the corrosion behavior of scratched polymer coatings, Prog. Org. Coat. 90 (2016) 296–303.

[9] I.A. Kartsonakis, E.P. Koumoulos, A.C. Balaskas, G.S. Pappas, C.A. Charitidis, G.C. Kordas, Hybrid organic–inorganic multilayer coatings including nanocontainers for corrosion protection of metal alloys, Corros. Sci. 57 (2012) 56–66.
[10] S. Amiri, A. Rahimi, Synthesis and characterization of supramolecular corrosion inhibitor nanocontainers for anticorrosion hybrid nanocomposite coatings, J. Polym. Res. 22 (5) (2015) 66.
[11] J.M. Yeh, H.Y. Huang, C.L. Chen, W.F. Su, Y.H. Yu, Siloxane-modified epoxy resin–clay nanocomposite coatings with advanced anticorrosive properties prepared by a solution dispersion approach, Surf. Coat. Technol. 200 (8) (2006) 2753–2763.
[12] L. Al-Naamani, S. Dobretsov, J. Dutta, J.G. Burgess, Chitosan-zinc oxide nanocomposite coatings for the prevention of marine biofouling, Chemosphere 168 (2017) 408–417.
[13] S.H. Yoo, C.K. Kim, Enhancement of the meltdown temperature of a lithium ion battery separator via a nanocomposite coating, Ind. Eng. Chem. Res. 48 (22) (2009) 9936–9941.
[14] H. Dong, P. Ye, M. Zhong, J. Pietrasik, R. Drumright, K. Matyjaszewski, Superhydrophilic surfaces via polymer−SiO_2 nanocomposites, Langmuir 26 (19) (2010) 15567–15573.
[15] W. Hou, Q. Wang, UV-driven reversible switching of a polystyrene/titania nanocomposite coating between superhydrophobicity and superhydrophilicity, Langmuir 25 (12) (2009) 6875–6879.
[16] A. Davis, Y.H. Yeong, A. Steele, I.S. Bayer, E. Loth, Superhydrophobic nanocomposite surface topography and ice adhesion, ACS Appl. Mater. Interfaces 6 (12) (2014) 9272–9279.
[17] S. Schilp, A. Rosenhahn, M.E. Pettitt, J. Bowen, M.E. Callow, J.A. Callow, M. Grunze, Physicochemical properties of (ethylene glycol)-containing self-assembled monolayers relevant for protein and algal cell resistance, Langmuir 25 (17) (2009) 10077–10082.
[18] A. Toor, H. So, A.P. Pisano, Improved dielectric properties of polyvinylidene fluoride nanocomposite embedded with poly (vinylpyrrolidone)-coated gold nanoparticles, ACS Appl. Mater. Interfaces 9 (7) (2017) 6369–6375.
[19] S. He, C. Lu, S. Zhang, Facile and efficient route to polyimide-TiO_2 nanocomposite coating onto carbon fiber, ACS Appl. Mater. Interfaces 3 (12) (2011) 4744–4750.
[20] E.B. Caldona, A.C.C. De Leon, P.G. Thomas, D.F. Naylor III, B.B. Pajarito, R.C. Advincula, Superhydrophobic rubber-modified polybenzoxazine/SiO_2 nanocomposite coating with anticorrosion, anti-ice, and superoleophilicity properties, Ind. Eng. Chem. Res. 56 (6) (2017) 1485–1497.
[21] A. Kaboorani, N. Auclair, B. Riedl, V. Landry, Mechanical properties of UV-cured cellulose nanocrystal (CNC) nanocomposite coating for wood furniture, Prog. Org. Coat. 104 (2017) 91–96.
[22] H. Gül, F. Kılıç, M. Uysal, S. Aslan, A. Alp, H. Akbulut, Effect of particle concentration on the structure and tribological properties of submicron particle SiC reinforced Ni metal matrix composite (MMC) coatings produced by electrodeposition, Appl. Surf. Sci. 258 (10) (2012) 4260–4267.
[23] M.Y. Zhang, C. Ye, U.J. Erasquin, T. Huynh, C. Cai, G.J. Cheng, Laser engineered multilayer coating of biphasic calcium phosphate/titanium nanocomposite on metal substrates, ACS Appl. Mater. Interfaces 3 (2) (2011) 339–350.
[24] V. Miikkulainen, M. Leskelä, M. Ritala, R.L. Puurunen, Crystallinity of inorganic films grown by atomic layer deposition: overview and general trends, J. Appl. Phys. 113 (2) (2013) 2.
[25] C. Sanchez, P. Belleville, M. Popall, L. Nicole, Hybrid materials themed issue, Chem. Soc. Rev. 40 (2011) 453–1152.

[26] S. Cho, K. Takagi, H. Kwon, D. Seo, K. Ogawa, K. Kikuchi, A. Kawasaki, Multi-walled carbon nanotube-reinforced copper nanocomposite coating fabricated by low-pressure cold spray process, Surf. Coat. Technol. 206 (16) (2012) 3488–3494.

[27] P. Nguyen-Tri, T.A. Nguyen, P. Carriere, C. Ngo Xuan, Nanocomposite coatings: preparation, characterization, properties, and applications, Int. J. Corros. 2018 (2018) 4749501.

[28] E.J.T. Pialago, C.W. Park, Cold spray deposition characteristics of mechanically alloyed Cu-CNT composite powders, Appl. Surf. Sci. 308 (2014) 63–74.

[29] H. Lee, M.Y. Song, J. Jurng, Y.K. Park, The synthesis and coating process of TiO_2 nanoparticles using CVD process, Powder Technol. 214 (1) (2011) 64–68.

[30] J.E. Lee, N. Lee, T. Kim, J. Kim, T. Hyeon, Multifunctional mesoporous silica nanocomposite nanoparticles for theranostic applications, Acc. Chem. Res. 44 (10) (2011) 893–902.

[31] Y. Wang, R.N. Dave, R. Pfeffer, Polymer coating/encapsulation of nanoparticles using a supercritical anti-solvent process, J. Supercrit. Fluids 28 (1) (2004) 85–99.

[32] A. Bu, J. Wang, J. Zhang, J. Bai, Z. Shi, Q. Liu, G. Ji, Corrosion behavior of ZrO_2–TiO_2 nanocomposite thin films coating on stainless steel through sol–gel method, J. Sol-Gel Sci. Technol. 81 (3) (2017) 633–638.

[33] N. Abu-Thabit, A.S.H. Makhlouf, Recent approaches for designing nanomaterials-based coatings for corrosion protection, in: M. Aliofkhazraei, A. Makhlouf (Eds.), Handbook of Nanoelectrochemistry, Springer, Cham, 2016.

[34] D.O. Grigoriev, M.F. Haase, N. Fandrich, A. Latnikova, D.G. Shchukin, Emulsion route in fabrication of micro and nanocontainers for biomimetic self-healing and self-protecting functional coatings, Bioinspired Biomim. Nanobiomater. 1 (2) (2012) 101–116.

[35] S.R. White, N.R. Sottos, P.H. Geubelle, J.S. Moore, M.R. Kessler, S.R. Sriram, S. Viswanathan, Autonomic healing of polymer composites, Nature 409 (6822) (2001) 794–797.

[36] M. Ferreira, M.F. Rubner, Molecular-level processing of conjugated polymers. 1. Layer-by-layer manipulation of conjugated polyions, Macromolecules 28 (21) (1995) 7107–7114.

[37] A. Mohammadzadeh, S.K.N. Zadeh, M.H. Saidi, M. Sharifzadeh, Mechanical engineering of solid oxide fuel cell systems: geometric design, mechanical configuration, and thermal analysis, in: Design and Operation of Solid Oxide Fuel Cells, Academic Press, 2020, pp. 85–130.

[38] X. Tang, X. Yan, Dip-coating for fibrous materials: mechanism, methods and applications, J. Sol-Gel Sci. Technol. 81 (2) (2017) 378–404.

[39] Y. Yin, C. Wang, Q. Shen, G. Zhang, C.M.A. Galib, Surface deposition on cellulose substrate via cationic SiO_2/TiO_2 hybrid sol for transfer printing using disperse dye, Ind. Eng. Chem. Res. 52 (31) (2013) 10656–10663.

[40] M. Guglielmi, P. Colombo, F. Peron, L.M. Degli Esposti, Dependence of thickness on the withdrawal speed for SiO_2 and TiO_2 coatings obtained by the dipping method, J. Mater. Sci. 27 (18) (1992) 5052–5056.

[41] W. Luther, A. Zweck, Innovationsbegleitung Nanotechnologie: Nanotechnologie in Architektur und Bauwesen. Zukünftige Technologien Nr. 62, 2006.

[42] S. Zhang, G. Sun, Y. He, R. Fu, Y. Gu, S. Chen, Preparation, characterization, and electrochromic properties of nanocellulose-based polyaniline nanocomposite films, ACS Appl. Mater. Interfaces 9 (19) (2017) 16426–16434.

[43] M. Shabani-Nooshabadi, S.M. Ghoreishi, Y. Jafari, N. Kashanizadeh, Electrodeposition of polyaniline-montmorrilonite nanocomposite coatings on 316L stainless steel for corrosion prevention, J. Polym. Res. 21 (4) (2014) 416.

[44] R.H. Fernando, Nanocomposite and nanostructured coatings: recent advancements, in: CS Symposium Series, American Chemical Society, 2009.
[45] F. Chen, P. Wan, H. Xu, X. Sun, Flexible transparent supercapacitors based on hierarchical nanocomposite films, ACS Appl. Mater. Interfaces 9 (21) (2017) 17865–17871.
[46] L. Deutsches, Brandschutz-Beschichtungen – 40 Minuten können Leben retten. Lack im Gespräch, Informationsdienst Deutsches Lackinstitut Nr. 112, Seite 5, 2012, Retrieved from: http://www.lacke-undfarben.de/fileadmin/templates/img/pdf/LIG_112.pdf.
[47] M. Steinfeldt, A.V. Gleich, U. Petschow, R. Haum, T. Chudoba, S. Haubold, Nachhaltigkeitseffekte durch Herstellung und Anwendung nanotechnologischer Produkte. Schriftenreihe des IÖW 17704. Berlin, 2004, Retrieved from: http://www.bmbf.de/pub/nano_nachhaltigkeit_ioew_endbericht.pdf.

Index

Note: Page numbers followed by *f* indicate figures, and *t* indicate tables.

D

E

F

G

H

I

L

M

T

U

W

Printed in the United States
by Baker & Taylor Publisher Services